Lecture Notes in Networks and Systems

Volume 12

Series editor

Janusz Kacprzyk, Polish Academy of Sciences, Warsaw, Poland
e-mail: kacprzyk@ibspan.waw.pl

The series "Lecture Notes in Networks and Systems" publishes the latest developments in Networks and Systems—quickly, informally and with high quality. Original research reported in proceedings and post-proceedings represents the core of LNNS.

Volumes published in LNNS embrace all aspects and subfields of, as well as new challenges in, Networks and Systems.

The series contains proceedings and edited volumes in systems and networks, spanning the areas of Cyber-Physical Systems, Autonomous Systems, Sensor Networks, Control Systems, Energy Systems, Automotive Systems, Biological Systems, Vehicular Networking and Connected Vehicles, Aerospace Systems, Automation, Manufacturing, Smart Grids, Nonlinear Systems, Power Systems, Robotics, Social Systems, Economic Systems and other. Of particular value to both the contributors and the readership are the short publication timeframe and the world-wide distribution and exposure which enable both a wide and rapid dissemination of research output.

The series covers the theory, applications, and perspectives on the state of the art and future developments relevant to systems and networks, decision making, control, complex processes and related areas, as embedded in the fields of interdisciplinary and applied sciences, engineering, computer science, physics, economics, social, and life sciences, as well as the paradigms and methodologies behind them.

More information about this series at http://www.springer.com/series/15179

H.R. Vishwakarma · Shyam Akashe
Editors

Computing and Network Sustainability

Proceedings of IRSCNS 2016

Springer

Editors
H.R. Vishwakarma
Department of Software and Systems Engineering
VIT University
Vellore, Tamil Nadu
India

Shyam Akashe
Department of Electronics and Communication Engineering
ITM University
Gwalior, Madhya Pradesh
India

ISSN 2367-3370 ISSN 2367-3389 (electronic)
Lecture Notes in Networks and Systems
ISBN 978-981-10-3934-8 ISBN 978-981-10-3935-5 (eBook)
DOI 10.1007/978-981-10-3935-5

Library of Congress Control Number: 2017933069

Printed on acid-free paper

This Springer imprint is published by Springer Nature
The registered company is Springer Nature Singapore Pte Ltd.
The registered company address is: 152 Beach Road, #21-01/04 Gateway East, Singapore 189721, Singapore

Preface

The volume contains the papers presented at the IRCNS 2016: International Symposium on Computing and Network Sustainability. The symposium was held during 1st and 2nd of July 2016 in Goa, India, and organized communally by Associated Chamber of Commerce and Industry of India, Computer Society of India Division IV and Global Research Foundation. It targets state of the art as well as emerging topics pertaining to ICT and effective strategies for its implementation for engineering and intelligent applications. The objective of this international conference is to provide opportunities for the researchers, academicians, industry persons, and students to interact, exchange ideas, and devise strategies for future studies in information and communication technologies. The IRSCNS 2016 proposes a set of technical presentations made by internationally recognized researchers and experts. Young researchers participating in this event experienced the opportunity to present their work and obtain feedback during dedicated sessions intended to further facilitate interactions and the exchange of ideas. The target audience of IRSCNS ranges from senior researchers to Ph.D. and masters' students. Researchers and engineers from companies also participated in the symposium. The symposium attracted a large number of high-quality submissions and provided a forum to researchers for insightful discussions. Industry and academia shared their experiences and discussed future solutions for design infrastructure for ICT. Research submissions in various advanced technology areas were received, and after a rigorous peer-reviewed process with the help of program committee members and external reviewer, 107 papers were accepted with an acceptance ratio of 0.56. The symposium featured many distinguished personalities like professors from BITS Pilani Goa, Goa University, Mr. Nitin from Goa Chamber of Commerce, Dr. Durgesh Kumar Mishra and Mr. Aninda Bose from Springer India Pvt. Limited, and also a wide panel of start-up entrepreneurs. Invited talks were organized in industrial and academia tracks on both days. We are indebted to CSI Goa Professional Chapter, CSI Division IV, Goa Chamber of Commerce, Goa

University, for their immense support to make this conference possible in such a grand scale. A total of three sessions were organized, and 29 papers were presented in the three technical sessions. The total number of accepted submissions was 43 with a focal point on computing and network sustainability.

Vellore, India
Gwalior, India
July 2016

H.R. Vishwakarma
Shyam Akashe

Conference Organizing Committee

Advisory Committee

Dr. Dharm Singh, Namibia University of Science and Technology, Namibia
Dr. Aynur Unal, Stanford University, USA
Mr. P.N. Jain, Add. Sec., R&D, Government of Gujarat, India
Prof. J. Andrew Clark, Computer Science University of York, UK
Dr. Anirban Basu, Vice President, CSI
Prof. Mustafizur Rahman, Endeavour Research Fellow, Australia
Dr. Malay Nayak, Director-IT, London
Mr. Chandrashekhar Sahasrabudhe, ACM India
Dr. Pawan Lingras, Saint Mary's University, Canada
Prof. (Dr.) P. Thrimurthy, Past President, CSI
Dr. Shayam Akashe, ITM, Gwalior, MP, India
Dr. Bhushan Trivedi, India
Prof. S.K. Sharma, Pacific University, Udaipur, India
Prof. H.R. Vishwakarma, VIT, Vellore, India
Dr. Tarun Shrimali, SGI, Udaipur, India
Mr. Mignesh Parekh, Ahmadabad, India
Mr. Sandeep Sharma, Joint CEO, SCOPE
Dr. J.P. Bhamu, Bikaner, India
Dr. Chandana Unnithan, Victoria University, Australia
Prof. Deva Ram Godara, Bikaner, India
Dr. Y.C. Bhatt, Chairman, CSI Udaipur Chapter
Dr. B.R. Ranwah, Past Chairman, CSI Udaipur Chapter
Dr. Arpan Kumar Kar, IIT Delhi, India

Organizing Committee

Organizing Chairs
Ms. Bhagyesh Soneji, Chairperson ASSOCHAM Western Region
Co-chair—Dr. S.C. Satapathy, ANITS, Visakhapatnam

Members
Shri. Bharat Patel, COO, Yudiz Solutions
Dr. Basant Tiwari, Bhopal
Dr. Rajveer Shekhawat, Manipal University, Jaipur
Dr. Nilesh Modi, Chairman, ACM Ahmedabad Chapter
Dr. Harshal Arolkar, Assoc. Prof., GLS Ahmadabad
Dr. G.N. Jani, Ahmadabad, India
Dr. Vimal Pandya, Ahmadabad, India
Mr. Vinod Thummar, SITG, Gujarat, India
Mr. Nilesh Vaghela, ElectroMech, Ahmedabad, India
Dr. Chirag Thaker, GEC, Bhavnagar, Gujarat, India
Mr. Maulik Patel, SITG, Gujarat, India
Mr. Nilesh Vaghela, ElectroMech Corporation, Ahmadabad, India
Dr. Savita Gandhi, GU, Ahmadabad, India
Mr. Nayan Patel, SITG, Gujarat, India
Dr. Jyoti Parikh, Professor, GU, Ahmadabad, India
Dr. Vipin Tyagi, Jaypee University, Guna, India
Prof. Sanjay Shah, GEC, Gandhinagar, India
Dr. Chirag Thaker, GEC, Bhavnagar, Gujarat, India
Mr. Mihir Chauhan, VICT, Gujarat, India
Mr. Chetan Patel, Gandhinagar, India

Program Committee

Program Chair
Dr. Durgesh Kumar Mishra, Chairman, Div IV, CSI

Members
Dr. Priyanks Sharma, RSU, Ahmedabad
Dr. Nitika Vats Doohan, Indore
Dr. Mukesh Sharma, SFSU, Jaipur
Dr. Manuj Joshi, SGI, Udaipur, India
Dr. Bharat Singh Deora, JRNRV University, Udaipur
Prof. D.A. Parikh, Head, CE, LDCE, Ahmedabad, India
Prof. L.C. Bishnoi, GPC, Kota, India
Mr. Alpesh Patel, SITG, Gujarat

Dr. Nisheeth Joshi, Banasthali University, Rajasthan, India
Dr. Vishal Gaur, Bikaner, India
Dr. Aditya Patel, Ahmedabad University, Gujarat, India
Mr. Ajay Choudhary, IIT Roorkee, India
Dr. Dinesh Goyal, Gyan Vihar, Jaipur, India
Dr. Devesh Shrivastava, Manipal University, Jaipur
Dr. Muneesh Trivedi, ABES, Ghaziabad, India
Prof. R.S. Rao, New Delhi, India
Dr. Dilip Kumar Sharma, Mathura, India
Prof. R.K. Banyal, RTU, Kota, India
Mr. Jeril Kuriakose, Manipal University, Jaipur, India
Dr. M. Sundaresan, Chairman, CSI Coimbatore Chapter
Prof. Jayshree Upadhyay, HOD-CE, VCIT, Gujarat
Dr. Sandeep Vasant, Ahmedabad University, Gujarat, India

Contents

Editors and Contributors

About the Editors

Prof. H.R. Vishwakarma an engineer turned academician, is serving VIT University, Vellore, since 2004. He has about 30 years of experience both in industry and in academia. He served R&D Division of ITI Limited (A Govt. of India Undertaking), Bangalore, during 1987–2001 and rose to the position of deputy chief engineer (R&D). Subsequently, he briefly worked in software industry at Bangalore and with IIITM Kerala, Trivandrum. After joining VIT University in 2004 as a professor of IT, Prof. Vishwakarma also became a founder member of TIFAC-CORE on Automotive Infotronics (sponsored by DST, Govt. of India). Later, he contributed to the formation of School of Computing Sciences and served till 2009 holding various leadership positions. He is currently with the School of Information Technology and Engineering as a senior professor. During his entire professional career, Prof. Vishwakarma has been striving to promote industry–academia interaction. He is a fellow of Computer Society of India (CSI) and active member of many professional societies including ACM and IEEE. He served CSI for over 15 years holding various positions including honorary secretary at national level during 2010–12. He also served as a president, Engineering Sciences Section of Indian Science Congress Association during 2011–12. Professor Vishwakarma has organized several conferences, published several technical papers and edited conference proceedings. He holds B.E. degree from Government Engineering College, Jabalpur and M.Tech. degree from IIT Bombay.

Dr. Shyam Akashe is a professor in ITM University, Gwalior, Madhya Pradesh, India. He did his Ph.D. at Thapar University, Punjab, and M.Tech. at the Institute of Technology and Management, Gwalior, in electronics and communication engineering. There are more than 190 publications to his credit including more than 50 papers in SCI-indexed journals. He has supervised approximately 50 PG students and 8 Ph.D. scholars till date. The main focus of his research is low-power system on chip (SoC) applications in which static random-access memories (SRAMs) are omnipresent. He has authored two books entitled 'Moore's Law Alive: Gate-All-Around (GAA) Next Generation

Transistor' published by Lambert Academic Publishing, Germany, and 'Low Power High Speed CMOS Multiplexer Design' published by Nova Science Publishers, Inc., New York, USA. He has also published over 120 papers in refereed journals of national and international repute. Dr. Akashe has participated in many national and international conferences and presented over 100 papers.

Contributors

B.V. Ajay Prakash Department of Computer Science and Engineering, SJBIT, Bangalore, India

A.S. Alvi Department of CSE, PRMIT & R, Badnera, Amravati, India

D.V. Ashoka Department of Computer Science and Engineering, JSSATE, Bangalore, India

Fakruddin Baig Department of Telecommunication Engineering, M.S. Ramaiah Institute of Technology, Bangalore, India

Jagdish Bakal Department of Information Technology, GH Raisoni College of Engineering, Nagpur, India

P.I. Basarkod Department of Electronics and Communication, REVA Institute of Technology and Management, VTU, Bengaluru, India

Shalini Batra CSED, Thapar University, Patiala, Punjab, India

Parag Bhalchandra School of Computational Sciences, S.R.T.M. University, Nanded, MS, India

Nikhil Bhat MSR College Road, MSR Nagar, Bengaluru, Karnataka, India

Krunal P. Bhatt MSR College Road, MSR Nagar, Bengaluru, Karnataka, India

Siddalingappagouda C. Biradar Department of Electronics and Communication Engineering, Don Bosco Institute of Technology, Bangalore, Karnataka, India

Inderveer Chana Computer Science and Engineering Department, Thapar University, Patiala, India

Anshul Chandel CSED, Thapar University, Patiala, Punjab, India

Karishma Chaudhary L.J. Institute of Engineering and Technology, Ahmedabad, Gujarat, India

Arnima Chaurasia Computer Science Department, Banasthali Vidyapith, Jaipur, India

Nekita Chavan Department of Information Technology, GH Raisoni College of Engineering, Nagpur, India

Ramesh Debur Department of Physiotherapy, M.S. Ramaiah Medical College, Bangalore, India

Meghana Deepak MS. Ramaiah Institute of Technology, Bangalore, India

Nilesh Deshmukh School of Computational Sciences, S.R.T.M. University, Nanded, MS, India

Kuntal Devi Department of Mathematics & Statistics, Manipal University Jaipur, Jaipur, Rajasthan, India

Rita Dewanjee MATS School of Information Technology, MATS University, Raipur, India

Uha Durbha Department of Telecommunication Engineering, M.S. Ramaiah Institute of Technology, Bangalore, India

Ajay Kumar Dwivedi Department of Electronics Communication, SIET, Allahabad, UP, India

Deepak Garg CSED, Thapar University, Patiala, Punjab, India

Shiwani Garg CSED, Thapar University, Patiala, Punjab, India

Praveen Gubbala Symbiosis Institute of Technology, Lavale, Pune, Maharashtra, India

Neeraj Gugnani DAV Dental College, Yamunanagar, India

Kailas Hambarde School of Computational Sciences, S.R.T.M. University, Nanded, MS, India

Konda Hari Krishna Department of Computer Science & Engineering, Lingaya's University, Faridabad, India; Bharat Institute of Engineering and Technology, Hyderabad, India

Vinayak Hegde Department of Computer Science, Amrita Vishwa Vidyapeetham Mysuru Campus, Amrita University, Mysuru, Karnataka, India

Shaikh Husen School of Computational Sciences, S.R.T.M. University, Nanded, MS, India

Payal Jain Department of Electronics & Communication, Geetanjali Institute of Technical Studies, Udaipur, Rajasthan, India

Fouzan Javeed Department of Telecommunication Engineering, M.S. Ramaiah Institute of Technology, Bangalore, India

Pramod Jayaram MS. Ramaiah Institute of Technology, Bangalore, India

Hitesh Jha St. John College of Engineering and Technology, Palghar, India

Rutvij Jhaveri Department of Computer Engineering, SVM Institute of Technology, Bharuch, Gujarat, India

Bhanu Prakash Joshi Amity School of Engineering & Technology, Amity University, Noida, India

Deepti Jyotiyana Department of Computer Science and Engineering, Government Women Engineering College, Ajmer, India

Vijendra Kamble School of Computational Sciences, S.R.T.M. University, Nanded, MS, India

P.T. Karule Yashwantrao Chavan College of Engineering, Nagpur, India

Gagandeep Kaur Department of Computer Science & Engineering, Chandigarh University, Gharuan, Punjab, India

Manjeet Kaur CSED, Thapar University, Patiala, Punjab, India

Manpreet Kaur Department of Computer Science & Engineering, Chandigarh University, Gharuan, Punjab, India

Anand Khandare Department of CSE, SGB Amravati University, Amravati, India

Anushri Kulkarni Zeal College of Engineering and Research, Pune, Maharashtra, India

Govind Kulkarni School of Computational Sciences, S.R.T.M. University, Nanded, MS, India

Lalit Kulkarni Department of Information Technology, Maharashtra Institute of Technology, Pune, India

Prahlad Kulkarni Department of Electronics and Telecommunication Engineering, Pune Institute of Computer Technology, Pune, Maharashtra, India

Ashish Kumar Department of Mathematics & Statistics, Manipal University Jaipur, Jaipur, Rajasthan, India

Divya Kumar Computer Science and Engineering Department, Motilal Nehru National Institute of Technology Allahabad, Allahabad, India

Indeewar Kumar Department of Automobile Engineering, Manipal University Jaipur, Jaipur, Rajasthan, India

Tapas Kumar Department of Computer Science & Engineering, Lingaya's University, Faridabad, India

V. Anand Prem Kumar Department of Computer Science and Engineering, National Institute of Technology, Tiruchirappalli, Tamil Nadu, India

Jeril Kuriakose St. John College of Engineering and Technology, Palghar, India

Kanika Lakhani Amity University, Noida, India

R. Madan Mohan Department of Computer Science & Engineering, Bharat Institute of Engineering and Technology, Hyderabad, India

Saurabh Maheshwari Department of Computer Science and Engineering, Government Women Engineering College Ajmer, Ajmer, India

V.N. Manjunath Aradya Department of Master of Computer Application, SJCE, Mysore, India

Rajeev Mathur Suresh Gyan Vihar University, Jaipur, India

Vivek Mehta Computer Science and Engineering Department, Thapar University, Patiala, India

Bhawna Minocha Amity University, Noida, India

Brijesh Mishra Department of Electronics and Communication, University of Allahabad, Allahabad, UP, India

Shivam Modi Department of Information Technology, Vellore Institute of Technology, Vellore, Tamil Nadu, India

Debajyoti Mukhopadhyay Department of Information Technology, Maharashtra Institute of Technology, Pune, India

G.M. Navami Patil Department of Electronics and Communication, REVA Institute of Technology and Management, VTU, Bengaluru, India

Swati Nigade Zeal College of Engineering and Research, Pune, Maharashtra, India

R.K. Nikshitha MS. Ramaiah Institute of Technology, Bangalore, India

Akhilesh Kumar Pandey Department of Electronics and Communication, University of Allahabad, Allahabad, UP, India

Gayatri Pandi (Jain) L.J. Institute of Engineering and Technology, Ahmedabad, Gujarat, India

Sonam Parekh Department of Electronics & Communication, Geetanjali Institute of Technical Studies, Udaipur, Rajasthan, India

Priya Patel Department of Computer Engineering, SVM Institute of Technology, Bharuch, Gujarat, India

Bibudhendu Pati Department of Computer Science and Engineering, C. V. Raman College of Engineering, Bhubaneswar, Odisha, India

Shantala Devi Patil Computer Science and Engineering, REVA Institute of Technology and Management, Bangalore, India

Binod Kumar Pattanayak Department of Computer Science and Engineering, Siksha ‘O’ Anusandhan University, Bhubaneswar, Odisha, India

Ritima Paul Amity School of Engineering & Technology, Amity University, Noida, India

Pooja Pawar St. John College of Engineering and Technology, Palghar, India

B. Prasanthi Department of CSE, MGIT, Hyderabad, Telangana, India

P. Prerna MS. Ramaiah Institute of Technology, Bangalore, India

S. Prithvi Alva MSR College Road, MSR Nagar, Bengaluru, Karnataka, India

P. Raghavendra Department of Telecommunication Engineering, M.S. Ramaiah Institute of Technology, Bangalore, India

B.M. Tanvi Raj MSR College Road, MSR Nagar, Bengaluru, Karnataka, India

N. Ramasubramanian Department of Computer Science and Engineering, National Institute of Technology, Tiruchirappalli, Tamil Nadu, India

Virender Ranga Department of Computer Engineering, National Institute of Technology, Kurukshetra, India

Mamata Rath C. V. Raman Computer Academy, Bhubaneswar, Odisha, India

Ranjana D. Raut S.G.B. University, Amravati, India

M. Sachin Department of Telecommunication Engineering, M.S. Ramaiah Institute of Technology, Bangalore, India

N. Sainath Department of Computer Science & Engineering, Bharat Institute of Engineering and Technology, Hyderabad, India

Er. Kamaljit Singh Saini Department of Computer Science & Engineering, Chandigarh University, Gharuan, Punjab, India

Monika Saini Department of Mathematics & Statistics, Manipal University Jaipur, Jaipur, Rajasthan, India

Vahab Samandi Department of Information Technology, Maharashtra Institute of Technology, Pune, India

Khan Samida Department of Telecommunication Engineering, M.S. Ramaiah Institute of Technology, Bangalore, India

R. Sanjeetha MSR College Road, MSR Nagar, Bengaluru, Karnataka, India

Azeem Sarwar Department of Electronics and Communication, University of Allahabad, Allahabad, UP, India

V. Satyanarayana Department of Computer Science & Engineering, Bharat Institute of Engineering and Technology, Hyderabad, India

Varun P. Saxena Department of Computer Science and Engineering, Government Women Engineering College, Ajmer, India

Nilay Shah Symbiosis Institute of Technology, Lavale, Pune, Maharashtra, India

Akanksha Sharma Department of Information Technology, Vellore Institute of Technology, Vellore, Tamil Nadu, India

Ramya Sharma Department of Computer Engineering, National Institute of Technology, Kurukshetra, India

Shivam Sharma Department of Information Technology, Vellore Institute of Technology, Vellore, Tamil Nadu, India

Vijay Kumar Sharma SCIT Manipal University, Jaipur, Rajasthan, India

Pradnya Shidhaye St. John College of Engineering and Technology, Palghar, India

Moirangthem Sailash Singh MS. Ramaiah Institute of Technology, Bangalore, India

Rajeev Singh Department of Electronics and Communication, University of Allahabad, Allahabad, UP, India

Shailendra Pratap Singh Computer Science and Engineering Department, Motilal Nehru National Institute of Technology Allahabad, Allahabad, India

Vivek Singh Department of Electronics and Communication, University of Allahabad, Allahabad, UP, India

Vinay Sridhar Department of Electronics and Communication, M.S. Ramaiah Institute of Technology, Bangalore, India

P.S. Srinivas Department of Telecommunication Engineering, M.S. Ramaiah Institute of Technology, Bangalore, India

Devesh Kumar Srivastava SCIT Manipal University, Jaipur, Rajasthan, India

Sanjeev Kumar Srivastava PIIT, Mumbai, India

Chitra Suman Department of Computer Science and Engineering, University College of Engineering, RTU Kota, Kota, India

Aishwarya Suresh Department of Computer Science, Amrita Vishwa Vidyapeetham Mysuru Campus, Amrita University, Mysuru, Karnataka, India

P. Suresh Department of IT, CBIT, Hyderabad, Telangana, India

Y. Suresh Babu P.G. Department of Computer Science, JKC College, Guntur, India

H.S. Sushma Rao Department of Computer Science, Amrita Vishwa Vidyapeetham Mysuru Campus, Amrita University, Mysuru, Karnataka, India

Viswanath Talasila Department of Telecommunication Engineering, M.S. Ramaiah Institute of Technology, Bangalore, India

Preetam Tamsekar School of Computational Sciences, S.R.T.M. University, Nanded, MS, India

Lokesh Tharani Department of Electronics & Communication Engineering, University College of Engineering, RTU Kota, Kota, India

Nimish Ukey Department of Information Technology, Maharashtra Institute of Technology, Pune, India

D. Vasumathi Department of CSE, JNTUH College of Engineering, Hyderabad, Telangana, India

D. Veeraiah Lakireddy Bali Reddy College of Engineering, Mylavaram, Andhra Pradesh, India

Anil Kumar Verma CSED, Thapar University, Patiala, Punjab, India

B.P. Vijayakumar Information Science and Engineering, M.S. Ramaiah Institute of Technology, Bangalore, India

Ranjana Vyas IIIT Allahabad, Allahabad, Uttar Pradesh, India

Pawan Wasnik School of Computational Sciences, S.R.T.M. University, Nanded, MS, India

Trust Model for Secure Routing and Localizing Malicious Attackers in WSN

G.M. Navami Patil and P.I. Basarkod

Abstract The principle venture resists through remote sensor systems is security. Acknowledge as valid with models had been nowadays guided as a productive security way for WSNs. In this errand, it prompts a trust model for secure directing and restricting malignant assailants in WSN. To start with, report conviction, vitality conviction, and data acknowledge as valid with are mulled over at some stage in the estimation of direct consideration. Moreover, if the source and destination hubs are far away, then exhortation and diagonal concur with are figured. Consider, consistency and consideration are characterized to reinforce the rightness of exhortation conviction. Malignant hubs might be related to low conviction values that is distinguished in direct and proposal concur with figuring. The proposed model can think about constancy of sensor hubs more prominent effectively and maintain a strategic distance from the security breaks additional accurately.

Keywords Security · Routing protocols · Belief levels

1 Introduction

WSN's are developing are day by day increasing and developing advancements that have been extensively used as a part of various congruity, for instance, crisis reaction, restorative administrations checking, fight zone recognition, environment watching, action association. The paper has inspected the sorts of hypothesis estimations which is used to deal with the strikes by checking firm activities of

G.M. Navami Patil (✉) · P.I. Basarkod (✉)
Department of Electronics and Communication,
REVA Institute of Technology and Management, VTU, Bengaluru, India
e-mail: navamipatil92@gmail.com

P.I. Basarkod
e-mail: basarkod@revainstitution.org

H.R. Vishwakarma and S. Akashe (eds.), *Computing and Network Sustainability*,
Lecture Notes in Networks and Systems 12, DOI 10.1007/978-981-10-3935-5_1

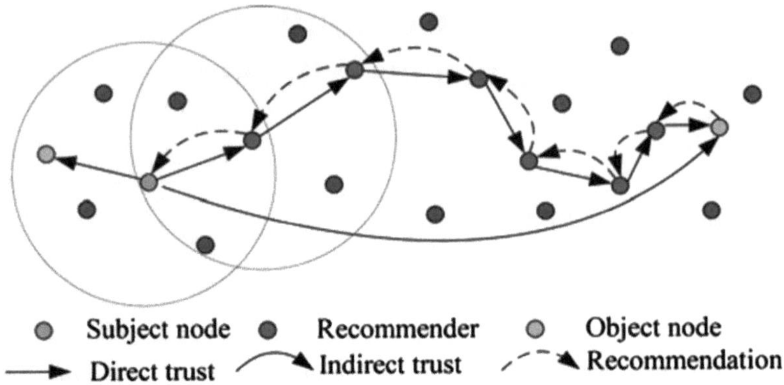

Fig. 1 Network structure

framework. The paper also tells methods for building conviction model. It moreover looks at present proposition models used as a piece of different fundamental initiative method of remote sensor frameworks.

1.1 Conceptual Diagram

See Fig. 1.

2 Paper Organization

This paper is organized into eight sections. Section 1 gives a general idea of Trust Model. Section 2 is about literature survey. Sections 3 and 4 describes the objective of the paper. Section 5 gives the scope of the work. Section 6 gives the methodology used to solve the problem. Section 7 gives derived results and followed by Section 8 including conclusion and future works.

3 Objective

To setup safe connections, we want to assure that all intertwining nodes are believed. This shows the reality that it is noteworthy to setup a belief model making a sensor node to deduce the reliability of other node.

4 Scope of the Work

The conviction structure has end up wide for horrendous center point's reputation in WSNs. It can control in piles of hindrances which join safe coordinating, secured substances add up to, and relied on key switch. Due to the remote natures of WSNs, it fancies a scattered trust adjustment with no center point, in which neighbor center points can check each other. Absolutely, a fit agree with model is fundamental to hold consider related in estimations in a shielded and persisting way.

5 Proposed Work

In this project, Firstly try to know all the trust values, and by adding those trust 48 values, finding the shortest path for the transaction to take place.

The communication trust is measured by,

$$T = \{b, d, u\}$$

The verbal exchange trust T_{com} is measured primarily based on a hit (s) and unsuccessful (f) verbal exchange packets:

$$T_{\text{com}} = \frac{2b+u}{2},$$
$$\text{where } b = \frac{s}{s+f+1},\ u = \frac{1}{s+f+1}.$$

The energy trust is measured by:

$$T_{\text{ene}} = \begin{cases} 1 - p_{\text{ene}} & \textit{if } E_{\text{res}} \geq \theta, \\ 0, & \textit{else}, \end{cases}$$

The data trust is measured by,

$$T_{\text{data}} = 2\left(0.5 - \int_{\mu}^{v_d} f(x)dx\right) = 2\int_{v_d}^{\infty} f(x)dx.$$

By combining all these trust values (Fig. 2),

$$T_{n-\text{direct}} = w_{\text{com}}T_{\text{com}} + w_{\text{ene}}T_{\text{ene}} + w_{\text{data}}T_{\text{data}},$$

As an enhancement, while finding the route for transaction by the belief values, it can also find the malicious node and can omit that. Means that node cannot be removed from the network, but it can be localized to the other nodes as malicious and can inform not to use that node for further process.

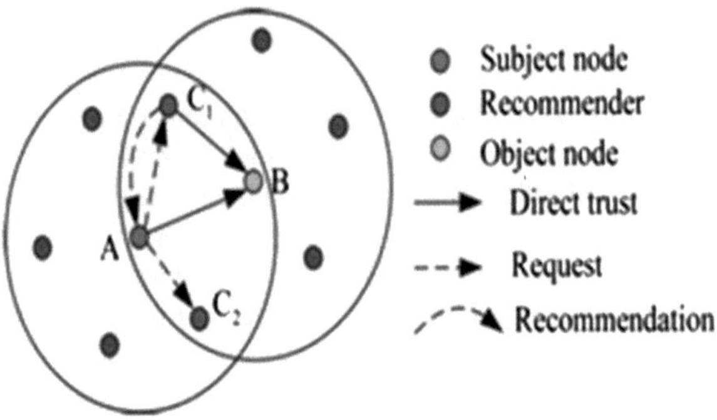

Fig. 2 Calculation of recommendation trust

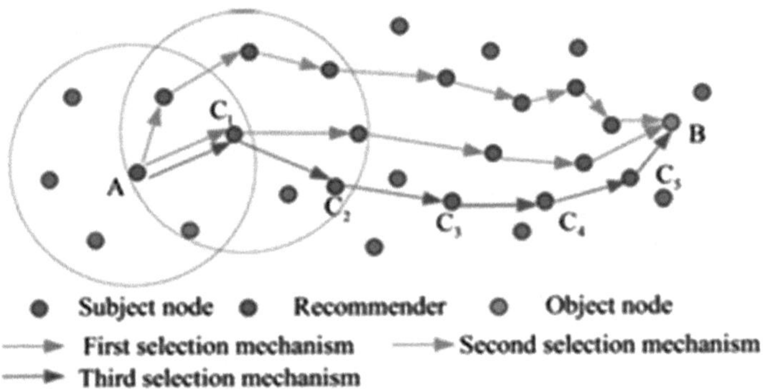

Fig. 3 Calculation of indirect trust

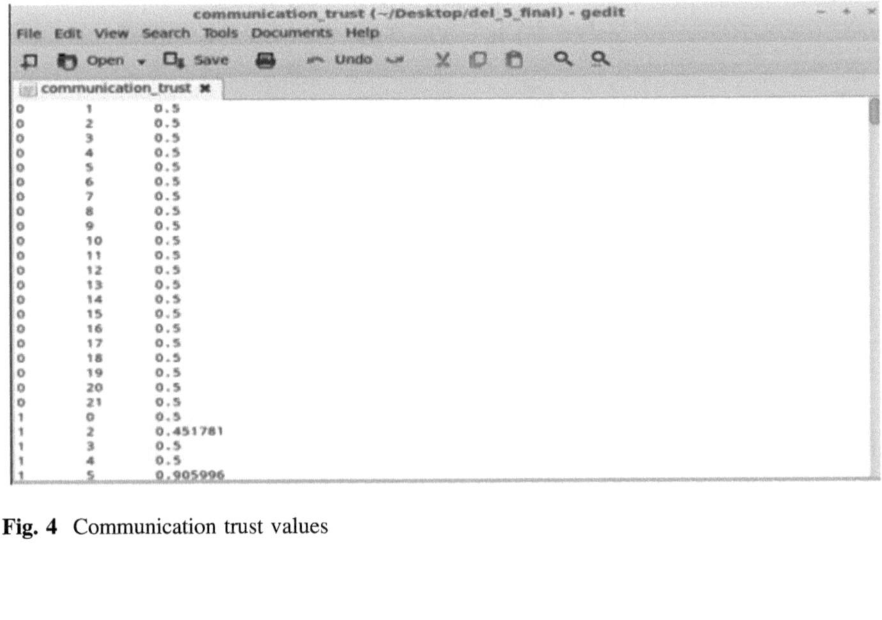

0	1	0.5
0	2	0.5
0	3	0.5
0	4	0.5
0	5	0.5
0	6	0.5
0	7	0.5
0	8	0.5
0	9	0.5
0	10	0.5
0	11	0.5
0	12	0.5
0	13	0.5
0	14	0.5
0	15	0.5
0	16	0.5
0	17	0.5
0	18	0.5
0	19	0.5
0	20	0.5
0	21	0.5
1	0	0.5
1	2	0.451781
1	3	0.5
1	4	0.5
1	5	0.905996

Fig. 4 Communication trust values

6 Result and Analysis

Figures 4, 5, 6, 7, 8, 9, and 10 show the direct, recommendation, and indirect trust values. Figure 11 shows the network setup and Fig. 12 shows the selection of route from the belief values measured.

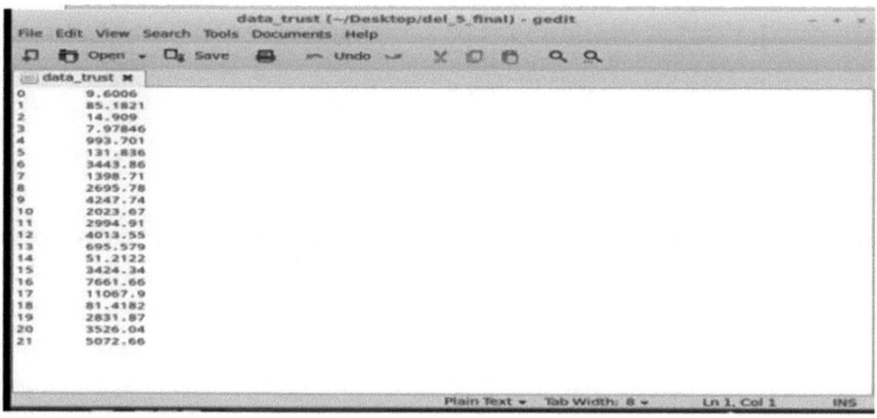

Fig. 5 Data trust values

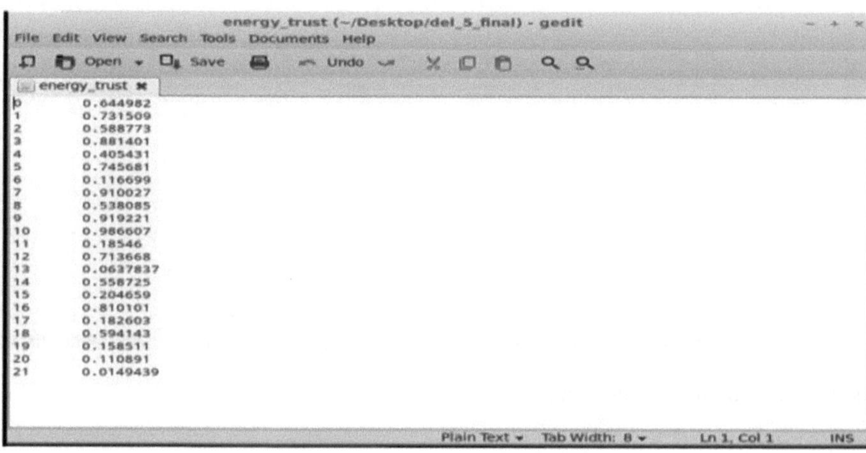

Fig. 6 Energy trust values

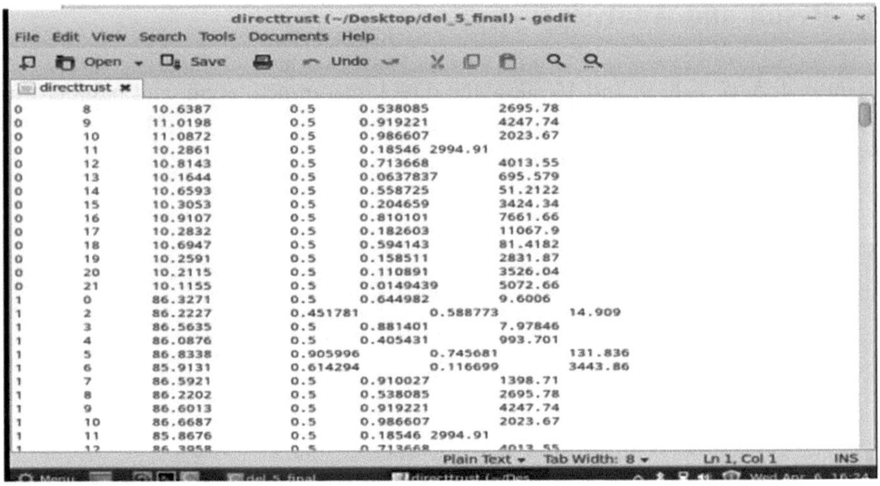

Fig. 7 Direct trust values

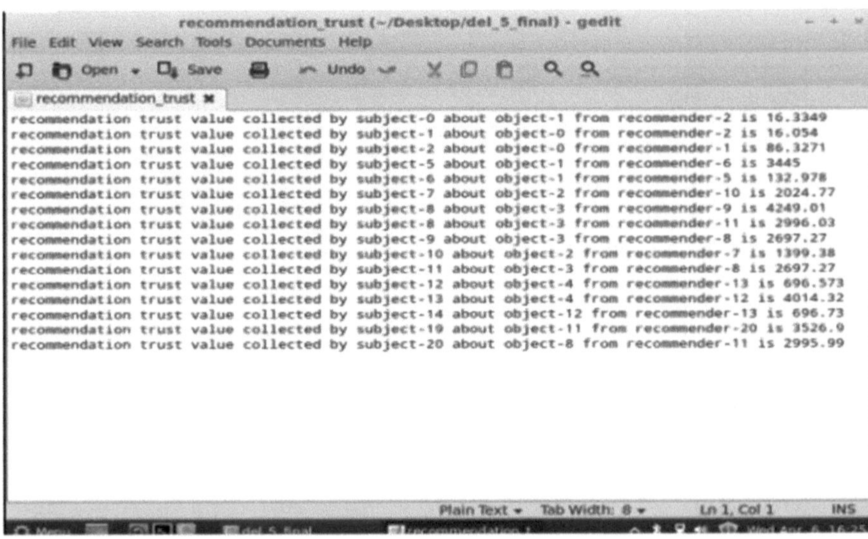

Fig. 8 Recommendation trust values

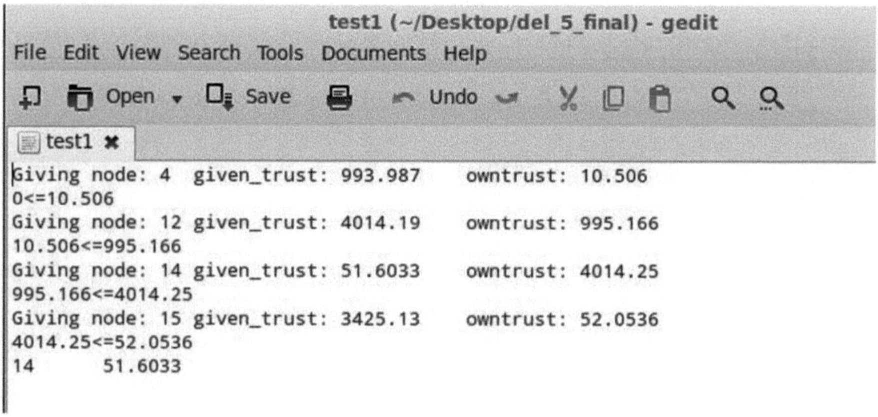

Fig. 9 Indirect trust values

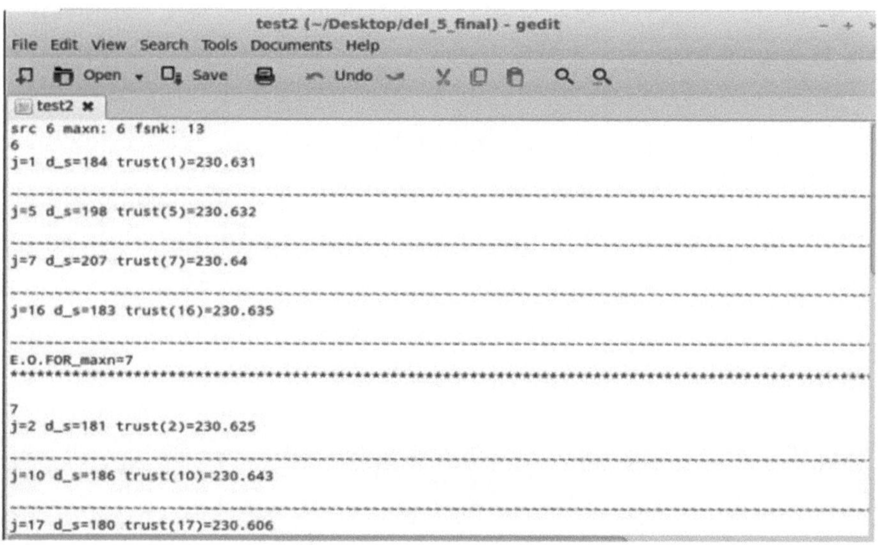

Fig. 10 Indirect trust calculation

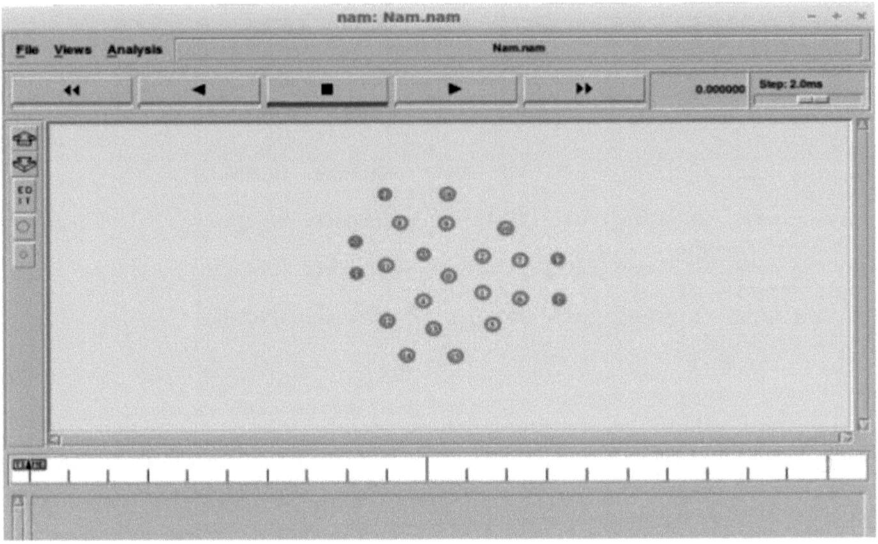

Fig. 11 Network setup

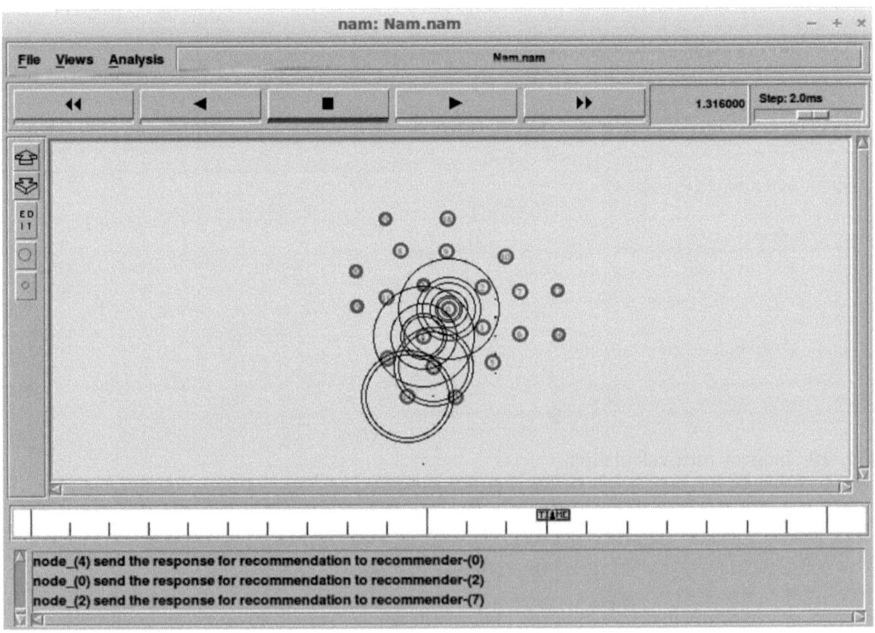

Fig. 12 Choosing for shortest path on the basis of trust values

7 Conclusion

The paper has built up Trust model for secure routing is built by the trust values and the path is traced. As an extension and enhancement detecting and localizing of malicious node is also analyzed and performed. The node which is having the lowest belief value is considered as the malicious node. The paper has chosen the recommenders based on their trust value and hence eliminate the malicious node taken as the recommender.

References

1. Chan H, Perrig A (2003) Security and privacy in sensor networks. Computer 36(10):103–105
2. Feng R, Xu X, Zhou X, Wan J (2011) A trust evaluation algorithm for wireless sensor networks based on node behaviors and d-s evidence theory. Sensors 11:1345–1360
3. Ganeriwal S, Balzano LK, Srivastava MB (2004) Reputation based framework for high integrity sensor networks. In: Proceedings of 2nd ACM workshop security ad hoc sensor networks, pp 66–77
4. Gungor VC, Bin L, Hancke GP (2010) Opportunities and challenges of wireless sensor networks in smart grid. IEEE Trans Ind Electron 57(10):3557–3564
5. Han G, Jiang J, Shu L, Niu J, Chao HC (2014) Managements and applications of trust in wireless sensor networks: a survey. J Comput Syst Sci 80(3):602–617
6. Huang YM, Hsieh MY, Chao HC, Hung SH, Park JH (2009) Pervasive, secure access to a hierarchical-based healthcare monitoring architecture in wireless heterogeneous sensor networks. IEEE J Sel Areas Commun 24(7):400–411
7. Jiang J, Han G, Wang F, Shu L, Guizani M (2015) An efficient distributed trust model for wireless sensor networks. IEEE Trans Parallel Distrib Syst 26(5)
8. Yao Z, Kim D, Doh Y (2008) PLUS: parameterized and localized trust management scheme for sensor networks security. In: Proceedings of IEEE international conference mobile adhoc sensor systems, pp 437–446

Probabilistic Analysis of Performance Measures of Redundant Systems Under Weibull Failure and Repair Laws

Indeewar Kumar, Ashish Kumar, Monika Saini and Kuntal Devi

Abstract The aim of the present paper is to analyze probabilistically various performance measures of two redundant systems under Weibull failure and repair activities using semi-Markov processes and regenerative point technique. Two stochastic models comprise of one original and one duplicate units are developed with the provision of a single repair facility and priority. All repairs and preventive maintenance, after a pre-specific time, are perfect. Recurrence relations of availability, mean time to system failure and profit function for both the models are derived. To highlight the importance of the system, numerical and graphical results for the difference of both models are obtained with respect to failure rate of original unit.

Keywords Weibull failure and repair laws · Redundant system · Preventive maintenance · Priority · Availability

1 Introduction

In the current age of science and technology, reliability, availability, and expected profit of a system play a very dominant role in the market of IT, communication, computers electrical, and manufacturing systems. During the last few decades, many reliability improvement techniques have been developed by researchers. Redundancy, provision of spare unit, is one technique that is used to enhance the reliability of the system. Standby redundant systems have been studied by many authors such as Cao and Wu [1], Goel and Sharma [3], Chandrasekhar et al. [2], Mahmoud and Moshref [9], Moghaddass et al. [10], Wu and Wu [11], and Kumar

I. Kumar
Department of Automobile Engineering, Manipal University Jaipur, Jaipur 303007, Rajasthan, India

A. Kumar (✉) · M. Saini · K. Devi
Department of Mathematics & Statistics, Manipal University Jaipur, Jaipur 303007, Rajasthan, India
e-mail: ashishbarak2020@gmail.com

H.R. Vishwakarma and S. Akashe (eds.), *Computing and Network Sustainability*, Lecture Notes in Networks and Systems 12, DOI 10.1007/978-981-10-3935-5_2

and Saini [6] discussed two-unit cold standby systems under different set of assumptions such as repair, replacement, random shocks, priority, preventive maintenance, inspection, and constant repair and failure distributions. The Weibull distribution has been used in many different fields with many applications. The hazard function of the Weibull distribution can be increasing, decreasing, or constant. It is a very attractive property of Weibull distribution. For many years, researchers Gupta et al. [4] suggest a reliability model for a cold standby system using Weibull distribution for repair and failure rates. Kishan and Jain [8] analyzed a two-unit parallel system by classical and Bayesian approach with Weibull failure and repair laws. Kumar and Malik [5] developed a reliability model for single unit systems in which all random variables are arbitrary distributed like Weibull distribution. Recently, Kumar et al. [7] analyzed a non-identical unit's redundant system using the concept of preventive maintenance, priority, and Weibull distribution for all-time variables.

The main objective of the present paper is to analyze two reliability models of two non-identical units by using Weibull distribution for all-time variables with different scale and equal shape parameters. For, concrete study, difference graphs are derived for various reliability characteristics of both models. The following reliability characteristics of interest to system developers by using regenerative point technique and semi-Markov technique are evaluated.

1. Steady state transition probabilities of different states for both reliability models
2. Mean sojourn times at different states for both reliability models,
3. Reliability and MTSF for both models,
4. Availability of both models,
5. Expected busy period of server due to repair and preventive maintenance for both models, and
6. Net profit of both the models.

2 System Model Description, Notations and States of the Model

The system consists of two units, one original and other duplicate, in cold standby redundancy pattern. In both models, system starts working in S0 state and upon failure of one unit system also remains operative and failed unit undergoes for repair or replacement. The system stops working upon failure of both units simultaneously in both models. In model-I all repair activities performed in FCFS manner while in model-II priority to preventive maintenance of original unit is given over repair of duplicate unit. All random variables associated with repair and failure activities are independent having Weibull density function with different scale parameters and common shape parameter as follows: failure times of the original and duplicate unit are denoted by $f(t) = \beta\eta t^{\eta-1}\exp(-\beta t^{\eta})$, $f_2(t) = h\eta t^{\eta-1}\exp(-ht^{\eta})$, $g(t) = \alpha\eta t^{\eta-1}\exp(-\alpha t^{\eta})$, $g_1(t) = \gamma\eta t^{\eta-1}\exp(-\gamma t^{\eta})$, $f_1(t) = k\eta t^{\eta-1}\exp(-kt^{\eta})$ and $f_3(t) = l\eta t^{\eta-1}\exp(-lt^{\eta})$ with $t \geq 0$ *and* $\theta, \eta, \alpha, \beta, h, k, l > 0$.

3 Notations

O	Original operative unit
DCs	Duplicate unit in standby
Do	Operative duplicative unit
Fur/FUR	Original unit failed and under repair/continuously under repair
DFur/DFUR	Duplicate unit failed and under repair/continuously under repair
DPm/DWPm	Duplicate unit under/waiting for preventive maintenance
Pm/WPm	Original unit under/waiting for preventive maintenance
PM/WPM	Original unit continuously under/waiting for preventive Maintenance
DPM/DWPM	Duplicate unit continuously under/waiting for preventive maintenance
Fwr/DFwr	Original unit after failure under/waiting for repair
FWR/DFWR	Duplicate unit after failure continuously under/waiting for repair

4 Various Transition States of the System Models

In view of the above notations and assumptions, the system may be in one of the following states:

Common in Model-I and II

$S_0 = (O, DCs)$, $S_1 = (Pm, Do)$, $S_2 = (Fur, Do)$, $S_3 = (O, DFur)$, $S_4 = (O, DPm)$, $S_5 = (DPM, Fwr)$ $S_6 = (FUR, DFwr)$, $S_7 = (FUR, DWPm)$, $S_8 = (DPM, WPm)$, $S_9 = (PM, DWPm)$, $S_{10} = (PM, DFwr)$, and $S_{11} = (Fwr, DFUR)$.

Distinct states in Model-I and II

$S_{12} = (DFUR, WPm)$, and $S_{12}(Pm, DFwr)$.

5 Transition Probabilities

By considering simple probabilistic arguments and using below mentioned formula, we easily obtained transition probabilities for all possible states of both models for a particular value of the shape parameter $\eta = 1$. The probabilities of model-I are given below and for model-II are derived in similar way by using the following formula:

$$p_{ij} = Q_{ij}(\infty) = \int q_{ij}(t)dt \tag{1}$$

For Model-I

$$\begin{aligned}
&p_{01} = \frac{\alpha}{\alpha+\beta},\ p_{02} = \frac{\beta}{\alpha+\beta},\ p_{27} = \frac{\alpha}{\alpha+h+k} = p_{24.7},\ p_{10} = \frac{\gamma}{\alpha+h+\gamma},\\
&p_{1.10} = \frac{h}{\alpha+\gamma+h} = p_{13.10},\ p_{19} = \frac{\alpha}{\alpha+\gamma+h} = p_{14.9},\ p_{3.12} = \frac{\alpha}{\alpha+\beta+l} = p_{31.12},\\
&p_{20} = \frac{k}{\alpha+k+h},\ p_{26} = \frac{h}{\alpha+h+k} = p_{23.6}, p_{30} = \frac{l}{l+\alpha+\beta},\ p_{3.11} = \frac{\beta}{\alpha+\beta+l} = p_{32.11},\\
&p_{40} = \frac{\gamma}{\alpha+\beta+\gamma},\ p_{45} = \frac{\beta}{\alpha+\beta+\gamma} = p_{42.5},\ p_{48} = \frac{\alpha}{\alpha+\beta+\gamma} = p_{41.8},\\
&p_{52} = p_{63} = p_{74} = p_{81} = p_{94} = p_{10.3} = p_{11.2} = p_{12.1} = 1
\end{aligned} \tag{2}$$

6 Mean Sojourn Times

The expected time to stay in any particular state by system prior visiting to any other position is called mean sojourn time. If T_i is sojourn time at state S_i, the mean sojourn time (ψ_i) is obtained by formula $\psi_i = \int P(T_i > t)dt$. Mean sojourn time of the model-I is given below for rest states we can obtained in similar way.

$$\psi_0 = \frac{\Gamma(1+1/\eta)}{(\alpha+\beta)^{1/\eta}},\ \psi_3 = \frac{\Gamma(1+1/\eta)}{(\alpha+\beta+l)^{1/\eta}},\ \psi_2 = \frac{\Gamma(1+1/\eta)}{(\alpha+k+h)^{1/\eta}},\ \psi_4 = \frac{\Gamma(1+1/\eta)}{(\alpha+\beta+\gamma)^{1/\eta}},\ \psi_1 = \frac{\Gamma(1+1/\eta)}{(\alpha+\gamma+h)^{1/\eta}}$$

7 Reliability Measures

7.1 Mean Time to System Failure

By using probability concepts, certain recursive relations for reliability analysis of a system model are obtained for cumulative density function between regenerative states where failed state is an absorbing state:

$$R_i(t) = \sum_j Q_{i,j}(t)\Theta R_j(t) + \sum_k Q_{i,k}(t), \quad \begin{cases} i=0,1,2,3,4 \; for\; Model-\text{I} \\ i=0,1,2,3,4 \; for\; Model-\text{II} \end{cases} \tag{3}$$

The mean time to system failure (MTSF) is obtained by taking LST of above relations and using the formula appended below

$$\lim_{s\to 0} \frac{1-V_0^{**}(s)}{s}.$$

7.2 Availability Analysis

Let $Av_i(t)$ denotes the up-state probability at regenerative state S_i at $t = 0$. The recursive relations for $Av_i(t)$ are given as

$$Av_i(t) = \sum_j q_{i,j}(t)\Theta Av_j(t) + M_i(t), \quad \begin{cases} i=0,1,2,3,4 \; for\; Model-\text{I} \\ i=0,1,2,3,4,12 \; for\; Model-\text{II} \end{cases} \tag{4}$$

The probability to stay operative in a particular regenerative state is denoted by $M_i(t)$ up to a time point without any transition. Taking LT from Eq. (4) and solving for $A_0^*(s)$, the steady state availability is given by

$$A_0(\infty) = \lim_{s\to 0} sA_0^*(s)$$

7.3 Some Other Reliability Characteristics

By using probabilistic arguments, recurrence relations for other reliability characteristics such as busy period of server due to preventive maintenance, repair, expected number of repairs, preventive maintenance, and expected number of visits by the server, respectively, denoted by $B_i^P(t)$, $B_i^R(t)$, $R_i^r(t)$, $R_i^p(t)$, and $N_i(t)$ are derived between regenerative states as follows:

$$B_i^p(t) = \sum_j q_{i,j}(t)\Theta B_j^p(t) + W_i(t), \quad \begin{cases} i=0,1,2,3,4 \; for\; Model-\text{I} \\ i=0,1,2,3,4,12 \; for\; Model-\text{II} \end{cases}$$

$$B_i^r(t) = \sum_j q_{i,j}(t)\Theta B_j^r(t) + W_i(t), \quad \begin{cases} i=0,1,2,3,4 \; for\; Model-\text{I} \\ i=0,1,2,3,4,12 \; for\; Model-\text{II} \end{cases}$$

$$R_i^{pm}(t) = \sum_j Q_{i,j}^{(n)}(t)\left[\delta_j + R_j^{pm}(t)\right], \quad \begin{cases} i=0,1,2,3,4 \; for\ Model-\mathrm{I} \\ i=0,1,2,3,4,12 \; for\ Model-\mathrm{II} \end{cases}$$

$$R_i^{r}(t) = \sum_j Q_{i,j}^{(n)}(t)\left[\delta_j + R_j^{r}(t)\right], \quad \begin{cases} i=0,1,2,3,4 \; for\ Model-\mathrm{I} \\ i=0,1,2,3,4,12 \; for\ Model-\mathrm{II} \end{cases}$$

$$N_i(t) = \sum_j Q_{i,j}^{(n)}(t)\left[\delta_j + N_i(t)\right], \quad \begin{cases} i=0,1,2,3,4 \; for\ Model-\mathrm{I} \\ i=0,1,2,3,4,12 \; for\ Model-\mathrm{II} \end{cases} \tag{5}$$

Taking LT of above relations and solving for $B_0^{*P}(\mathrm{s})$, $B_0^{*R}(\mathrm{s})$, $R_i^{*r}(s)$, $R_i^{*pm}(s)$, and $N_i^{**}(t)$. The time for which server is busy due to preventive maintenance, repair, expected number of preventive maintenance, repairs, and expected number of visits is given by

$$B_0^P = \lim_{s\to 0} sB_0^{*P}(s),\; B_0^R = \lim_{s\to 0} sB_0^{*R}(s),\; N_0(\infty) = \lim_{s\to 0} s\tilde{N}_0(s)$$
$$R_0^r(\infty) = \lim_{s\to 0} s\tilde{R}_0^r(s),\; R_0^P(\infty) = \lim_{s\to 0} s\tilde{R}_0^P(s)$$

8 Profit Analysis

The net expected profit incurred to the system model by defining various costs as K_i in steady state can be obtained for both models as follows:

$$P = K_0 Av_0 - K_1 B_0^R - K_2 B_0^P - K_3 R_0^R - K_4 R_0^P - K_5 N_0$$

8.1 Comparative Study

The availability and profit function for both models are obtained for a set of particular values $\alpha = 2$, $\eta = 0.5$, $\gamma = 5$, $k = 1.5$, $h = 0.009$, and $l = 1.4$. The model-II is more available and profitable than that of the model-I for all cases in which shape parameter $\eta = 0.5, 1, 2$. The availability and profit of both models are increased by increasing the value of $\gamma = 5$ to $\gamma = 7$. But, all reliability characteristics decrease with the increase in shape parameter. Hence, we can concluded on the basis of present study that the concept of priority to preventive maintenance of original unit is given over repair of duplicate unit is more profitable.

9 Graphical Results

See Figs. 1 and 2.

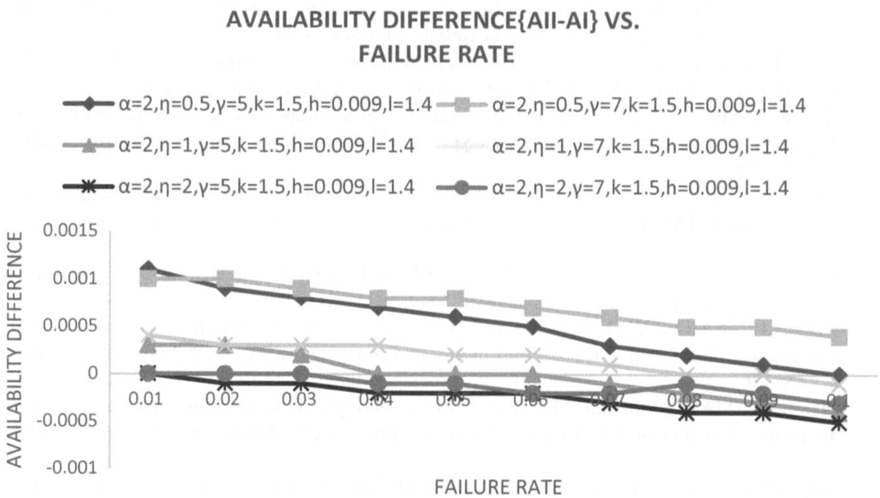

Fig. 1 Availability (AI-AII) versus failure rate (β) for shape parameter η = 0.5, 1, 2

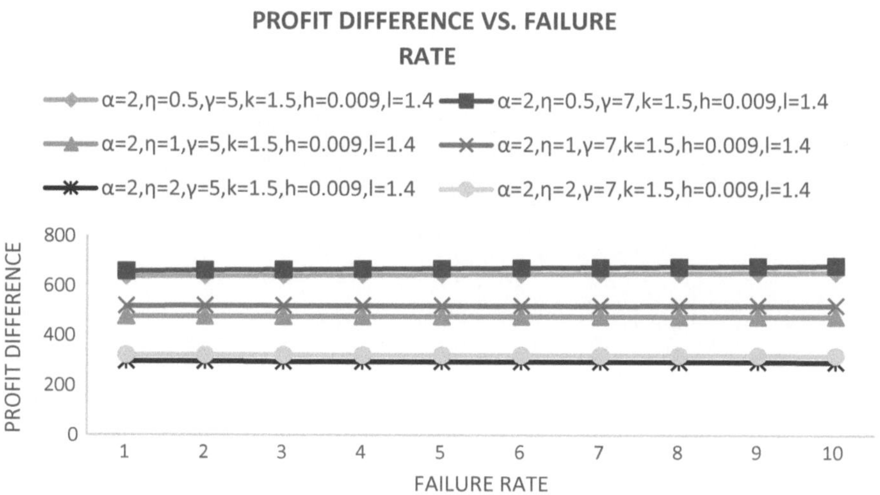

Fig. 2 Profit (PI-PII) versus failure rate (β) for shape parameter η = 0.5, 1, 2

References

1. Cao J, Wu Y (1989) Reliability analysis of a two-unit cold standby system with a replaceable repair facility. Microelectron Reliab 29(2):145–150
2. Chandrasekhar P, Natarajan R, Yadavalli VSS (2004) A study on a two unit standby system with Erlangian repair time. Asia-Pac J Oper Res 21(03):271–277
3. Goel LR, Sharma SC (1989) Stochastic analysis of a 2-unit standby system with two failure modes and slow switch. Microelectron Reliab 29(4):493–498
4. Gupta R, Kumar P, Gupta A (2013) Cost-benefit analysis of a two dissimilar unit cold standby system with Weibull failure and repair laws. Int J Syst Assur Eng Manage 4(4):327–334
5. Kumar A, Malik SC (2012) Reliability modeling of a computer system with priority to s/w replacement over h/w replacement subject to MOT and MRT. Int J Pure Appl Math 80 (5):693–709
6. Kumar A, Saini M (2014) Cost-benefit analysis of a single-unit system with preventive maintenance and Weibull distribution for failure and repair activities. J Appl Math Stat Inform 10(2):5–19
7. Kumar A, Saini M, Devi K (2016) Analysis of a redundant system with priority and Weibull distribution for failure and repair. Cogent Math 3(1):1–11
8. Kishan R, Jain D (2014) Classical and Bayesian analysis of reliability characteristics of a two-unit parallel system with Weibull failure and repair laws. Int J Syst Assur Eng Manage 5 (3):252–261
9. Mahmoud MAW, Moshref ME (2010) On a two-unit cold standby system considering hardware, human error failures and preventive maintenance. Math Comput Model 51(5):736–745
10. Moghaddass R, Zuo MJ, Qu J (2011) Reliability and availability analysis of a repairable-out-of-system with repairmen subject to shut-off rules. IEEE Trans Reliab 60 (3):658–666
11. Wu Q, Wu S (2011) Reliability analysis of two-unit cold standby repairable systems under Poisson shocks. Appl Math Comput 218(1):171–182

Blockage With in Wi-Fi Sensor Networks in Addition to Systems Regarding Controlling Congestion

Konda Hari Krishna, Tapas Kumar, Y. Suresh Babu, R. Madan Mohan, N. Sainath and V. Satyanarayana

Abstract The intriguing characteristics of remote sensor systems, for instance, mindful nature of action to base station that happens through its various to-one topology and effect in physical channel are essential reasons of stop up in remote sensor frameworks. Also when sensor center points imbue material data into framework the block is possible. Blockage impacts the predictable stream of data, loss of information, deferral in the arrival of data to the destination and undesirable use of gigantic measure to a great degree obliged measure of imperativeness in the hubs. Along these lines Clog in remote sensor frameworks ought to be controlled with a particular deciding objective to draw out structure lifetime improve sensibility, high essentialness capability, and upgrade nature of organization. This broadsheet has generally depicted the trademark and the substance of stop up controlling remote sensor establish and surveys the inspection related to the block control traditions for remote sensor systems.

K. Hari Krishna (✉) · T. Kumar
Department of Computer Science & Engineering, Lingaya's University, Faridabad, India
e-mail: kharikrishna396@gmail.com

T. Kumar
e-mail: Kumartapus534@gmail.Com

K. Hari Krishna
Bharat Institute of Engineering and Technology, Hyderabad, India

Y. Suresh Babu
P.G. Department of Computer Science, JKC College, Guntur, India
e-mail: yalavarthi_s@yahoo.com

R. Madan Mohan · N. Sainath · V. Satyanarayana
Department of Computer Science & Engineering,
Bharat Institute of Engineering and Technology, Hyderabad, India
e-mail: rmmnaidu@gmail.com

N. Sainath
e-mail: natukulasainath@gmail.com

V. Satyanarayana
e-mail: satyav@biet.ac.in

H.R. Vishwakarma and S. Akashe (eds.), *Computing and Network Sustainability*, Lecture Notes in Networks and Systems 12, DOI 10.1007/978-981-10-3935-5_3

Keywords Wireless sensor networks · Hub · Congestion control · Conventions

1 Introduction

Wireless frameworks are generally comprised of one or more sink nodes and a large number of sensor center points scattered over a physical space. Using a blend of identified information, the sensors identify physical information, process poor/useless information, and report the required information to the sink node. These sensors are often very small in size and can sense, measure, and collect information from nature in addition to transmitting the required data to the customer through decision making programing. The standard task of a sensor center point is to assemble information from the scene of an event and send the data to a sink node. Figure 1 exhibits the standard remote sensor setup involving various numbers of sensor centers and one sink node to which data is sent from the surrounding environment. Remote frameworks can be used for many applications, e.g., live observations of a particular area, security observations, industrial machine monitoring, less invasive analysis of systems etc. Remote sensor frameworking is cross-disciplinary, requiring a solid understanding of framework communication, and is a front line tool for global research.

2 Bottlenecks in Wireless Sensor Networks

Remote sensor framework requisition requires that readings are alternately recognized and assembled and directed appropriately. Blockages can occur in these frameworks when gathering and sending date to the sink. Blockages happen mostly

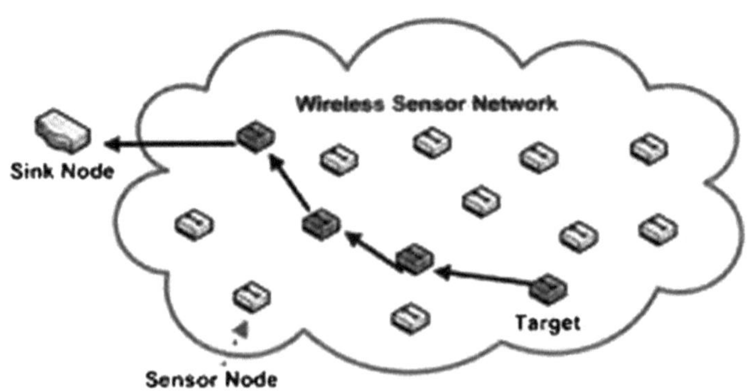

Fig. 1 A wireless sensor network

in the sensor-to-sink. Using a system to monitor blockages can impact upon the framework's execution and objective, i.e., changing data transfer time. The purpose behind wireless sensor network blockage control is to upgrade the sensor framework's throughput, decreasing the deferral of transmitting data. This results in an improved framework efficiency, improved data flow to nodes, and improved interchanges in transmitted information. Upgrading some remote sensor frameworks can be achieved through the assemblies outline.

3 Types of Blockages in Wireless Systems

a. **Hub-level blockage**: In this type of blockage the central level of the system becomes obstructed and is therefore unable to transmit data further—this represents the most common occurrence in standard frameworks. It can be caused by data general damage to sensors.
b. **Transmitter-level blockages**: Over a particular region, data crashes may occur when various dynamic sensor hubs malfunction while attempting to transmit data. This type of blockage diminishes both the use of the system as a whole and general throughput of data (Fig. 2).

4 Review of Congestion Control Protocols in Wireless Sensor Networks

There has been some research into this subject with claims of blockage control. Below are some methodologies.

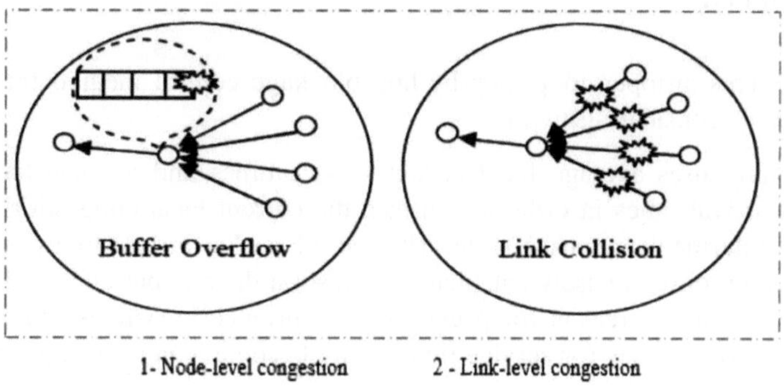

Fig. 2 Congestion types in wireless sensor networks

4.1 Clog Avoidance and Detection

This methodology recognizes obstructions. In this method a sensor center educates its neighbor in order to decrease its transmission rate. Transmitting data means that the sensor becomes detached from its sensing role for certain periods of time. One can anticipate these busy channels over pre-characterized time periods, thus allowing preparation for such blockages.

4.2 Fairness Aware Congestion Control

This methodology uses packs of sensors. At the point where a blockage is experienced, it illuminates those downstream hubs which would be affected and attempts to find an alternative data route to the sink node. Intermediate relaying sensor nodes are re-categorized as near-source nodes and near-sink nodes. Near-source nodes maintain a per-flow state and allocate an approximately fair rate to each passing flow. On the other hand, near-sink nodes do not need to maintain a per-flow state and use a lightweight probabilistic dropping algorithm based on queue occupancy and hit frequency. This provides good performance in terms of throughput, packet loss, energy efficiency, and fairness.

4.3 Versatile Rate Control

Each sensor center appraises the amount of transmission for upstream hubs and sets a limit based on the past. The ensuing transmission ability allocated to each center is in this way harshly controlled.

4.4 SenTCP

This represents an open-loop, hop-by-hop blockage control method for upstream traffic control considering two cases.

(1) It jointly uses average local packet service times and average local packet inter-arrival times in order to estimate the current local congestion degree in each intermediate sensor node. The use of packet arrival times and service times not only precisely calculates congestion degree, but effectively helps to differentiate the reason for packet loss occurrence in wireless environments, since arrival times (or service times) may become small (or large) if congestion occurs.

(2) It uses hop-by-hop congestion control. In SenTCP, each intermediate sensor node will issue a feedback signal backwards and hop-by-hop. The feedback signal, which carries local congestion degree and buffer occupancy ratio, is used for neighboring sensor nodes in order for them to adjust their sending rate in the transport layer. The use of hop-by-hop feedback control can remove congestion quickly and reduces packet dropping, which in turn conserves energy. SenTCP realizes higher throughput and good energy efficiency since it obviously reduces packet dropping; however, SenTCP copes with only congestion and guarantees no reliability.

4.5 Reasonable Awareness Congestion Control

This is an obstruction control system, which controls blockages and accomplishes a sensible transmission limit for each stream. This method distinguishes the obstruction by analyzing the packet drop rate towards the sink center. Hubs are divided under two classes: close to the sink center and end wellspring center in their perspective. When a packet is lost, those end sink hubs send a "warning message" to the closest hotspot center which in turn sends a "control message" to the sourball center. In turn the sourball hubs adjust their sending rate in terms of movement on the channel thus introducing a new sending rate.

4.6 Priority-Based Congestion Control Protocol (PCCP)

This system prevents upstream congestion in a wireless sensor network. The PCCP creates a priority table based on the importance of each node, and then sends this information to all the nodes within the network. The major function of PCCP is to measure congestion levels in the network using the ratio of packet inter-arrival times against packet service time. PCCP is used to control upstream congestion and degree of congestion. PCCP is a hop-by-hop upstream congestion control protocol which works under both single path and multipath routing. PCCP ensures reduced packet loss as well as delay, the result being that each node can avoid unfairness and achieve flexible throughput.

4.7 Stream

Stream's key objective is basic: a spot sensor transmits code metadata even though it need not be listened to by other sensors. However, since several other sensors are transmitting the same thing, this permits a constant trickle of data. There are two possible outcomes with this trickle of data. Possibly each bit that hears those messages will be a beneficiary of an update.

For instance, if a telecasts that it needs code φ, yet b also needs code $\varphi + 1$, then b understands that a prerequisite is an upgrade. Also, if b reveals that it needs $\varphi + 1$, it also understands that it needs an overhaul. If b reveals it requires upgrades, its neighbors could receive them without highlighting their need for it. In addition, a portion of sensors claiming these updates may not even have listened to a's transmission. "Stream" therefore allows exchange of code metadata with its framework neighbors.

4.8 Siphon

Siphon aims to control congestion as well as handle the funneling effect, which is where events generated under various workloads move quickly towards one or more sink nodes. This increases traffic at the sink which leads to packet loss. Virtual sinks are randomly distributed across the sensor network which takes the traffic load off the already loaded sensor node. In siphon what happens is that there is an initial virtual sink discovery undertaken. After congestion detection traffic is redirected from the overloaded physical sink to virtual sinks. It is done by setting a redirection bit in the network layer header.

4.9 Prioritized Heterogeneous Traffic-Oriented Congestion Control Protocol (PHTCCP)

This performs hop-by-hop rate adjustment controlling congestion and ensures an efficient rate for prioritized diverse traffic. PHTCCP is an efficient congestion control protocol for handling diverse data with different priorities within a single node. The PHTCCP module works by interacting with the MAC layer to perform congestion control functions. In this protocol, congestion can be controlled by ensuring there are adjustment transmission rates for the different types of data that have been generated by the sensors with various priorities. The sink node assigns an individual priority for each type of sensed data and each node has n number of equally sized priority queues for n types of sensed data. Heterogeneous applications can reflect the number of queues in a node. In this congestion detection method, congestion levels at each sensor node are presented by packet service ratio.

4.10 Learning Automata-Based Congestion Avoidance Algorithm in Sensor Networks (LACAS)

In LACAS the problem of congestion control in sensor nodes is dealt with utilizing an adaptive approach based on learning automata. This protocol causes the rate of processing (rate of entry of data) in nodes to be equivalent to the rate of transmission in them, so that the occurrence of congestion gradually decreases. An automaton is placed in each node which has the ability to learn. In fact it can be considered as a small piece of code that interacts with its environment and makes decisions based on the characteristics it finds.

5 Conclusion

Generally there is a need to create an interest in remote sensor networks. The impact from claiming remote sensor frameworks around our commonplace life might make ideally contrasted for the thing that web need carried out with us. Both those variables of obstruct control Furthermore reliability aides over lessening package misfortune, which acquires something like a vitality profitable operation of the system, which will be a magic variable for stretching those lifetime of the sensor framework. In turn part on be taken under record by those vehicle assemblies will be those compelled possessions of the center gadgets. Despite the way that these obstruct control frameworks would guaranteeing there are at present there are various challenges should get it to remote sensor framework on handle blockage control proficiently. What's more, All the more Look into endeavors would required to continue upgrading obstruct control on remote sensor networks.

References

1. Akyildiz F, Su W, Sankarasubramaniam Y, Cayirci E (2002) Wireless sensor network: a survey. Comput Netw 38(4):393–422
2. Congestion control and fairness for many- to-one routing in sensor networks. In: Proceedings of 2nd international conference on embedded networked sensor systems, pp 148–161
3. Congestion control and fairness for many-to-one routing in sensor networks. In: Proceeding ACM sensys'04, 2004
4. Congestion detection and avoidance in sensor networks. In: Proceedings of the first international conference on embedded networked sensor systems (SenSys'03), Los Angeles, CA, USA, 2003, pp 266–279
5. Fang W, Chen J, Shu L, Chu T, Qian D (2009) Congestion avoidance, detection and alleviation in wireless sensor networks. J Univ Sci 11:63–73
6. Heikalabad SR, Ghaffari A, Hadian MA, Rasouli H (2011) DPCC dynamic predictive congestion control in wireless sensor networks. Int J Comput Sci (IJCSI) 8(1):1694–0814
7. Hull B, Jamieson K, Balakrishnan H (2004) Mitigating congestion in wireless sensor networks. In: 2nd international conference on embedded networked sensor systems, Maryland

8. Jones J, Atiquzzaman M. Transport protocols for wireless sensor networks: state-of-the-art and future directions
9. Pang Q, Wong VWS, Leung VCM (2008) Reliable data transport and congestion control in wireless sensor networks. Int J Sens Netw 3(1):16–24
10. Trickle: a selfregulating logarithm for code propagation and maintenance in wireless sensor networks. In: Proceedings of first symposium networked system design and implementation (NSDI)
11. Wang C, Li B, Sohraby K, Daneshmand M, Hu Y (2007) Upstream congestion control in wireless sensor networks through cross-layer optimization. IEEE J Sel Areas Commun 25 (4):786–795
12. Wang C, Sohraby K, Li B (2005) SenTCP: a hop-by-hop congestion control protocol for wireless sensor networks. In: Proceedings of IEEE INFOCOM 2005 (Poster paper), Miami, Florida, USA
13. Wang C, Sohraby K, Lawrence V, Li B (2006) Priority based congestion control in wireless sensor networks. In: IEEE international conference on sensor networks, ubiquitous and trustworthy computing, Taiwan, pp 22–31
14. Woo A, Culle D (2001) A transmission control scheme for media access in sensor networks. In: Seventh annual international conference on mobile computing and networking, pp 221–235
15. Yin X, Zhou X, Li Z, Li S (2009) A novel congestion control scheme in wireless sensor networks. In: 5th international conference on mobile ad-hoc and sensor networks, Fujian, pp 381–387

Author Biographies

Konda Hari krishna received his **M.TECH.** in **computer science** from **Jawaharlal Nehru Technological University,** Kakinada & A.P. and pursuing **Ph.D. in LINGAYA's University, Faridabad.** He is working as an **Assistant Professor** in **Bharat Institute of Engineering & Technology** in **Dept. of Computer Science & Engineering**. He published **18 Research Papers** in Various **International Journals** of **Reputed** and His **Research Area** is **Mining of applications in Wireless Sensor Networks**. He is a **good researcher** & who has worked mostly on **Wireless Sensor networks**, **Ad hoc Networks, Network security** and **Data mining.**

Dr. Tapas Kumar Working as a **Professor, Dean & H.O.D in School of Computer Science & Engineering, Lingaya's University, Faridabad.** He holds a **Doctorate in Computer Science & Engineering.** He has more than experience of **15 years in Academics & Administration.** He has **published various Research papers in various National & International Journals of Reputed.**

Dr. Y. Suresh Babu Working as a **Professor** in **Dept of Computer Science, JKC COLLEGE, GUNTUR.** He holds a **Doctorate in Computer Science & Engg, Image processing as specialization with a combined experience of 23 years in Academics & Administration.** He has **published nearly 45 research papers in various National and International Journals of reputed.**

R. Madan Mohan Associate Professor at **Bharat Institute of Engineering & Technology.** He has more than experience of **10 years in Academics & Administration.** He has **published various Research papers in various National & International Journals of Reputed.**

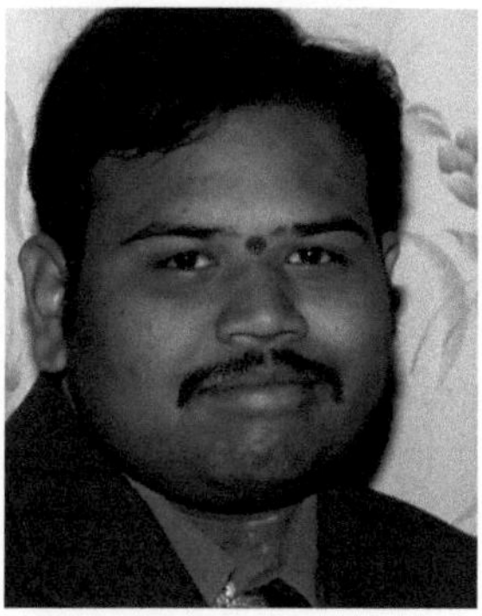

N. Sainath B.Tech. from JayaPrakash Narayana College of Engineering & **M.Tech.** SE from Srinidhi Institute of Technology, Pursuing **Ph.D.** from **JNTU Hyderabad** currently he is working as Associate Professor at **Bharat Institute of Engineering & Technology.** His **areas of interest** include Data mining, Network Security, Software Engineering, Sensor Networks, and Cloud Computing. He is enrolled for the memberships of IEEE, CSI, and ISTE. He has Published 17 papers in International Journals and has 11 International conference Proceedings and attended 15 workshops and 10 National conferences.

V. Satyanarayana Associate Professor at Bharat Institute of Engineering & Technology. He has more than experience of 10 years in Academics & Administration. He has published various Research papers in various National & International Journals of Reputed.

Privacy Preserving Using Video Encryption Technique—the Hybrid Approach

Karishma Chaudhary and Gayatri Pandi (Jain)

Abstract In this epoch, Web and network applications are new quick. With the rapid development of various multimedia technologies, thousands of multimedia expertise are created and transmitted within the administration places of work (CID, FBI), study organization, E-trade, and navy fields, so the necessities to ensure such purposes are expanded. Video encryption calculations have turned into a vital discipline of study nowadays. As the cost of making use of video is getting increased, the security of video expertise seems to be extra essential. Hence, there is an incredible interest for storing and transmission methods for secured information. In the course of the most recent couple of years, numerous encryption algorithms have offered to secure constant video transmission, while countless encryption techniques have been proposed and some have been utilized as a part of real time. In this work, we introduce another relative study between IDEA and RSA encryption algorithms.

Keywords Video encryption · IDEA · RSA · Public key · Private key · Hybrid

1 Introduction

The World Wide Web is a vast collection of different forms of data. Those data are divided into so many different forms of type. There are mainly divided into four kinds of data: text, audio, video, and image. Nowadays, Web uses most of multimedia messages such as image message, and audio and video forms of messages.

As the utilization of the Web is turning out to be broadly acknowledged nowadays, it is a pattern of transmitting information or data over the Web. Be that

K. Chaudhary (✉) · G. Pandi (Jain)
L.J. Institute of Engineering and Technology, Ahmedabad, Gujarat, India
e-mail: krizchaudhary@gmail.com

G. Pandi (Jain)
e-mail: priyal.kartik@gmail.com

H.R. Vishwakarma and S. Akashe (eds.), *Computing and Network Sustainability*,
Lecture Notes in Networks and Systems 12, DOI 10.1007/978-981-10-3935-5_4

as it may, individuals have found strategies to break the security of the data, which ought to be sent over the Internet. Safety is turning into a rising trepidation in a revolutionary multimedia outlined world. The later rise of embedded interactive media applications, for example, versatile television, video informing, and telemedicine, has expanded the impact of interactive media and its security of our individual lives. For instance, a tremendous expansion within the software of distributed video surveillance science to monitor site visitors and public locations has raised distresses concerning the privateness and safety of the targeted subjects [1].

2 Cryptography Information

Cryptography is the method by which encryption and decryption of data is carried out using a secret key such that it cannot be detected. Cryptography creates figure message, which may incite the intruder [2]. Cryptography is an art of changing the information into an illegible and untraceable format which is known as cipher textual content. Only the person who owns the secret key can decipher it or decode the original message into the unique form. Cryptography supplies the means by which one can send and share the data or information in an ease and secret means. For the period of the process of cryptography, the data by all accounts seem like garbage value to an individual and it becomes practically tough to find the data content which actually exists under the picture or a video file [3]. Cryptography is one of the security procedures which make the data/information transfer secure and protected (Fig. 1).

3 Video Encryption Algorithm Approach

IDEA (International Data Encryption Technique) is used to transform 64-bit plain textual content to 64-bit cipher text utilizing a single key of 128 bits for encryption and decryption. RSA is used an asymmetric key to ciphering and deciphering data.

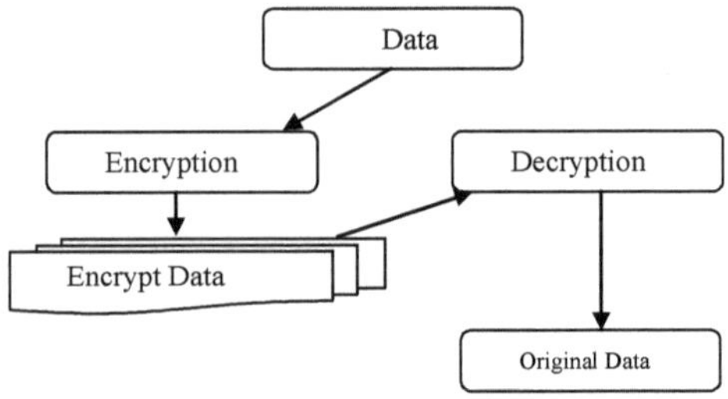

Fig. 1 Cryptography cycle

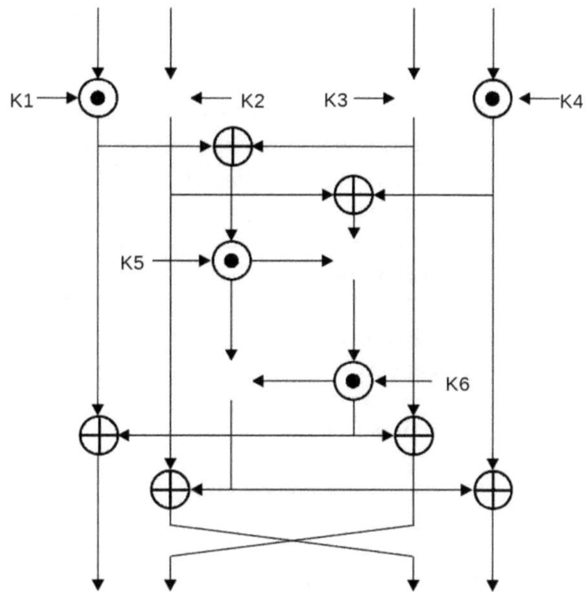

Fig. 2 Encryption round of idea using six intermediate keys [4]

1. **IDEA (International Data Encryption Technique)**

IDEA which stands for "The International Data Encryption Algorithm" is a block cipher designed in 1991 by James Massey and Xuejia Lai. It was once supposed to be a substitute for the data encryption standard (Fig. 2).

IDEA works on 64-bit blocks, making use of a 128-bit key. It includes an advancement of eight indistinct transformations (rounds) and one yield transformation (the half-round). It does this for a total of 8.5 rounds. The strategies for encryption and decryption are comparative.

IDEA infers fairly a little bit of its protection by means of interleaving operations from more than a few gathering modular addition and multiplication, and bitwise exclusive OR (XOR) which are selected to be "algebraically incompatible" [4].

Each and every of the eight round utilizations 6 sub-keys, whilst the half round utilizations four; for a sum of fifty two sub-keys. Each sub-key has a length of 16 bits. The initial 8 sub-keys are extricated chiefly from the 128-bit key, with K1 being probably the most minimal sixteen bits and K8 is the highest sixteen bits; a different group of eight keys is made by way of pivoting the predominant key left twenty-five bits after the generation of the previous group; six rotations create all sub-keys.

2. **RSA Algorithm**

The algorithm named "RSA" is normally used by nowadays computers to scramble and decode messages. It is an algorithm which acts as non-symmetric cryptographic method of protecting data. Non-symmetric means that there are 2 exceptional keys. Additionally known as open key cryptography, because one of all of them can be given to each individual. The second key need to be stored secretly. It relies on the actuality that searching the factor of an integer is hard [5].

INPUT: Needed modulus bit length, k.

OUTPUT: ((N, e), d) is an RSA key pair, wherein N is the modulus, N = pq should not exceed k bits in size (p and q are prime); e stands for public exponent, a number lower than and coprime to $(p-1)(q-1)$; and d is known as private exponent wherein $ed \equiv 1 \pmod{(p-1)(q-1)}$.

Pseudo code [4]

1) Choose a value of e from (3, 5, 17, 257, 65537)
2) **repeat**
3) p ← genprime(k/2)
4) **unless** (p mod e) ≠ 1
5) **repeat**
6) q ← genprime(k - k/2)
7) **unless** (q mod e) ≠ 1
8) N ← pq
9) L ← (p-1)(q-1)
10) d ← modinv(e, L)
11) **return** (N, e, d)

4 Proposed Methodology for Video Encryption

There are basic four phases of this proposed approach. In the first phase, data are encrypted using IDEA algorithm after received decrypted data used as second-phase input data and once again data are encrypted using RSA there after got receive hybrid encrypted data. Then, in third phase decrypted by RSA and after receiving decrypted data which is used as four-phase input and finally we decrypt this data using IDEA and got our original data.

First Phase: Data Encryption by way of IDEA

T raw data (input data)
k 128-bit encrypted key
C cipher data

$$\mathbf{C = T + k\ (128\,bit\,key)}$$

Here, basic IDEA encryption of 64-bit plain text block is split into 16-bit intermediate blocks: x1, x2, x3, x4. Then, we need to add 128-bit key split with 16-bit block, which becomes z1, z2, z3, z4, z5, z6, z7, and z8. Here, first 6 intermediate keys are used in first attempt and remaining are used in second attempt.

Approach to the encryption of IDEA [4, 6]

Step 1 w = x1 * z1
Step 2 z2 = w + x2
Step 3 z3 = z2 + x3
Step 4 z4 = z3 * x4
Step 5 add Step 1 and Step 3 using XOR

$$w1 = y \oplus z3$$

Step 6 add Step 2 and Step 4 using XOR

$$w2 = z2 \oplus z4$$

Step 7 w3 = w1 * z5
Step 8 w4 = w2 + w3
Step 9 w5 = w4 * z6
Step 10 Sum steps 7 and 9

$$w6 = w3 + w5$$

Step 11 Bit-wise operation

$$w7 = w + w5$$

Step 12 w8 = z3 + y5
Step 13 w9 = z2 + y6
Step 14 w10 = z4 + y6

Here, w, w1, w2, …, w10 is assumed variable which is used to store operation value for further reference for data. After this operation, we got final encrypted data. We can say it is **C** (**cipher data**) (Fig. 3).

Second Phase: Data Encryption by way of RSA [5, 6]

Received cipher data used as input and encrypt data once again.

Key generation for RSA

Step 1 Select 2 different prime numbers randomly, "p" and "q", such that both are of same length.
Step 2 Use the formula n = p * q to calculate "n". This "n" serves as the modulus for both private and public keys.
Step 3 Calculate φ(n) = (p − 1) (q − 1). φ(n) is known as the Euler's totient function, i.e., positive integer lower than or equal to are prime to "n".
Step 4 Select an integer "e" such that 1 < e < φ(n). And, gcd (e, φ(n)) = 1. Here, "e" and "φ(n)" are primes. This "e" will be act as PUBLIC KEY EXPONENT.

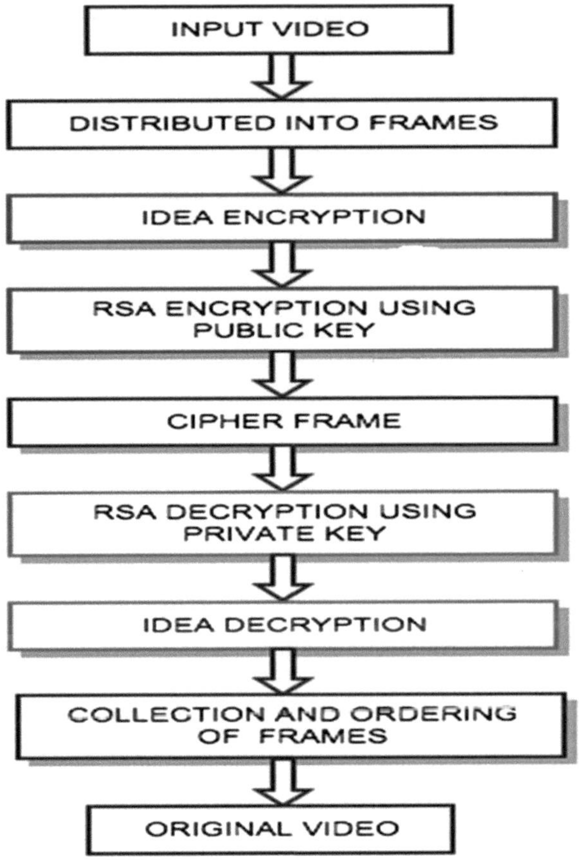

Fig. 3 Proposed system flowchart

Step 5 Calculate multiplicative inverse of e.

$$d - 1 = e(\bmod \varphi(n))$$

This "d" will be act as PRIVATE KEY EXPONENT.
After this, we concluded that

Public Key = n + e
Private Key = n + d

Then, the final encrypted data C1 (cipher data) are

$$\mathbf{C1 = C + k1\ (512\ bit\ RSA\ key)}$$

Third Phase: Decryption the way of RSA

Encrypted data C1 are now secure with hybrid approach, but now it is time to decrypt data. Using RSA, data are decrypted as follows:

$$\mathbf{C = C1 - k1\,(512\,bit\,RSA\,key)}$$

Fourth Phase: Decryption the way of IDEA

Encrypted data C1 are now decrypted in the third phase, and now we have C cipher data which is decrypted by IDEA. After performing this, we will be getting our original data.

$$\mathbf{C = T - k\,(128\,bit\,key)}$$

T original data
C cipher data
k 128-bit IDEA key.

5 Pros and Cons of Proposed Methodology

The key advantage of this proposed system is developing a cryptostructure, which takes out the brittle key idea, which presented another bit of key length in this system. Additionally, this gives enhancement in the privacy of IDEA.

The bottleneck of this system includes a huge execution of encryption processing, and hence there arise a need of additional actions and time.

6 Simulation Study

Here, we have taken an original video image (Figs. 4, 5, and 6).

Original video is converted into number of image frames which is encrypted using RSA and IDEA. After encryption did the same process of decryption to

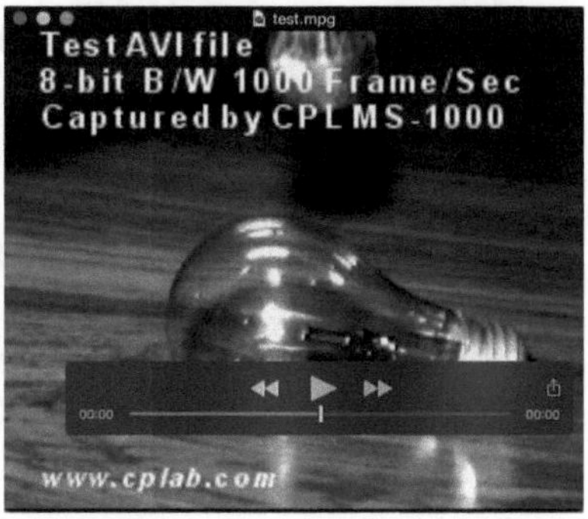

Fig. 4 Original video

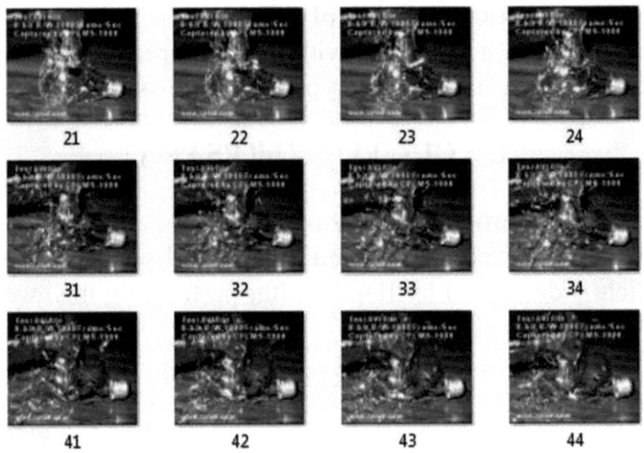

Fig. 5 Video converted into frames

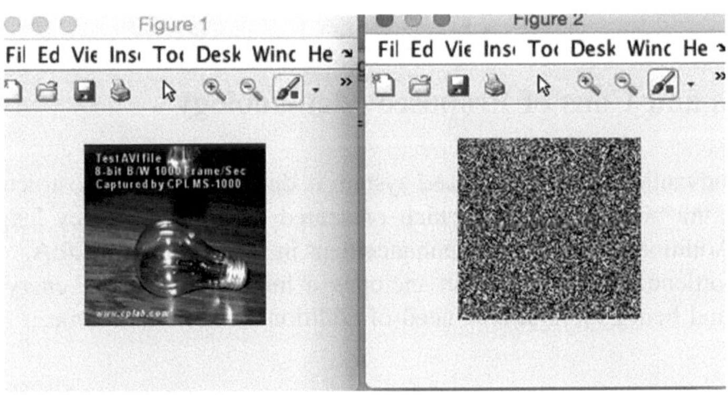

Fig. 6 Encryption using IDEA + RSA

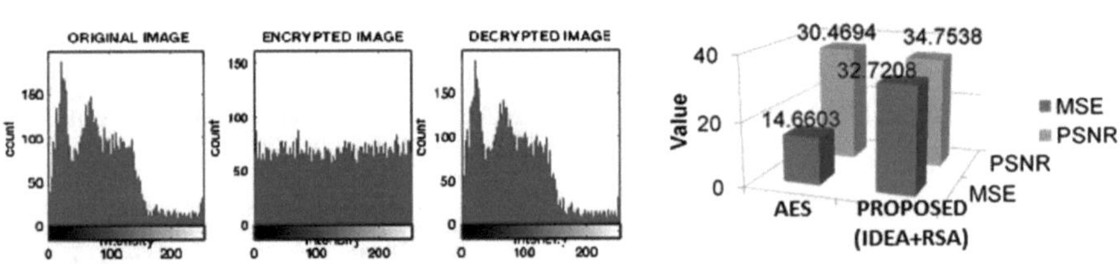

Fig. 7 Histogram and result analysis

get the original image. On the table, the encrypted and decrypted data results are given.

Fig. 7 shows that when we encrypt original image and we can see the changes in the original image, encrypted image histogram graph changes. Finally, when we decrypt the image, we can see that the graph, which would be the same like the original image. Graph shows the comparison of existing system and proposed system.

7 Conclusion

Cryptography is an important field of study in today's world, whether in our daily lives or in national security; here, we were learning about different encryption algorithms and techniques to secure the video transmission. IDEA had weak key categories that hindered in the perfect safety of the information; there was a necessity of this sort of cipher, which does not has vulnerable keys. Proposed procedure created it viable to fade away the vulnerable keys from the cipher consequently improved the data protection. It uses RSA algorithm furthermore to the suggestion cipher making an improvement in it. Now the keys are not able to be detected effectively. So, knowledge healing is not possible. The addition of RSA cipher has incorporated the thought of two special keys each and every for encryption and decryption, which was once just one key in IDEA.

References

1. Dumbere DM, Janwe NJ (2014) Video encryption using AES algorithm. In: Current trends in engineering and technology, ICCTET' 2014, pp 332–337
2. Gesu S, Vasudeva S, Bhatt S, Santhosh B (2015) A novel technique for secure data transmission using audio/video files. Int J Eng Res Technol 912–916
3. Viral GM, Jain DK, Ravin S (2014) A real time approach for secure text transmission using video cryptography. In: 2014 fourth international conference on communication systems and network technologies, Bhopal. IEEE, pp 635–638. Print ISBN:978-1-4799-3069-2, INSPEC Accession Number: 14348660. doi:10.1109/CSNT.2014.133
4. Chang H-S (2004) International data encryption algorithm. http://www.googleusercontent.com, http://www.users.cs.jmu.edu
5. Rivest RL, Shamir A, Adleman L. A method for obtaining digital signatures and public-key cryptosystems, vol 02139. Laboratory for Computer Science, Massachusetts Institute of Technology, Cambridge, MA, pp 1–15
6. Bhatnagar A, Pangaria M, Shrivastava V (2013) Enhancement of security in international data encryption algorithm (idea) by increasing its key length. Int J Adv Res Comput Commun Eng 2 (10):3869–3871
7. Ren YJ, O'Gorman L, Chang F, Wood TL, Zhang JR (2013) Authenticating lossy surveillance video. In: IEEE Trans Inf Forensics Secur 1678–1687. Date of publication: 23 August 2013, Date of current version: 10 September 2013, Issue date: Oct. 2013, Sponsored by: IEEE Signal Processing Society, ISSN: 1556-6013, INSPEC Accession Number: 13747543. doi:10.1109/TIFS.2013.2279542
8. Cao X, Liu N, Du L, Li C (2014) Preserving privacy for video surveillance via visual cryptography. In: Signal and information processing (ChinaSIP), 2014 IEEE China summit and international conference, Xi'an. Print ISBN: 978-1-4799-5401-8, INSPEC Accession Number: 14563468, Date of conference: 9–13 July 2014, pp 607–61. doi:10.1109/ChinaSIP.2014.6889315

Performance Analysis of Improved AODV Routing Protocol in Mobile Ad hoc Network

Sanjeev Kumar Srivastava, Ranjana D. Raut and P.T. Karule

Abstract In this paper, it has been discussed improved AODV routing protocol for mobile ad hoc network. Later, we compared performance of AODV, improved AODV (AODV+), and DSR routing protocols in mobile and non-mobile scenarios. From thorough simulation analysis, we conclude that our improved AODV protocol can improve the average throughput by an average of 30% and energy consumption by an average of 16% as compared to existing AODV protocol.

Keywords AODV · AODV+ · DSR · Performance analysis · Ad hoc networks · Average throughput · Average energy consumption

1 Introduction

Ad hoc wireless networks communicate along with another node that is immediately within or outside their radio ranges without any fixed infrastructure or base station. Its characteristics contain various types of controls of the network which are distributed among the node.

The main issue is finding the route of packet from source to destination [1]. Route can be either static table without much change in the networks, or dynamic one with updating automatically in the network.

Therefore, the research contribution of this work has been surveyed of existing work in detailed way and has studied about their proposals of limitations which do not exist in current literature survey as per best of knowledge. It has done extensive

S.K. Srivastava (✉)
PIIT, Mumbai, India
e-mail: sanjeevkumar.srivastava1@gmail.com; ssrivastav@mes.ac.in; sanju_iway97@rediffmail.com

R.D. Raut
S.G.B. University, Amravati, India

P.T. Karule
Yashwantrao Chavan College of Engineering, Nagpur, India

H.R. Vishwakarma and S. Akashe (eds.), *Computing and Network Sustainability*, Lecture Notes in Networks and Systems 12, DOI 10.1007/978-981-10-3935-5_5

simulation survey to compare existing routing protocols for detailed study [2]. It has also checked on proposed idea with detailed study of routing protocols to see the presence in different conditions. It has done the performance analysis of proposed idea with mobility and without mobility. It has proposed a modified AODV (i.e., AODV+) routing protocol to improve the performance of average throughput and average energy consumption by verifying on various parameters in ad hoc wireless networks. It has compared the large scalability till nodes of 100 which has been addressed in very few researcher articles to the best of knowledge.

2 Proposed Idea Description

An idea has been proposed about the general way to design routing protocols in the existing wireless network [1]. To do so, three basic steps have been discussed which are route discovery, route path formation, and route maintenance. In route discovery procedures, when the source node sends route information to some destination node without any valid route, it initiates this process in order to find the location of other node. It broadcasts a route request (RREQ) packet to its neighbors, which then forward the request to their neighbors, and so on, until either the destination or an intermediate node with a new route to the destination is located as shown in Fig. 1.

In the route path formation, it consists of route request (RREQ) in the form of Hello messages and route reply (REPLY). Figure 2 shows the route formation of request messages. Hello messages contain the information related to SCR (Sourcc) ID, DEST (Destination) ID, and packet ID, etc. for broadcast query. SCR ID is used for packet forwarding, and DEST ID is used for router identification. Type indicates about the packet priority of current service. Hop count shows the address/identifiers field of the query packet in order to broadcast to its neighbors in header port.

This field provides the services of packet acknowledgment, routing, etc. Route link quality provides the information about the quality of packets transmitted to its neighbors in header port (i.e., average, good, and excellent). Sequence number prevents duplicate packets and to provide unique address of packets in header port. CRC (cyclic redundancy code) is used to detect the error in the incoming packets to

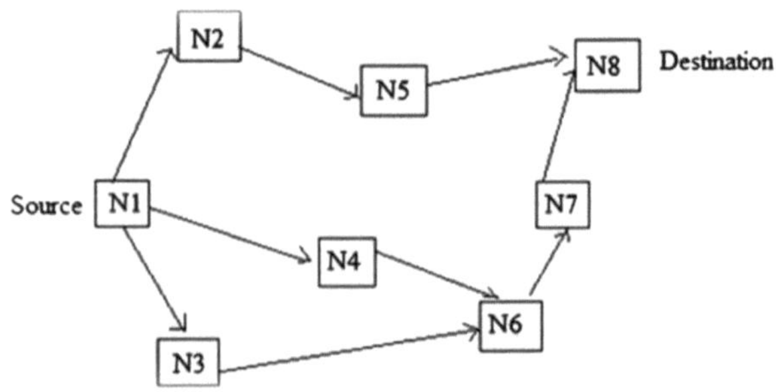

Fig. 1 Route discovery phase

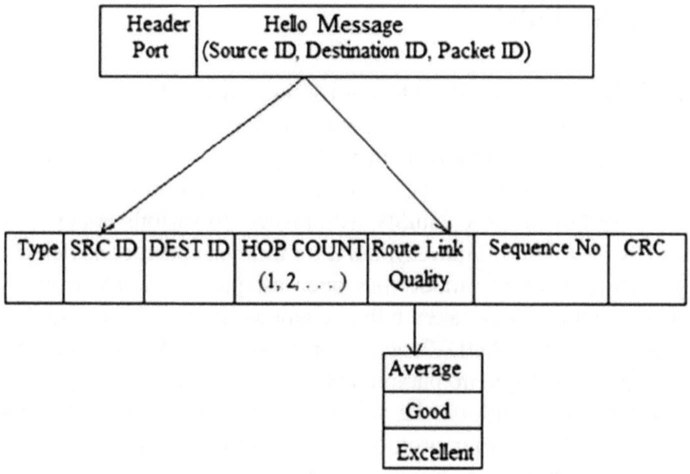

Fig. 2 Route request (RREQ) message

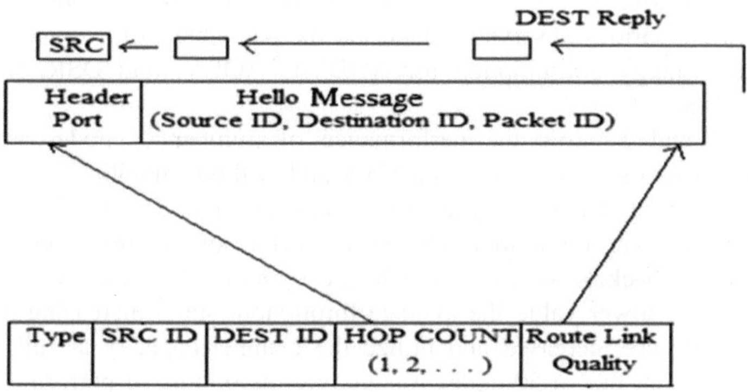

Fig. 3 Route reply (RREP) message

the header. After receiving the first broadcast query, DEST knows all the possible routes and their qualities (i.e., average, good, and excellent) as shown in Fig. 3.

Hence, it marks the valid route to DEST. It means that all other possible route is inactive and will not relay packet destined for DEST, even if they hear transmission. This therefore avoids duplicated packets from arriving at DEST. In the route maintenance, if a source node finds any error in that route, it has to reinitiate the route discovery protocol to find a new route to the destination.

To improve the network performance, it has been made modification in existing AODV protocol. When protocol is making the routing table, it is suggested the link state in consideration. Therefore, it is made sure that routing table is created based on link state so it can get a better quality path for packet forwarding. This assumption is that link quality will not change on frequent basis. This idea has been implemented in this simulation experiment and compared with existing protocols.

3 Simulation Model

To improve the performance of AODV+ routing protocol, there will be change in routing scheme by considering link between different nodes (i.e., 100 nodes). Nodes are assumed to be random with mobility and without mobility. By considering area of $500 \times 500\ m^2$, 4–5 sources are taken into account. It is very important to contribute high scalability with respect to various nodes. Traffic model is taken as CBR (constant bit rate). Simulation time has been considered as 150 and 250 s, respectively. Mobility model has been assumed as random way point, and MAC layer protocol has been taken into account as IEEE 802.11 with 2.4 GHz. For enhanced performance of AODV routing protocol, there has been a change in the simulation results on different parameters, i.e., average throughput and average energy consumption. Therefore, the performance of existing AODV routing protocol along with DSR routing protocol and modified AODV (AODV+) routing protocol simulation results has been compared.

This section is about simulating and discussing the performance of existing AODV, DSR, and AODV+ (modified) on different parameters, i.e., average throughput and average energy consumption with and without mobility.

It shows the various observation tables at the parameters of average throughput and average energy consumption for AODV+, AODV, and DSR, respectively (Tables 1, 2, and 3).

Figures 4 and 5 show the performance of number of nodes vs. average throughput in two scenarios, with mobility and without mobility, respectively. It has been observed that initially at lower number of nodes, DSR, AODV, and AODV+ show lower throughput value as network resources are underutilized, i.e., less number of packets are generated in the systems. As numbers of nodes are increased above lower value the average throughput starts increasing as network resources/traffic are increased and utilize the channel capacity at full. From the graph, it is evident that at 50 nodes it is getting the optimized performance and as number of nodes is increasing performance starts degrading because collision and competition to get channel access becomes very severe. From the graph, it can be observed that AODV+ is giving better results in mobility scenario as frequently updating routing table with link state is easy compared to traditional DSR and AODV protocols. However, in non-mobility scenario, DSR and AODV protocols

Table 1 Observation table of AODV+ with mobility and non-mobility

Nodes	Mobility		Non-mobility	
	Average throughput	Average energy consumption	Average throughput	Average energy consumption
25	11335.7	0.244328	131475	0.285821
50	20374.6	0.502504	254881	0.492444
75	20050.1	0.00107704	40902.3	0.00582103
100	21092.7	0.00464658	25263.2	0.00381687

Table 2 Observation table of AODV with mobility and non-mobility

Nodes	Mobility		Non-mobility	
	Average throughput	Average energy consumption	Average throughput	Average energy consumption
25	11564.8	0.95285	184090	1.18223
50	26894.1	1.82229	338093	1.94491
75	30686.9	0.0325869	33715.4	0.0206947
100	36185.1	0.0613873	32288	0.0257826

Table 3 Observation table of DSR with mobility and non-mobility

Nodes	Mobility		Non-mobility	
	Average throughput	Average energy consumption	Average throughput	Average energy consumption
25	18366.9	1.50317	184593	1.26221
50	34303.1	2.63418	337091	2.45817
75	44440.5	0.0748171	35771.4	0.0249182
100	36208.1	0.0317686	31485.7	0.0313606

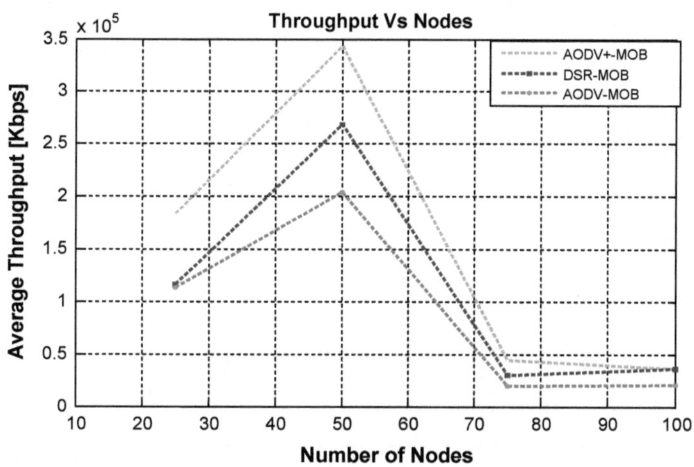

Fig. 4 Average throughput with mobility

show better performance compared to AODV+ till medium number of nodes because stability of their routing table is better. As nodes are not mobile, their routing table is better and static as compared AODV+.

By referring the observation table of AODV+, AODV, and DSR with mobility and without mobility and their respective graphs for average energy consumption as

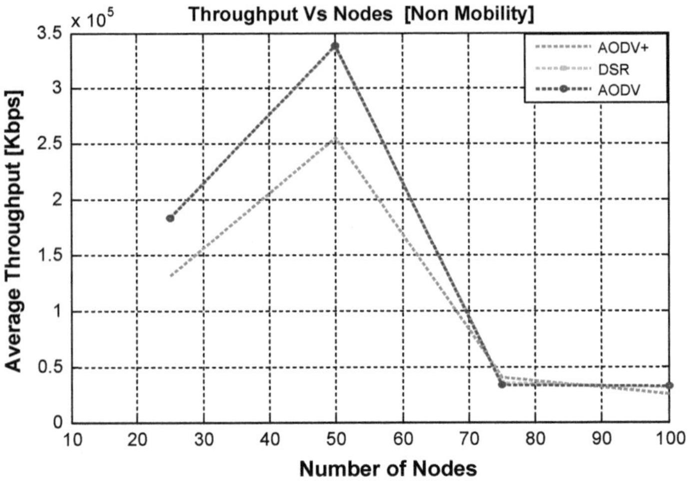

Fig. 5 Average throughput without mobility

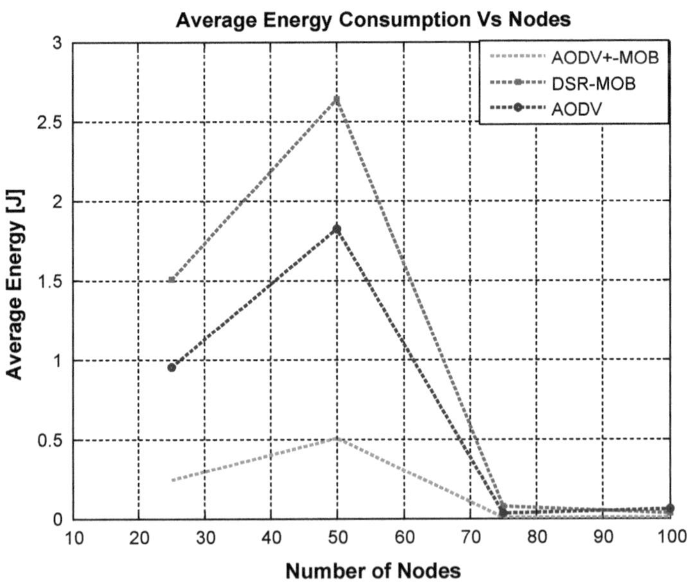

Fig. 6 Average energy consumption with mobility

shown in Figs. 6 and 7, it is observed that as the nodes are increasing, the performance of AODV+ routing protocol, DSR routing protocol, and AODV routing protocol results is almost consistent with this analysis. But overall performance of

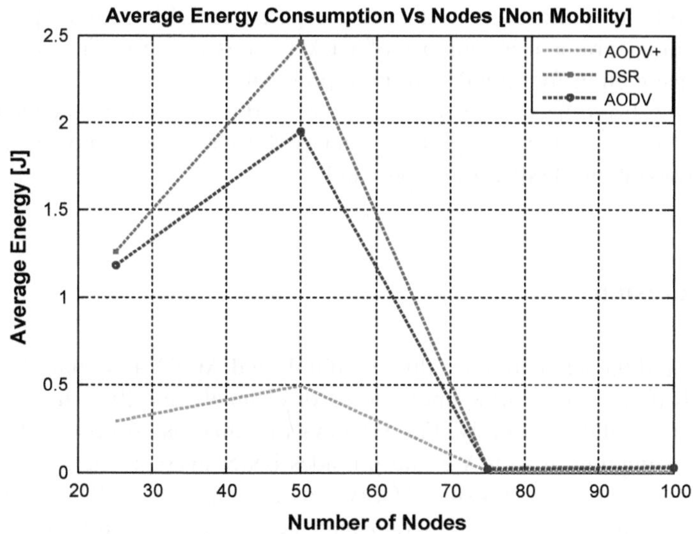

Fig. 7 Average energy consumption without mobility

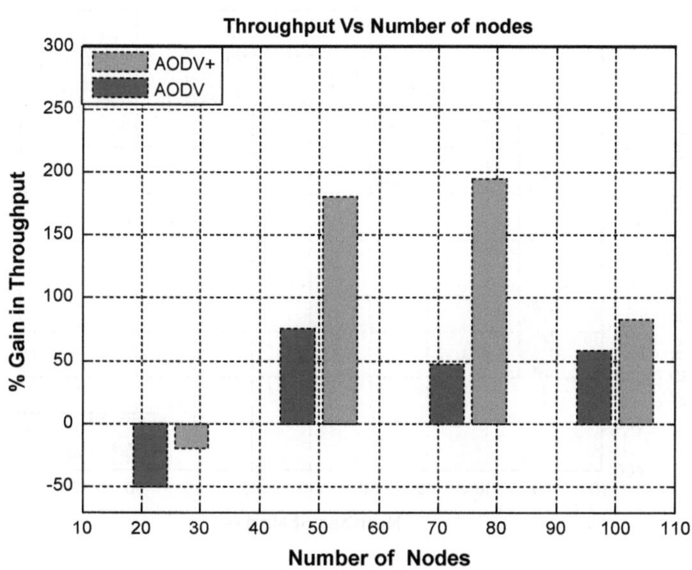

Fig. 8 Comparison of AODV and AODV+ of average throughput

AODV+ routing protocol is better as compared to DSR routing protocol and AODV routing protocol as number of packet drops are less. Therefore, average energy consumption for AODV+ routing protocol is less.

Therefore, from the overall observation of various graphs, it is concluded that AODV+ routing protocol is giving better performance as compared to AODV routing protocol and DSR routing protocol.

4 Conclusion

It is compared the simulation results for AODV and AODV+. Figure 8 shows that there is 30 times better throughput gained of channel capacity using AODV+ as compared to AODV. Similarly, Fig. 9 shows that there is 16 times better average energy consumption in AODV+ and provides better average energy consumption gained of channel capacity using AODV+ as compared to AODV. In future, we would like to include few more parameters such as QoS (Quality of Service) and priority to different load and data traffic.

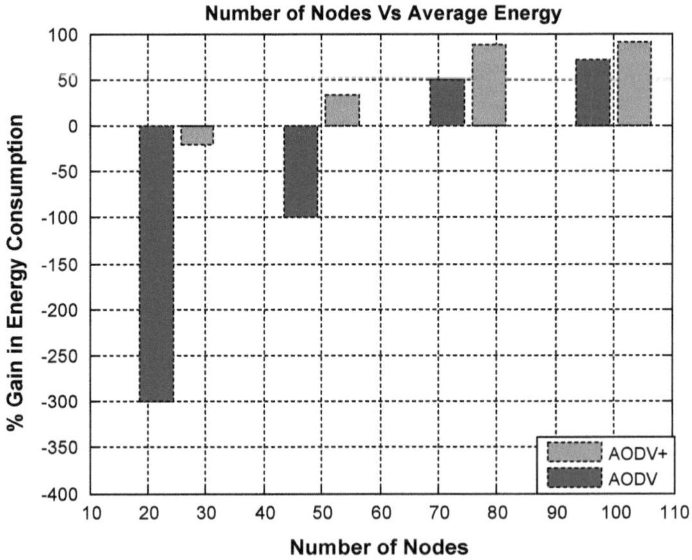

Fig. 9 Comparison of AODV and AODV+ of average energy consumption

References

1. Toh C-K (2013) Ad hoc mobile wireless networks: protocols and systems. Pearson Education
2. Li J, Blake C, De Couto DSJ, Lee HI, Morris R (2002) Capacity of ad hoc wireless networks. IEEE
3. Dhokal D, Gautam K (2013) Performance comparison of AODV and DSR routing protocols in mobile ad-hoc networks: a survey. Int J Eng Sci Innov Technol (IJESIT) 2(3):258–265
4. Aarti, Tyagi SS (2013) Study of MANET: characteristics, challenges, applications and security attacks. IJARCSSE (Int J Adv Res Comput Sci Softw Eng 3(5):252–257
5. Bhimla S, Yadav N (2012) Comparison between AODV protocol and DSR protocol in MANET. Int J Adv Eng Res Stud (IJAERS)
6. Amnai M, Fakhri Y, Abouchabaka J (2011) Impact of mobility on delay-throughput performance in multi-service mobile ad-hoc networks. Int J Commun Netw Syst Serv 4:395–402
7. Srivastava SK, Raut RD, Karule PT (2015) Evaluation of performance comparison of DSR and AODV routing protocols in mobile ad hoc networks. Int J Electron Commun Soft Comput Sci Eng (IJECSCSE). In: 46th midterm IETE symposium, vol 50, pp 258–261. doi:10.1109/ICCN
8. Jha RK, Kharga P (2015) A comparative performance analysis of routing protocols in MANET using NS-3 simulator. Int J Comput Netw Inf Secur 4:62–68
9. Srivastava SK, Raut RD. Study and analysis of ad-hoc wireless network system to maximize the performance of throughput capacity, Thesis technical report, Nagpur University (For internal reference only)

Slots Loaded Multilayered Circular Patch Antenna for Wi-Fi/WLAN Applications

Brijesh Mishra, Vivek Singh, Ajay Kumar Dwivedi, Akhilesh Kumar Pandey, Azeem Sarwar and Rajeev Singh

Abstract In the present work, a circular dual-band patch antenna structure with four rectangular slots is proposed and fabricated for wireless applications. The performance of patch antenna has been analyzed in terms of antenna parameters return loss, VSWR, radiation pattern, and gain. A maximum gain of 3.93 dB at lower resonant frequency of 3.71 GHz and a gain of 3.09 dB at upper resonant frequency of 5.51 GHz are achieved. E-field beam width of 230° and 205° and H-field beam width of 250° and 165° are observed for lower and upper resonance frequencies at −3 dB. The experimental results of return loss and VSWR of the proposed structure are presented along with the simulated results.

Keywords Microstrip patch antenna · HFSS · Radiation pattern · Beam width · VSWR

B. Mishra · V. Singh · A.K. Pandey · A. Sarwar · R. Singh (✉)
Department of Electronics and Communication, University of Allahabad, Allahabad, UP, India
e-mail: rsingh68@allduniv.ac.in

B. Mishra
e-mail: brijesh.mishra0933@gmail.com

V. Singh
e-mail: vivek.10singh@gmail.com

A.K. Pandey
e-mail: akhileshjkit@gmail.com

A. Sarwar
e-mail: azeembuitec11@gmail.com

A.K. Dwivedi
Department of Electronics Communication, SIET, Allahabad, UP, India
e-mail: er.ajaydwivedi@yahoo.in

H.R. Vishwakarma and S. Akashe (eds.), *Computing and Network Sustainability*, Lecture Notes in Networks and Systems 12, DOI 10.1007/978-981-10-3935-5_6

1 Introduction

Microstrip patch antennas are structures having low volume, economical to fabricate, and can be easily integrated with devices; therefore, it is used in many wireless applications, such as radar, aircraft, missiles, satellite communications, biomedical, telemetry and remote sensing. [1]. Dual- or multiband operation on a single microstrip antenna has replaced the use of many antennas operating at different frequencies in order to cope with the limited frequency spectra available for wireless communication systems [2]. Narrow bandwidth, low gain, and low radiation efficiency restrict the use of patch antennas in wireless applications.

Application such as SAR (synthetic aperture radar) requires large bandwidth which motivated the scientists and researchers to evolve techniques and methods to enhance the bandwidth. Bandwidth enhancement is accomplished by forming patches on multiple substrates, using different structural configurations of patch, etching slots on the patch, and by means of aperture coupled and probe feeding techniques [3–7]. Loading of notches and slots on a single radiating surface increases the surface current path which reduces the patch size for desired frequency of operation. Frequency ratio and bandwidth of the antenna can be slightly varied with the variation in length and width of the notch [8, 9]. Shorting pin and shorted wall reduce the patch size and produce multiband operation in microstrip antennas [8–10].

Broadening of bandwidth can also be done by increasing substrate thickness, by laminating various patches on same or different layers of antenna [11–13], and by adding parasitic elements to the antenna structure. Addition of parasitic elements increases bandwidth by reducing impedance variation of antenna with frequency. Several papers have been reported [12, 14] wherein enhancement in bandwidth and gain has been observed by using techniques such as multilayered structures and parasitic elements.

The newly introduced antenna design for this paper consists of symmetrical slots of different sizes on a circular radiating patch along with two FR4-epoxy substrate layers with air of height 2 mm in between them to obtain dual-band operation with improved bandwidth and gain. The design was simulated by using electromagnetic tool and high-frequency structure simulator (HFSS). Antenna is fabricated in printed circuit board (PCB) laboratory using FR4-epoxy as substrate, and results are measured by vector network analyzer E5071C. Four different antenna structures as shown in Fig. 2 have been simulated to observe the effect of return loss by varying its slot thickness ‘g’, vertical slot length ‘L_s’, and horizontal slot length ‘W_s’. The final and proposed structure was chosen for fabrication and for experimental verification.

2 Antenna Design

The proposed antenna structure consists of two dielectric substrates (FR4-epoxy) with $\varepsilon_r = 4.4$ and $h_1 = h_3 = 1.6$ mm. The overall dimensions of the two substrates are $36 \times 36 \times 1.6$ mm^3 and are separated by an air gap of 2 mm. A circular

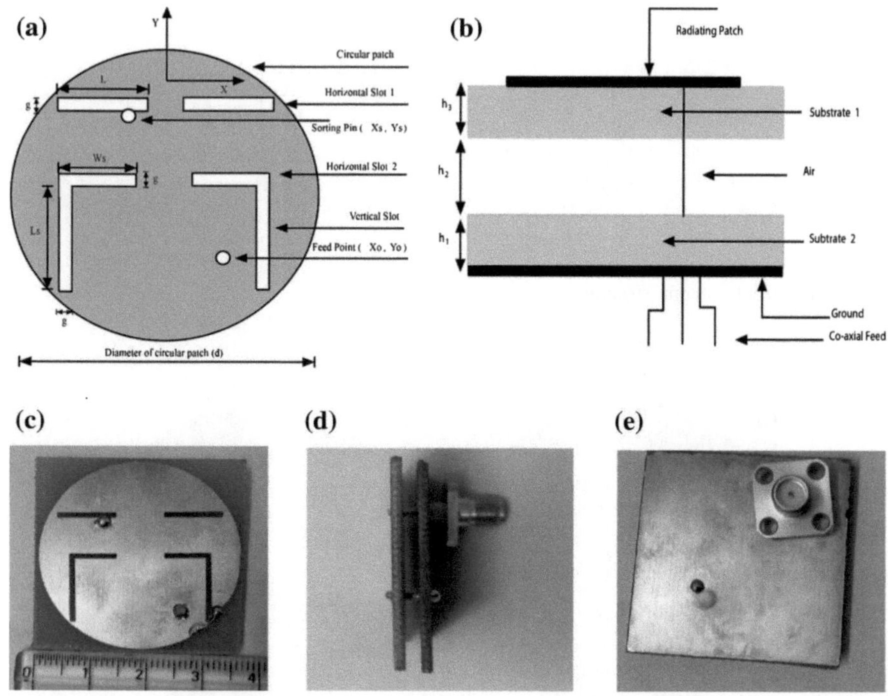

Fig. 1 Geometrical configuration and fabricated photograph of patch antenna **a** front look, **b** side look, **c** front look (fabricated), **d** side look (fabricated), and **e** back look (fabricated)

conducting patch of diameter ‘d’ was left on top upper substrate, and rest is etched out. The bottom lower substrate acts as a ground. In order to obtain dual-band resonating structure, different slots have been cut out on circular patch. Two vertical cuts of dimension ($L_s \times g$) mm are made along y-axis, and two horizontal slots of dimension ($L \times g$) mm and two horizontal slots of dimension ($W_s \times g$) mm are made along x-axis. The coaxial feed of 50 Ω with 1 mm diameter is used to excite the patch. The front view and the side view geometry of patch antenna are shown in Fig. 1a and b, respectively, and related parameter values are presented in Table 1.

Table 1 Design specifications of the proposed antenna with dimensions

Substrate material used	FR4-epoxy and air
Relative permittivity (ε_r) of FR4 and air	4.4 and 1
Dimension of ground	(36×36) mm^2
Thickness of substrate material (h_1, h_2, h_3)	(1.6, 2, 1.6) mm
Dimension of horizontal slot 1 ($L \times g$)	(10×1) mm^2
Dimension of horizontal slot 2 ($Ws \times g$)	(8×1) mm^2
Dimension of vertical slots ($Ls \times g$)	(12×1) mm^2
Diameter of circular patch (d) and thickness of slots (g)	34 mm and 1 mm
Location of feed (X_O, Y_O) and sorting pin (X_S, Y_S)	(7, −10) mm and (−6, 6) mm

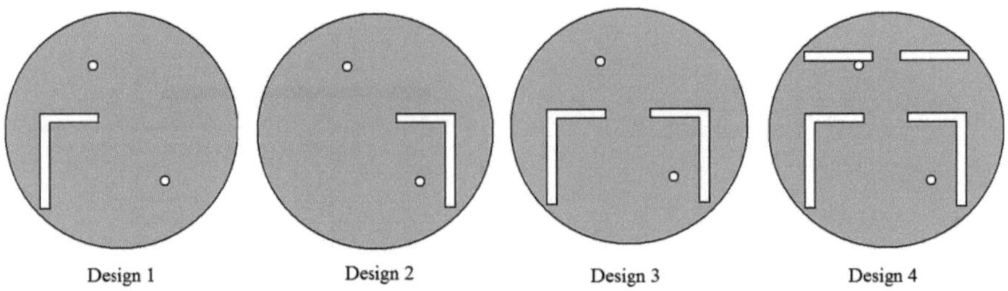

Fig. 2 Different antenna structures

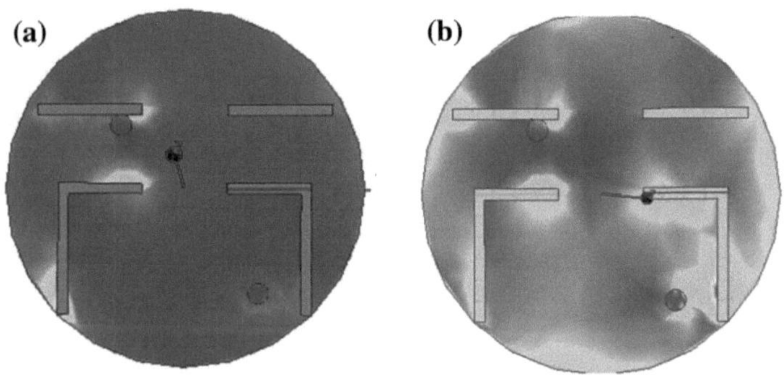

Fig. 3 Current density distribution on the surface at **a** 3.7 GHz and **b** 5.5 GHz

Figure 1c–e shows the photograph of front, side, and back view of fabricated antenna, respectively.

Figure 2 shows four different antenna design structures which have been simulated. Design 1 has an inverted L shape on the left side. Design 2 has an inverted L shape on the right side, and design 3 is the combination of design 1 and design 2. Design 4 is the proposed structure which incorporates inverted L shapes from design 3 in left and right along with two horizontal slots on top of them and a sorting pin on top. All the four structures have been simulated, and the effect of return loss by varying its slot thickness 'g', vertical slot length 'L_s', and horizontal slot length 'W_s' has been observed. Figure 3a and b shows the current density distribution on the surface of the antenna at 3.7 GHz and 5.5 GHz, respectively, for the proposed and fabricated antenna. It is observed that whole patch is radiating at higher resonant frequency, while only inverted L shape on the left side and horizontal slot on left top of it are responsible to resonate the antenna at lower resonant frequency.

3 Result and Discussion

The proposed antenna has been simulated using 3D electromagnetic structure solver tool HFSS, and fabricated antenna has been experimentally tested for its desired performance. The antenna behavior in terms of return loss, radiation pattern, gain, VSWR, and beam width is analyzed and discussed in this section. A comparison of return loss of the four antenna designs has been shown in Fig. 4. From Fig. 4, it is clear that the resonant frequency of antenna depends on its configuration. Design 1 resonates only at single frequency, whereas dual band is achieved in designs 2, 3, and 4. The configuration of design 4 provides best return loss and gains at both the resonant frequencies; therefore, it is chosen as proposed design for fabrication and experimental verification. The change in return loss with frequency for different slot thickness 'g' is shown in Fig. 5, and the value of return loss increases with the increase in the slot thickness. Dual bands are observed for values of g $\geq$ 1 mm. The antenna having values of g = 1.5 mm also provides dual-band operation, but the higher resonant frequency is shifted toward higher frequency region having a resonant frequency of 5.8 GHz.

Return loss variation with respect to frequency for different lengths of horizontal slot 2 (Ws) and vertical slots (Ls) is shown in Figs. 6 and 7, respectively. As the value of Ws increases, the resonant frequencies are shifted toward lower frequency region. The length of horizontal slot 2 in the proposed antenna is 8 mm. As observed from Fig. 7, it is clear that the antenna resonates at dual frequency only when Ls $\geq$ 10 mm. For Ls = 12 mm, optimal value of return loss is observed for the proposed antenna design

Fig. 4 Return loss variation for different antenna structures

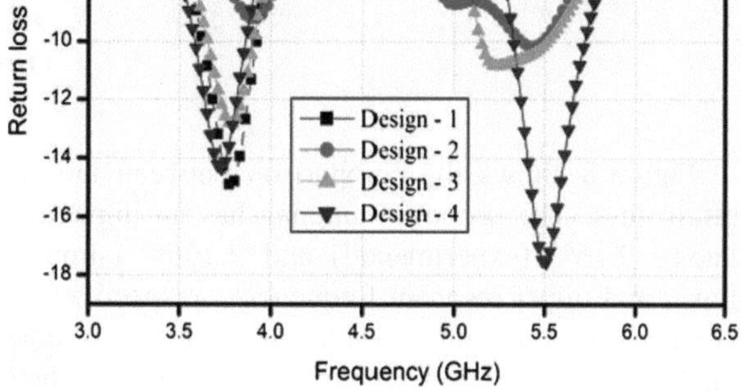

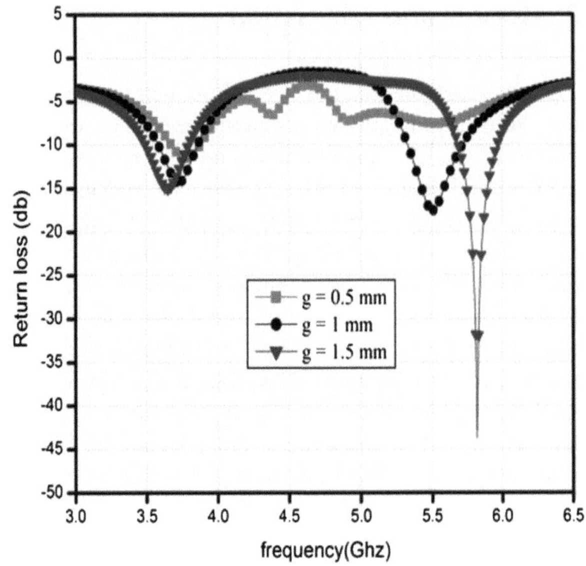

Fig. 5 Return loss variation with frequency for different values of slot thickness 'g'

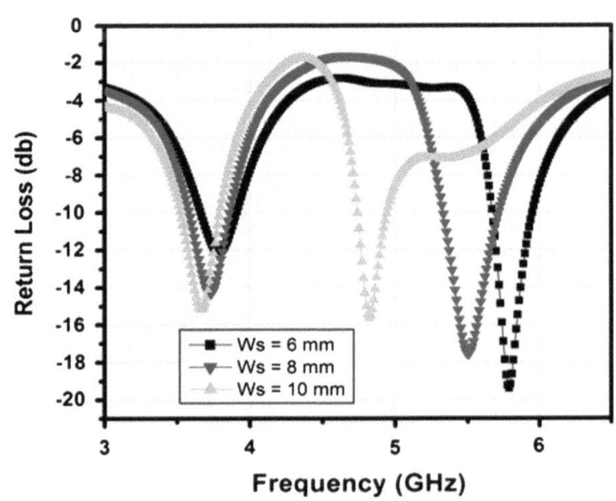

Fig. 6 Return loss variation with frequency for different lengths of horizontal slot 2 (Ws)

Figure 8 shows the comparison between the simulated and the experimental return loss. The proposed antenna has an impedance bandwidth of 7.48% (simulated), 9.16% (experimental) and 7.106% (simulated), 8.89% (experimental) at lower and upper resonant frequencies, respectively. The discrepancies between the simulated and measured results are attributed due to mathematical approximations made in the electromagnetic tool used for antenna simulation and analysis,

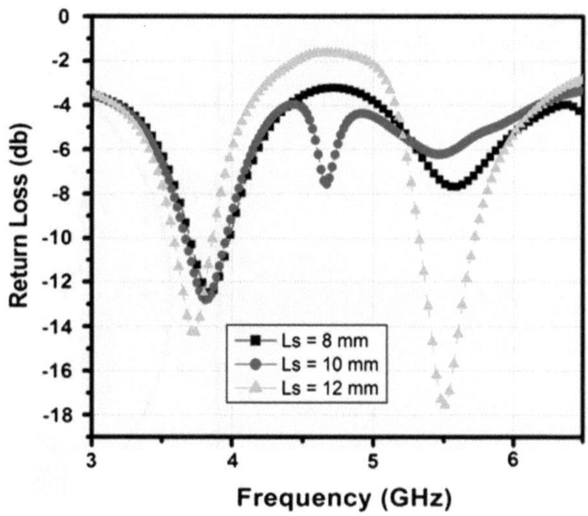

Fig. 7 Return loss variation with frequency for different values of slot length (Ls)

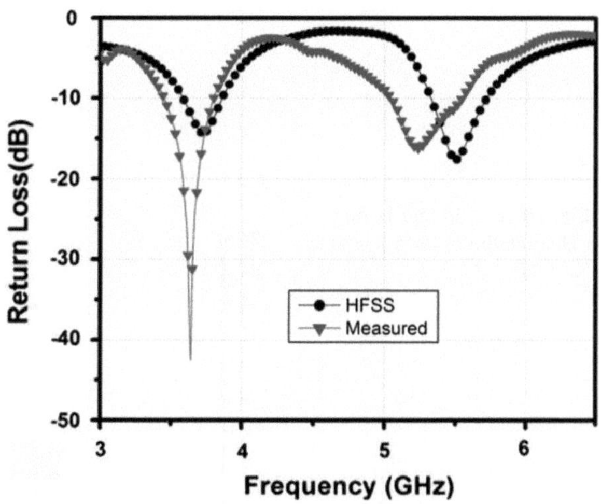

Fig. 8 Comparative plot of simulated and measured return loss for the proposed antenna

assumptions made in calculation of effective dielectric constant, mechanical couplings, fringe capacitances, etc.

The electric and magnetic field radiation patterns at lower and upper resonating frequency are shown in Figs. 9 and 10, respectively. From Figs. 9 and 10, it is clear that the maximum power is radiated in the broadside direction within the operating band.

A gain of 3.93 dB and 3.09 dB is observed at 3.7 GHz and 5.5 GHz resonant frequencies, respectively, as shown in Figs. 11 and 12. E-field beam width is found to be 230° and 205° and H-field beam width is 250° and 165° at lower and upper

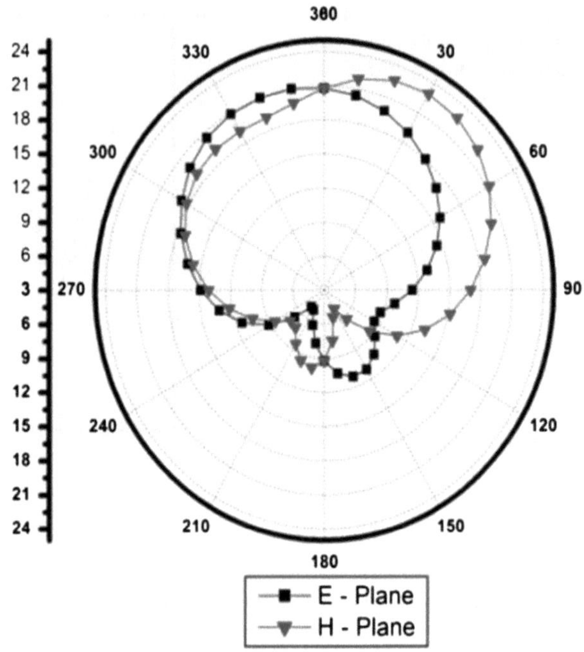

Fig. 9 E-field and H-field radiation plot of the antenna at 3.7 GHz

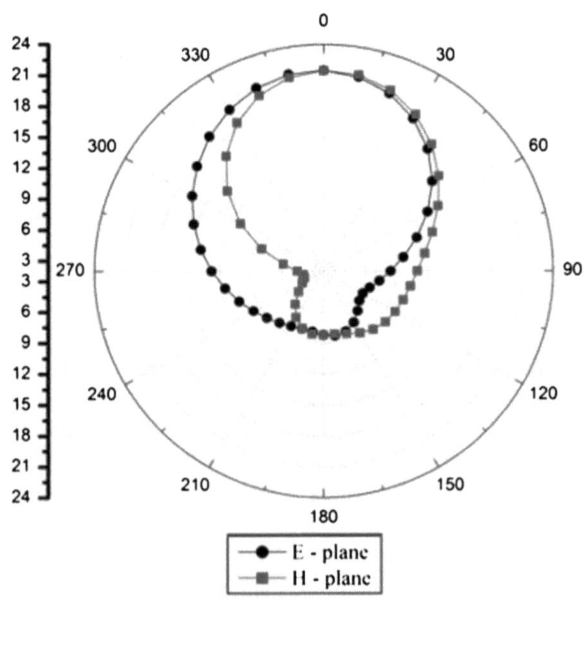

Fig. 10 E-field and H-field radiation plot of the antenna at 5.5 GHz

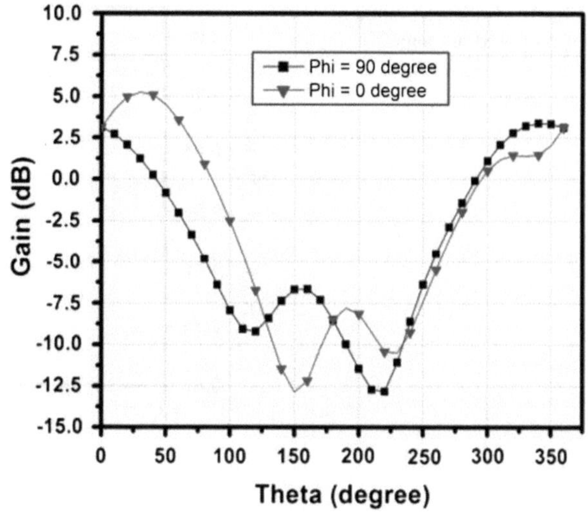

Fig. 11 Gain versus theta plot at 3.7 GHz of proposed antenna

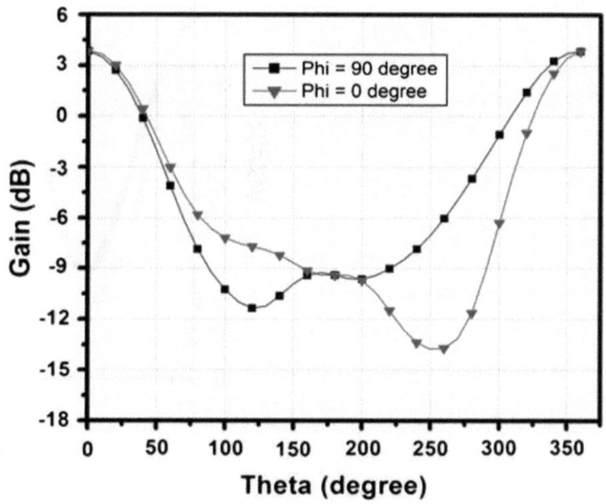

Fig. 12 Gain versus theta plot at 5.5 GHz of proposed antenna

resonant frequency, respectively, at −3 dB. Maximum radiation efficiency is observed (cf. Fig. 13) to be 92.6% for lower band and 88.3% for upper band.

From Fig. 14, it is observed that the simulated and measured values of VSWR are 1.49 and 1.28 at lower resonant frequency, whereas it is 1.85 (simulated) and 1.79 (experimental) at upper resonant frequency. The experimental and simulated values are fairly close and also in good range (i.e., <2).

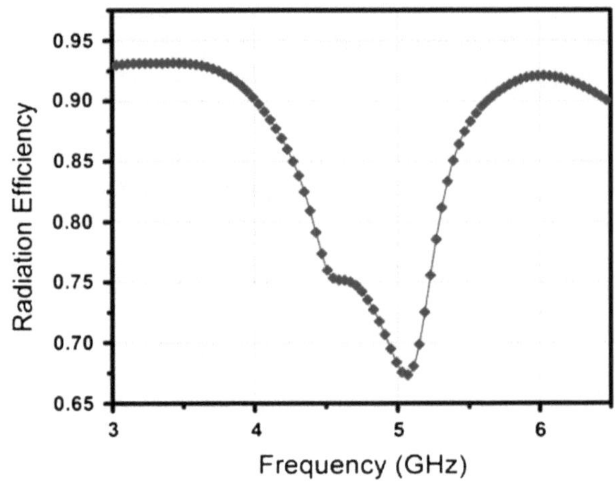

Fig. 13 Radiation efficiency of proposed antenna

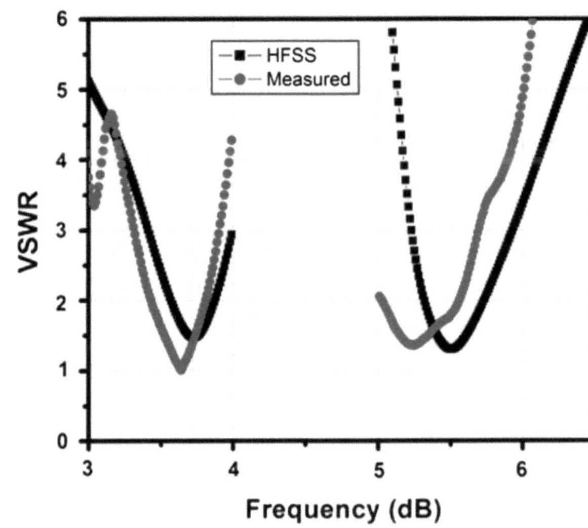

Fig. 14 VSWR versus frequency plot of proposed antenna

4 Conclusions

A compact, lightweight novel design has been presented for Wi-Fi/WLAN applications. The different antenna parameters and effect of slot dimensions on return loss are investigated. Separation between resonant frequencies depends upon slot dimensions, number of horizontal and vertical cuts, and multiple layers of substrates. The antenna introduced here resonates at 3.7 GHz and 5.5 GHz with a bandwidth of 7.48% (simulated), 9.16% (experimental) and 7.106% (simulated), 8.89% (experimental) at lower and upper resonant frequencies, respectively.

Maximum gain and efficiency obtained at lower frequency are 3.93 dB and 92.6%, and at higher resonant frequency, it is 3.09 dB and 88.3%. The results represent good agreement between measured and simulated result for proposed design.

References

1. Garg R, Bhartia P, Bahl I, Ittipiboon A (2001) Microstrip antenna design handbook. Artech House, Norwood, MA
2. Maci S, Gentilli GB (1997) Dual frequency patch antenna. IEEE Antennas Propog Mag 39 (6):13–20
3. Ansari JA, Mishra A, Yadav NP, Singh P (2010) Dual-band slot loaded circular disk patch antenna for WLAN applications. Int J Microw Opt Technol 5(3):13–20
4. Mishra A, Ansari JA, Kamakshi K, Singh A, Aneesh Md, Vishwakarma BR (2014) Compact dual-band rectangular microstrip patch antenna for 2.4/5.12 GHz wireless applications. Wireless Netw 21(2):347–355
5. Mak CL, Luk KM (2000) Experimental study of a microstrip patch antenna with an L shaped probe. IEEE Trans Antennas Propag 48(5):777–783
6. Wang Z, Fang S, Fu S, Lu S (2009) Dual-band probe fed stacked patch antenna for GNSS applications. IEEE Antenna Wireless Propag Lett 8:100–103
7. Ghassemi N, Mohassel JR, Mohanna Sh, Moradi Gh (2009) A wideband aperture coupled microstrip patch antenna for S and C bands. Microw Opt Technol Lett 51(8):1807–1809
8. Shivnarayan, Vishvakarma BR (2006) Analysis of notch loaded patch for dual operation. Indian J Radio Space Phys 35:435–442
9. Mishra A, Singh P, Yadav NP, Ansari JA, Vishvakarma BR (2009) Compact shorted microstrip patch antenna. Prog Electromagn Res C 9:171–182
10. Yoon C, Choi S-H, Lee H-C, Park H-D (2008) Small microstrip patch antennas with short-pin using a dual-band operation. Microw Opt Technol Lett 50(2):367–371
11. Ooi BL, Qin S, Keong MS (2002) Novel design of broad band stacked patch antenna. IEEE Trans Antenna Propag 50(10):1391–1395
12. Ronglin L, DeJean G, Maeng M, Lim K, Pinel S, Tentzeris MM (2004) Design of compact stacked patch antennas in LTCC multilayer packaging modules of wireless applications. IEEE Trans Adv Packag 27(4):581–589
13. Singh VK, Ali Z, Singh AK, Ayub S (2013) Dual-band triangular slotted stacked microstrip antenna for wireless applications. Cent Eur J Eng 3(2):221–225
14. Ansari JA, Singh P, Yadav NP (2009) Analysis of wideband multilayer patch antenna with two parasitic elements. Microw Opt Technol Lett 51(6):1397–1401

Triggering a Functional Electrical Stimulator Based on Gesture for Stroke-Induced Movement Disorder

P. Raghavendra, Viswanath Talasila, Vinay Sridhar and Ramesh Debur

Abstract This paper presents the design of a smart intent-based triggering system to trigger a functional electric stimulator, based on an electromyography (EMG) measurement of voluntary muscle activity and IMU measurement of gait. Thus, an atrophied/weakened muscle is electrically stimulated by identifying pre-defined gestures and EMG signal strength of the active muscle bundle, thereby restoring the movements of the affected muscle. This smart FES triggering system is designed as a strap on module and deployed at the site of the affected muscle.

Keywords Electromyography · Functional electrical stimulator · Muscle activity · Inertial sensing

1 Introduction

An EMG signal is generated when a motor neuron action potential from the spinal cord arrives at a neuromuscular junction. Electromyography is a technique to measure myoelectric signals, which are generated by physiological variations in the state of the muscle fibre membranes [1, 2]. Loosely speaking, this means that

P. Raghavendra (✉) · V. Talasila
Department of Telecommunication Engineering, M.S. Ramaiah Institute of Technology, Bangalore, India
e-mail: raghavendra.karnad@gmail.com

V. Talasila
e-mail: viswanath.talasila@gmail.com

V. Sridhar
Department of Electronics and Communication, M.S. Ramaiah Institute of Technology, Bangalore, India
e-mail: vinay.sridhar33@gmail.com

R. Debur
Department of Physiotherapy, M.S. Ramaiah Medical College, Bangalore, India
e-mail: rameshdebur@gmail.com

H.R. Vishwakarma and S. Akashe (eds.), *Computing and Network Sustainability*, Lecture Notes in Networks and Systems 12, DOI 10.1007/978-981-10-3935-5_7

electromyography is a technique in which the electrical activity of a muscle is measured. Measured EMG potentials ranges between less than 50 V and up to few millivolts, depending on the muscle under observation. This paper aimed at developing a low-cost reliable embedded system that used the muscle action to trigger a functional electrical stimulator. Electrical muscle stimulation is a technique in which the paralyzed muscle is made to contract via electrical impulses delivered through the surface electrodes placed on the skin. This can be a rehabilitation tool for patients with motor dysfunctions (e.g. after spinal cord injury or stroke) [3].

The major challenge is to fire at the muscle to be activated without activating the neighbouring muscle bundle. This can be done by choosing an electrode with suitable form factor and placing them precisely [4]. Studies show that the use of multi-pad electrodes can improve the selectivity during stimulation, thus causing less fatigue to the muscle [5]. There are two major classes of FES devices, from a control viewpoint: open loop and closed loop. By pressing the buttons on a rolling frame, elbow frame-walker, or crutches, the user can trigger stimulation sequences for standing up, stepping forward, or sitting down, as well as increase/decrease the stimulation intensity [6, 7]. An example is shown below for triggering the FES in a frame walker (Fig. 1).

The other class of FES is the closed-loop systems, where walking has been used to generate controlled and target-directed movements [8, 9].

In this paper, we adopt a slightly different philosophy. We design a closed-loop control (the trigger to the FES), but instead of pressing a series of buttons on a walker (or a FES strap), we design a gesture recognition system which requires minimal voluntary action from the patient in order to trigger the FES. The key novelty is the use of minimal gestures in order to trigger the FES as shown in the Fig. 2. (Note that the action of pressing a series of buttons is a fairly complex psychological function).

There are many variants of muscle stimulator available in the market. The most common is a pocket-sized device usually used in sports training and rehabilitation centres. The triggering of the FES happens at regular intervals of time with a

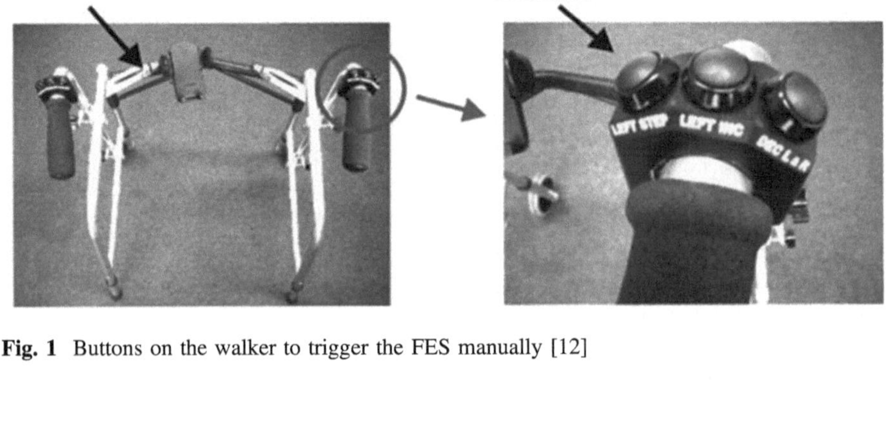

Fig. 1 Buttons on the walker to trigger the FES manually [12]

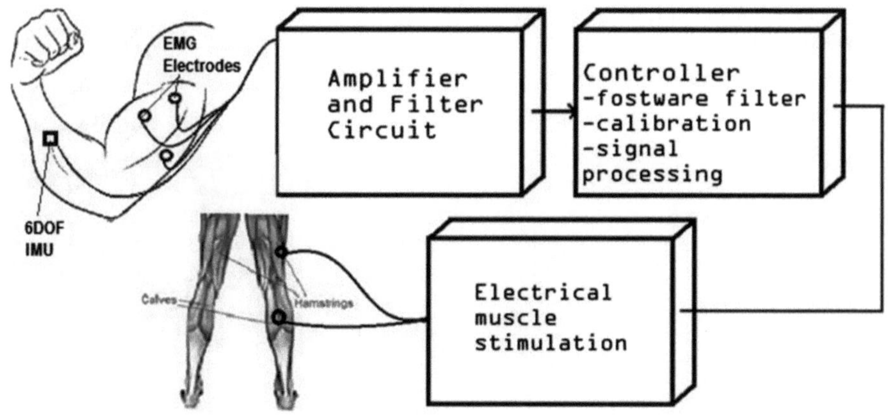

Fig. 2 Conceptual idea of biceps being used to stimulate the hamstring muscle

predefined duty cycle. FES-a [10] by Tecnalia Research and Innovation makes use of multi-pad electrodes and the target muscle(s) can be fired very accurately [11]. The hand worn module receives commands from a computer wirelessly.

These systems are typically very expensive and usually unaffordable in developing countries like India. While the actual FES is itself a fairly standard and inexpensive device, the associated software and automation drive up the cost. For example, some FES devices comprise a wireless hand-held control unit, which stores important information about system usage, stimulation intensity levels. This is backed up by software that allows for data analysis. It is the additional modules that significantly increase the costs of modern FES devices.

Further, most FES products are designed in such a way that a trigger is provided which can be used to activate the FES placed on (for example) a hand. To activate, the trigger itself requires the use of another hand. This limits the degrees of freedom available when a triggering action is required. In other words, the subject must interrupt the actions of the (functioning) hand in order to trigger the FES. From a patient viewpoint, this can often be psychologically frustrating and can be an obstacle for using the FES.

The goal of our work is to design an automated low-cost, portable, lightweight FES system. The goal of this specific paper is to focus on the automated triggering of the FES by designing and prototyping an intent-based trigger. The idea is to trigger the FES through muscular activity which does not interfere with normal task-oriented functioning. The FES systems are used in contraction of muscles to produce functions such as rasping, walking, and standing. A person re-learns how to execute impaired functions, instead of depending on prosthesis. The Fig. 3 shows a single-channel functional electrical stimulator stimulating the muscle group near the wrist. This device stimulates the muscle group to cause a muscle contraction and perform a motor function.

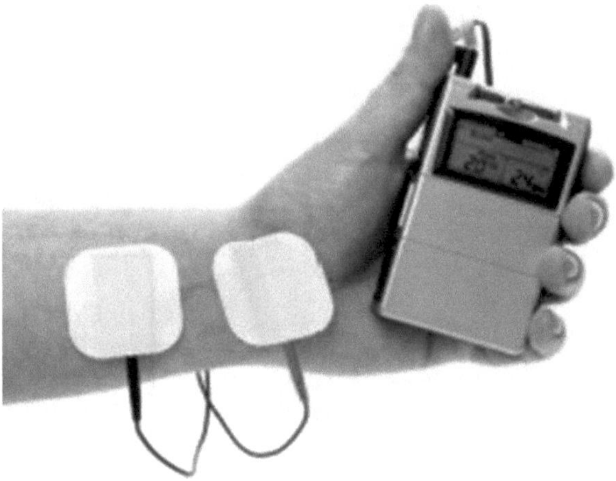

Fig. 3 Functional electrical stimulator [13]

2 Experimental Setup

2.1 Electromyography

Figure 4 shows the amplifier and filter circuit diagram used in the first stage. There are 3 electrodes, of which the first goes to the mid of the muscle of observation, the second goes to the end of that muscle, and the third electrode is the reference ground. In the first stage, the EMG signals are picked up by the electrodes and amplified by the instrumentation amplifier, INA129. The signal is then passed through a first order low-pass filter with cut-off frequency 90 Hz and a gain of 15. The amplified signal is then passed through a first order high-pass filter with a cut-off frequency of 10 Hz and a gain equal to 5. Thus, the overall gain of the circuit is 50 * 15 * 5 = 3,750.

In the second stage, the amplified output signal is sampled at 2 kHz using a microcontroller (Teensy 3.2 ARM Cortex-M4 92 MHz). The controller has a

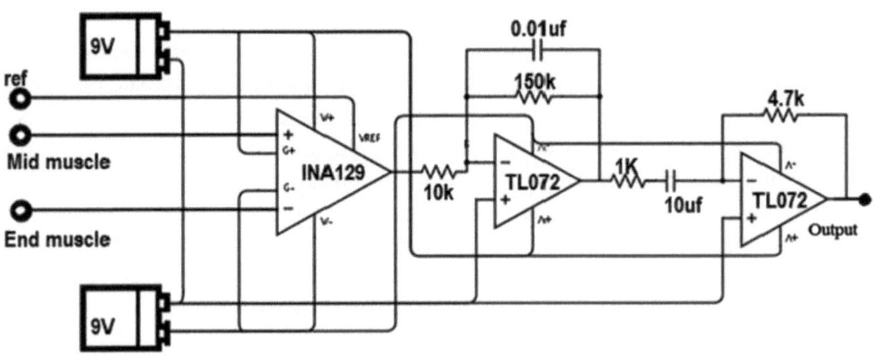

Fig. 4 Amplifier and filter circuit

software notch filter to eliminate the 50 Hz noise due to the transmission line interference. The controller calibrates itself by taking continuous samples of EMG signal for 10 s and records the maximum and minimum reading in the interval. The output of the controller is made high if the voltage crosses the threshold level thereby triggering the FES. Therefore the output updates five times in every second. This refresh rate can be varied as per the requirement.

2.2 *Attitude Estimate*

A 6DOF IMU is used to know the attitude of the hand. A quaternion-based complementary filter is used to estimate the orientation and thereby avoiding false triggering of the FES and prevent turning off of the FES in the middle of a gesture which is explained in the next section. MEMS gyroscope is used as a primary sensor as it can capture high frequency dynamics while the MEMS accelerometer is used to correct the gyroscope drift generated due to integrating noise. This is the idea of a complementary filter in which the gyroscope data is passed through a high-pass filter and the accelerometer data is passed through a low-pass filter, and the results are combined to obtain the attitudes.

Figure 5 shows block diagram of 6dof quaternion-based complementary filter. Data from the gyroscope along with the attitude correction vector is integrated to

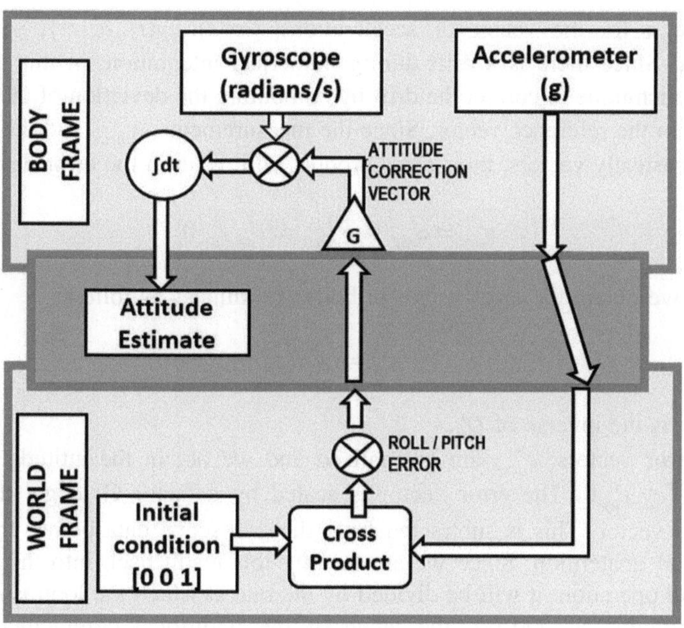

Fig. 5 Block diagram of 6dof attitude estimation algorithm

form the incremental quaternion. The main quaternion which related the inertial frame and body frame is the cumulative sum of incremental quaternions.

Accelerometer outputs (ax, ay, az); similarly, gyroscope outputs (wx, wy, wz); we define the reference 3axis vector of accelerations in inertial frame as

$$\mathbf{a}^i_{ref} = [0\,0\,1] \tag{1}$$

let $\Delta q = q_w + q_x i + q_y j + q_z k$ be an incremental quaternion. Converting the axial rotation from the gyroscope to quaternion form, we have

$$\begin{aligned} q_w &= \cos(\theta/2) \\ q_x &= U_x \sin(\theta/2) \\ q_y &= U_y \sin(\theta/2) \\ q_z &= U_z \sin(\theta/2) \end{aligned}$$

where, $\vec{U}$ is the normalized gyroscope data, $\vec{\omega}$ and θ are amount rotations around the given unit vector.

Let Q^i_b denotes the quaternion that relates body frame to inertial frame. Given $ax^b(t)$, $ay^b(t)$, $az^b(t)$ be the accelerometer values in body coordinates, let $ax^i(t)$, $ay^i(t)$, $az^i(t)$ be the representation in inertial coordinates, with

$$Q^i_b\colon \{ax^b(t),\, ay^b(t),\, az^b(t)\} \quad \mapsto \quad \{ax^i(t),\, ay^i(t),\, az^i(t)\}$$

Denote $\mathbf{a}$ for the vector of accelerations $\{ax(t), ay(t), az(t)\}$, so we have $Q^i_b\colon \mathbf{a}^b \mapsto \mathbf{a}^i$ Since there is a drift during numerical integration, in this section we present a technique to correct the drift by computing the deviation of the measured vector from the reference vector. Since the measurements, $\mathbf{a}^i_{meas}$ and the reference, $\mathbf{a}^i_{ref}$, are basically vectors, their cross product will result in the error deviation, i.e.

$$\varepsilon^i_{acc} = \mathbf{a}^i_{ref} \times \mathbf{a}^i_{meas} = [\varepsilon^i_{ax}\,\varepsilon^i_{ay}\,0] \tag{2}$$

Then, we obtain the error vector in body coordinates as follows

$$\varepsilon^b_{acc} = Q^b_i\,\varepsilon^i_{acc} \tag{3}$$

where Q^b_i is the inverse of Q^i_b.

The error vectors, ε^b_{acc}, are normalized and we obtain the attitude correction factor as $[\varepsilon^i_{ax}\,\varepsilon^i_{ay}\,0]$ The error vector is scaled by a factor G to get the attitude correction vector. This is subtracted from the gyroscope data before forming the incremental quaternion. Since we are feeding this result back into the gyroscope integration operation, it will be divided by the rate at which the loop is running. If loop is running at 500 Hz, the value of the traditional complementary filter

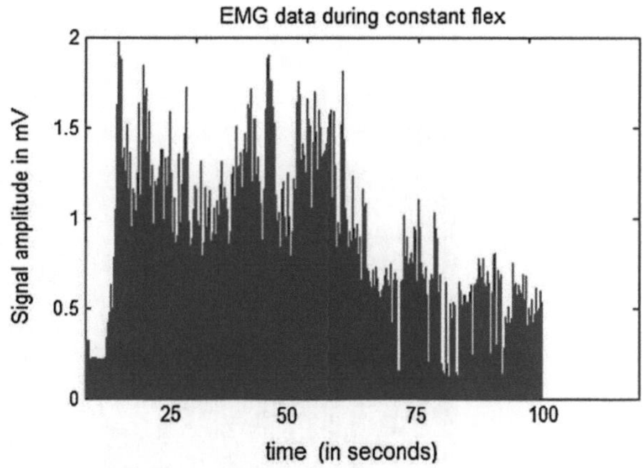

Fig. 6 Drop in the EMG signal strength during constant flex

coefficient (α) is 0.002 * G. The value of G is the trade-off between amount of noise let in from the sensors and the rate at which it corrects the error.

2.3 Trigger Mechanism

When the muscle is flexed for a long time, the EMG signal amplitude gradually decreases as shown in the figure above. This is because the muscle gets tired and central nervous system looks for an alternate/neighbouring muscle to do the same hand movement/flex. The graph shown in Fig. 6 was obtained by plotting 2,00,000 samples for EMG signal recorded at 2 kHz. When the EMG signal strength drops close to the threshold level, it crosses the threshold level rapidly due to the small noise present, thereby triggering the FES on and off rapidly which is dangerous as it may lead to muscle fatigue. Therefore, we use two levels of thresholds say Th1 and Th2 (Th2 > Th1).

The FES is turned on when the EMG signal strength is greater than TH2 and is kept on until the signal strength drops to less than TH1. This overcomes the rapid triggering of the FES when the signal strength decreases (Fig. 7).

3 Results

Figure 8 is a comparison between the pitch angles from the gyroscope and complementary filter. During numerical integration, the noise and bias get added and over a period of time the pitch value drifts. The plot shows data samples taken over 60 s. The gyroscope data approximately drifts at about 0.633°/s. This applies even to the roll angle.

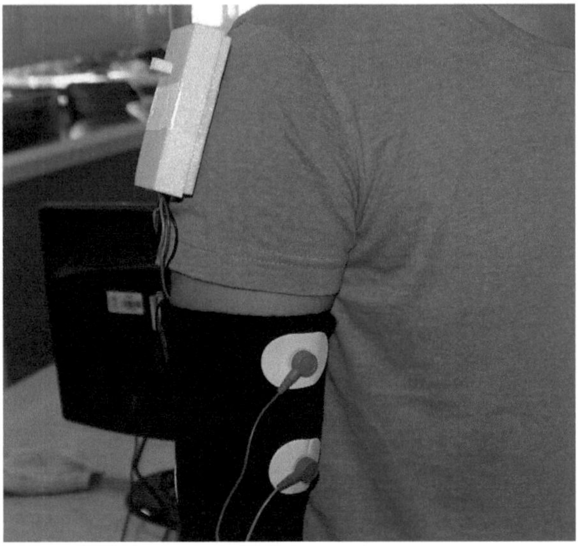

Fig. 7 Illustration on electrodes and IMU attached to the patient collecting limb rotation and electrical activity

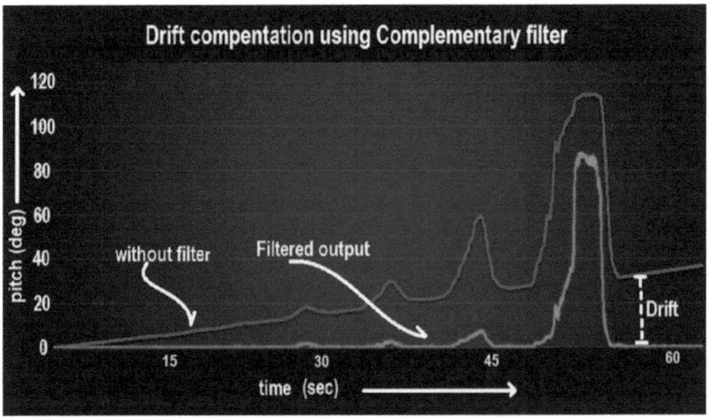

Fig. 8 The pitch angle with and without complementary filter

In a task involving the gripping of a cup (of coffee), assuming that the cup is placed in a way that it would involve the extension of the arm (at the elbow joint), with a (half) pronation action, followed by opening the wrist and gripping the cup handle and bringing it towards the patient's body through a combination of a supination and flexion (at elbow) actions. For a patient with wrist drop to lift the cup, he extends his hand towards the cup and flexes his biceps. The IMU monitors his gait and when the hand is in predefined position, FES is triggered enabling him

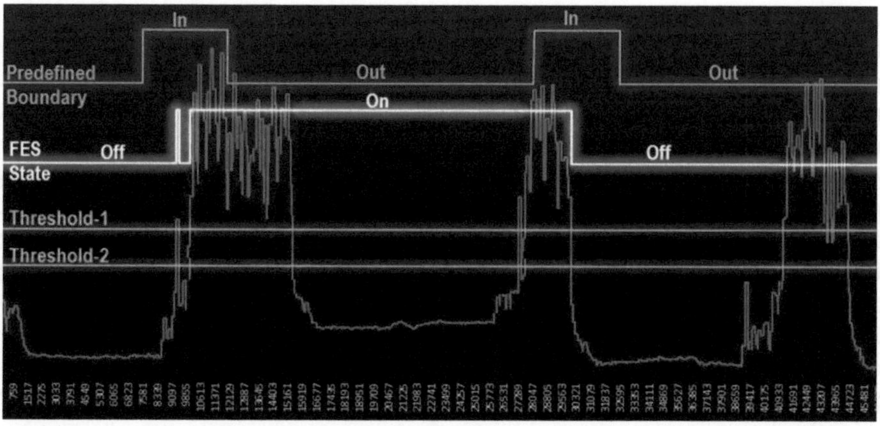

Fig. 9 Pictorial representation of the system in action

to grip the cup. Now, when the patient lifts the cup and relaxes this biceps, the IMU ensures the FES does not go off when he is still holding the cup to his mouth. Only when he places the object back and relaxes the biceps, the FES is turned off.

In the timing diagram as shown in Fig. 9, the patient tries to extend his hand towards the cup and the orientation estimation algorithm shows that his hand is in

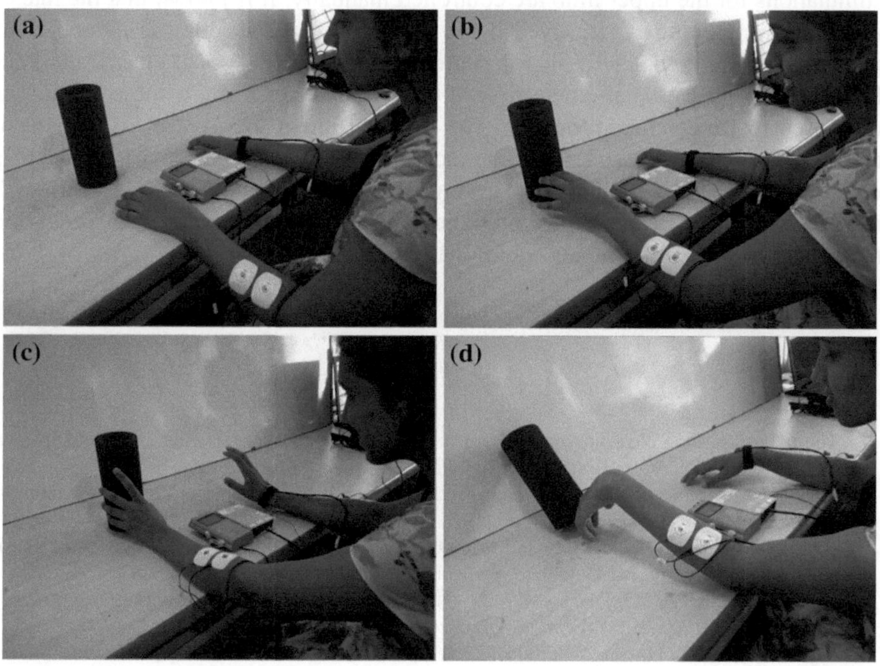

Fig. 10 Experiment done to show the intent-based triggering system

the position. Now, he flexes his biceps in order to grab the cup and the FES is turned on. As he moves the cup towards his mouth, the orientation is out of the predefined boundary. Even the patient relaxes his biceps in the middle of the task, the FES is still kept on. When the patient comes back to the initial position and relaxes his biceps, the FES is turned off.

The same experiment is shown visually, where the patient tries to grab a cup placed on a table. In Fig. 10a, the object placed on the table is seen by the patient. In Fig. 10b, the patient tries to reach out to the cup. If the stimulation using the FES is not applied to the hand of the patient, the patient will not be able to grab the cup and hence may accidentally drops the cup as shown in Fig. 10c. With the stimulator being triggered by our device and stimulation given to the muscle, the patient is able to grab the cup, as shown in Fig. 10d, and stimulator is on till the IMU is present in the boundary conditions. Only a small voluntary action is required to start this stimulation.

4 Conclusion

Muscular electrical stimulation is used in neuro-rehabilitation to re-learn basic motor tasks and to improve muscular strength/coordination, all of which are affected by specific neurological damage to the brain. Currently, most electrical stimulations for the upper limb are controlled manually. It is known that the rate of neurological learning is improved if the learning of motor action is natural. This paper presents a complete design and prototyping of a functional electrical stimulator through the automatic detection of a movement intent by a subject which is then used to trigger the stimulator. The detected intent involves measurement of complex limb motion and muscular activity. The wearable, smart device has undergone successful preliminary testing in a clinical setting. Future work will involve the testing of the device on patients with specific movement disorder and the detection of more complex motor activity.

References

1. The use of surface electromyography in biomechanics. In: Wartenweiler memorial lecture. International Society of Biomechanics, 5 July 1993
2. The H200 for hand paralysis, 4th edn. Product information available at http://www.bioness.com/NewsMedia/MediaGallery/H200.php
3. Westerveld A (2011) Selectivity and resolution of electrical stimulation of finger muscles for grasping support. In: 3rd DutchBio-medical engineering conference, 3rd edn
4. Elsaify A, Fothergil JC, Peasgood W (2004) Portable FES system optimizes electrode array using twitch response. In: IFESS conference
5. Fuji T, Seki K, Handa Y (2004) Development of a new FES system with trained supermultichannel surface electrodes. In: IFESS conference

6. Hermie J (2000) Hermens development of recommendations for SEMG sensors and sensor placement procedures. J Electromyogr Kinesiol 10(2000):361–374
7. Chizeck HJ, Kobetic R, Marsolais EB, Abbas JJ, Donner IH, Simon E (1988) Control of functional neuromuscular stimulation systems for standing and locomotion in paraplegics. Proc IEEE 76:1155–1165
8. Jezernik S, Wassink RGV, Keller T, Glen M (2004) Davis sliding mode closedloop control of FES: controlling the shank movement. IEEE Trans Biomed Eng 51:263–272
9. The ABC of EMG a practical introduction to kinesiological electromyography. Noraxon, April 2005
10. FES-a Tecnalia Research and Innovations: http://www.tecnalia.com/en/
11. Velik R (2011) INTFES: a multi-pad electrode system for selective transcutaneous electrical muscle stimulation. In: 16th annual conference of the international functional electrical stimulation society
12. Neopraxis Exostim: walking frame with buttons to evoke functional electrical stimulation mobility tasks, Retrieved from http://www.web.calstatela.edu/faculty/dwon/JClub/Braz09ReviewFESinSCIFeedback.pdf
13. Electric muscle stimulator tens unit: Retrieved from http://www.electricmusclestimulators.com/wpcontent/uploads/2011/05/electricmusclestimulator21.jpg
14. Braz GP, Russold M, Davis GM (2009) Functional electrical stimulation control of standing and stepping after spinal cord injury, vol 12, Number 3. doi:10.1111/j.1525-1403.2009.00213

Academic Dashboard—Prediction of Institutional Student Dropout Numbers Using a Naïve Bayesian Algorithm

Aishwarya Suresh, H.S. Sushma Rao and Vinayak Hegde

Abstract Every year, many students enroll themselves on various courses offered by institutions. In that bundle of admissions, a few tend to fall out of their academic programs. Students drop out of their courses due to varied reasons. Analyzing these reasons in order to predict the dropout rate of an institution is of interest. In this research chapter, we are considering a few reasons such as student attendance, educational history, medical history, family background, disciplinary issues, attendance, etc. as factors to compute and predict future dropout rates of registered courses at institutions. To compute and predict dropout rate, a pre-survey and post-survey is conducted. By applying a Naïve Bayesian classifier we predict the probability of students dropping out. Early prediction of student dropout rates, will help to improve the performance of an organization, both professionally and economically.

Keywords Naive Bayesian · Dropout · Analysis · Prediction · Probability

1 Introduction

One of the mottos of every institution is to have a low number of dropouts. When a student joins an institution they have no idea whether they are likely to drop out of their academic programs during their educational journey. When a student drops out of a program or institution, it is normal for their institution to fail to record the reason why. Dropout numbers differ each year due to different reasons. Recording

A. Suresh (✉) · H.S. Sushma Rao · V. Hegde
Department of Computer Science, Amrita Vishwa Vidyapeetham Mysuru Campus,
Amrita University, Mysuru, Karnataka, India
e-mail: aishwarya3939@gmail.com

H.S. Sushma Rao
e-mail: sushmarao.1396@gmail.com

V. Hegde
e-mail: vinayakhegde92@gmail.com

H.R. Vishwakarma and S. Akashe (eds.), *Computing and Network Sustainability*,
Lecture Notes in Networks and Systems 12, DOI 10.1007/978-981-10-3935-5_8

these reasons helps a institution to know where they are lacking and also helps them to understand student mentality. The objective of this chapter is to analyze, and predict, the number of dropouts based on the parameters like attendance, educational history, medical history, family background, subject backlogs, etc. This may help an institution to take corrective measures to improve, and reduce, student dropout rates. Data for this chapter was gathered using survey forms. Two surveys were performed, i.e., a pre-survey and post-survey. The data was pre-processed by converting text into zeros and ones as well as finding and updating missing/null values. With use of the data a Naïve Bayesian classifier was used to predict the output.

The survey forms were based on the following research questions.

1. What makes a student drop out of an organization? Is there any student related character, academic factor, family background which could be linked to the dropping out?
2. How does dropping out affect an organization's growth?
3. Under which circumstances do students feel they need/must drop the course?
4. What impact does academic performance have on dropping out?

The Naïve Bayesian algorithm was used with the help of WEKA to discover and extract results. The rest of the chapter is organized as follows: Sect. 2 describes the literature survey; Sect. 3 describes the methodology used to carry out the research; Sect. 4 looks at research ethics; Sect. 5 describes experimental results; and Sect. 6 concludes the chapter.

2 Related Work

Sweta Rai, proposed a prototype machine learning tool based on classification which automatically recognizes whether a student will continue their study or drop out of their course based on certain factors commonly considered as being responsible for dropouts. The technique used here is based on a decision tree and extraction of hidden knowledge from a large dataset. Considering various factors, a discriminant analysis was used to extract frequent patterns and correlations from the dataset [1]. An association rule for mining was applied to the dataset. The association rule and the decision tree was carried out using the WEKA data-mining tool. The results gathered from these rules supported the fact that 0.68% of dropouts were down to personal problems. The main reason for dropouts was mostly linked to sickness (home sickness), adapting to new courses, and poor hostel facilities, along with simply not adjusting to the campus environment as well as low placement rates.

Abu-oda and El-halees (2015) made use of different data-mining approaches to predicting student dropouts from different courses. The data was collected historically from the first two years of study. To classify and predict the datasets, different classifiers were used such as decision tree and Naïve Bayes. These methods were tested using 10-fold cross validation [2]. The result of the accuracy of the classifiers was 98.14% and 96.86%, respectively. From the result, it was also observed that one of the reasons for dropping out was relationships between students, which were not outwardly obvious. Yathongchai et al. (n.d.). considered three issues affecting student dropout rates. The factors were grouped thus: conditions related to students before admission, during study periods at university, and all the other factors [3]. Jadri (2010) concentrated on enhancing the efficiency of studying to understand the dropout problem. Statistical data processing was performed with data-mining methods. The first segment presents basic information on the structure of the student. The second segment presents an analysis using logistic regression, decision trees, and neural networks [4]. Models were identified according to SEMMA and were compared in order to select best at predicting. The attributes and parameters were listed and selected for on the basis of the analysis needed. The missing value replacement by most frequent value of neural network. Bayer et al. (n.d.) focused on school failure regarding dropouts due to social behavior and student records. This novel method for student failure prediction reduced the number of incorrectly classified dropouts. The classifier created, using only social behavior, had a data accuracy which did not exceed 69% [5]. By adding attributes describing social behavior this increased by 11%. The highest accuracy was obtained by PART, True Positive (TP). The best results were obtained using the decision tree learner, 82.53% and TP, 78.50%. This supports the hypothesis that four semesters represents a period model which can predict dropping out with high probability.

3 Methodology

The objective of this chapter is to find dropout rates of students from institutions. A predictive student dropout model may help institutions recognize/anticipate early dropouts. Two survey were conducted, i.e., one during student admission (pre-survey) and the other during the middle term of the course (post-survey), in order to collect data and predict which students are likely to drop out. Consideration was given to parameters like parental income, student medical history, student educational history, legal issues, and so on. The data gathered from the surveys were pre-processed by converting texts to zeros and ones. Using these sets of zeros and ones, an algorithm was used to compute and results. According to past results the Naïve Bayesian algorithm forms a better model than any other.

The Naïve Bayesian classifier provides a method for computing the probability using independent assumptions of predictors. The Bayesian classifier helps predict values from datasets.

3.1 Survey Form During Student Admission

At the time of student admission, a survey form was given to students. This form helped collect data, such as, student educational history, family background, parents' educational background, parents' economic background, student health, student disciplinary issues, previous failure in courses, etc. These were collected and saved in a database.

3.2 Survey Form for the Middle of Term

During the middle term of the course, another survey was issued, based on whether the student was adjusting to their course? Is the student settled their peers? Are they facing any difficulties from the institution or fellow students? Are they facing any difficulties in their hostel? Are they home sick? What is their attendance status? These and many more questions were posed on the survey form.

The data (categorical) from the survey was stored in the form of zeros and one in a database. Later, the date was retrieved and computed using the Naïve Bayesian algorithm in order to predict which student was likely drop out.

3.3 Work Methodology

See (Fig. 1).

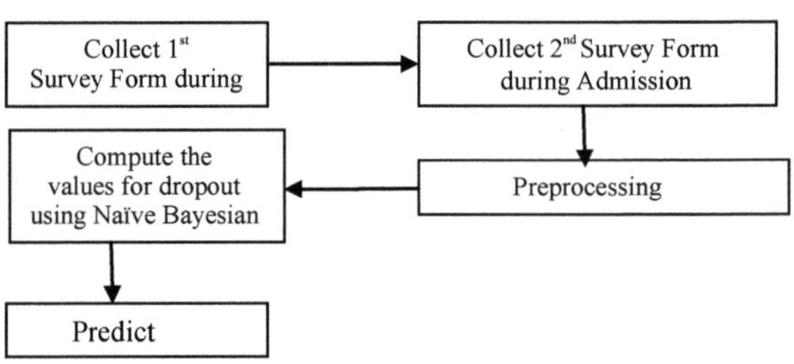

Fig. 1 Methodology undertaken during research. The data is collected via survey forms. Later, the data is pre-processed and a Naïve Bayesian classification is used to compute and predict the drop out values

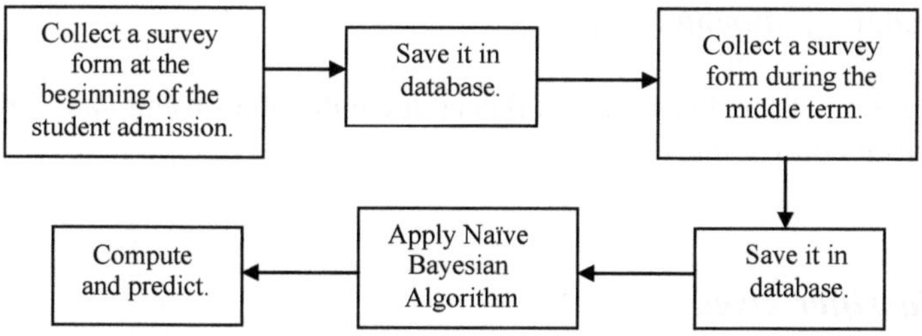

Fig. 2 A flow chart for the research

3.4 Predictions of Dropping Out

See (Fig. 2).

4 Research Ethics

Before conducting the survey, permission was requested from the head of the institution. Student drop out data was gathered through observation and interaction. Consent given by the institution was relayed to the student. The students understood they were participating in research, the confidentiality of research, what information was to be used, and the possible outcomes of the research. All relevant information was explained to the participants.

5 Experimental Results

5.1 Data Pre-processing

5.1.1 Converting Text to Zeros and Ones

The data gathered from the pre-survey and post-survey was in the form of text. This texts was converted into zeros and ones and stored in a database.

5.1.2 Finding Null Values and Separating Them

In the dataset collected, using this method, we separated the null values from the dataset.

5.1.3 Updating Missing Values

After separation of the datasets, we updated the null values using values relevant to other corresponding values.

5.2 *Classifier Used*

5.2.1 The Algorithm

The Naive Bayesian classifier uses the concept of independent assumptions existing between predictors, based on the Bayes' Theorem. A Naive Bayesian classifier uses the fact that the value of a predictor (*x*) on a given class (*c*) is independent of the values of other predictors. This assumption is called class conditional independence.

$$P(c/x) = P(x/c)P(c)/P(x) \tag{1}$$

P(c|x)	is the posterior probability of a class (target) given predictor (attribute)
P(c)	is the prior probability of the class
P(x|c)	is the likelihood of the probability of a predictor given class
P(x)	is the prior probability of the predictor

5.3 *Dataset*

The datasets are collected using surveys. Two survey forms were designed, i.e., for post-survey and pre-survey. Pre-survey was completed during the admission of the student. Post-survey was completed during the middle term of the course. The questions were framed in a way to obtain a YES or NO answer, which will then be pre-processed into zeros and ones. The frequency count of these is considered and, based on the survey's questions, categorization is undertaken, which helps understand which category is the main reason for a student dropping out. The summarized data is categorized thus: lack of attendance, number of failing subjects, satisfactory rating of hostel facilities, involvement in disciplinary issues, financial status, confidence in English speaking and writing.

5.4 *Evaluation and Measurements*

See (Tables 1, 2).

Table 1 Dropout predictions and probability distribution

— Prediction on test split —

Institution	Actual	Predicted	Error	Probability distribution
1	1:No	1:No	*0.834	0.166
2	2:Yes	2:Yes	0.028	*0.972
3	1:No	1:No	0.99	0.01
4	2:Yes	2:Yes	0.129	*0.871
5	1:No	2:Yes	0.338	*0.662
6	1:No	1:No	*0.608	0.392
7	1:No	1:No	*0.864	0.136
8	2:Yes	1:No	*0.781	0.219
9	2:Yes	1:No	*0.563	0.437
10	2:Yes	2:Yes	0.223	*0.777

Table 2 Detailed accuracy of the dataset

== Detailed accuracy analysis by class ==

TP rate		FP rate	Precision	Recall	F-measure	ROC area	Class
	0.902	0.128	0.881	0.902	0.892	0.964	No
	0.872	0.098	0.895	0.872	0.883	0.964	Yes
Weighted average	0.888	0.113	0.888	0.888	0.887	0.964	

=== Confusion matrix ===

a b < – classified as

37 4 | a = No

5 34 | b = Yes

=== Evaluation of test set ===

=== Summary ==

Correctly classified instances	71	88.75%
Incorrectly classified instances	9	11.25%
Kappa statistic	0.7747; agreement of prediction with true class	
Mean absolute error	0.19; not squared before averaging	
Root mean squared error	0.288; squared before averaging, so large errors have more influence	
Relative absolute error	38.0192%; relative values are ratios, and have no units	
Root relative squared error	57.6279%; total number of instances equals 80	

Table 3 Summary of student responses to survey questions

No	Label	Count
1	No	39
2	Yes	37

5.5 *Results and Discussions*

See (Table 3).

The total data collected was 76 of which 39 counts were NO and 37 counts were YES. We came to the conclusion that students who said No represented a negative indication of satisfaction towards academia. We observed through graphical representation that all students who had low attendance, a higher number of course failures, and had no chance of attending an exam because of a lack of attendance would go on to drop their course between terms.

5.6 *Visualization*

See (Figs. 3, 4, 5).

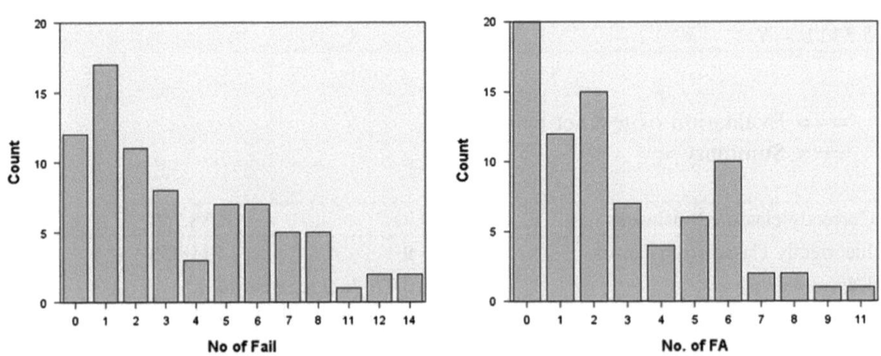

Fig. 3 Graphical representation of both academically failed students and lack of student attendance

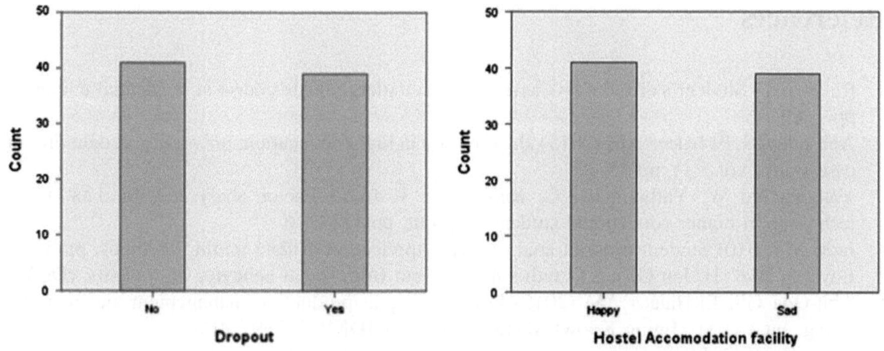

Fig. 4 Representation of willingness to take a decision to drop a course and/or continue a course

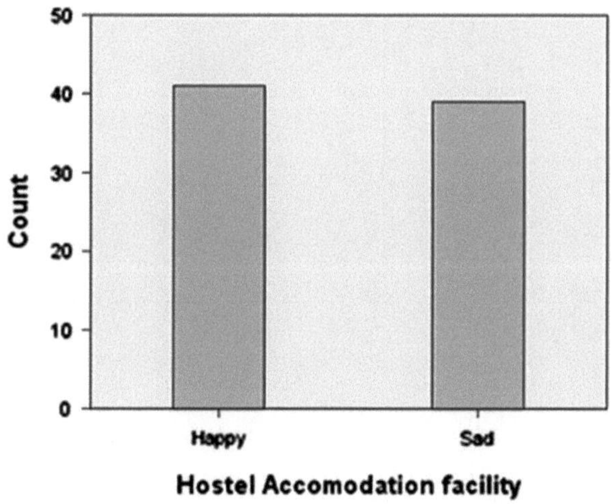

Fig. 5 Representation of individual opinions about residence in student hostels

6 Conclusion

The research was undertaken not only to identify potential dropouts, but also to help institutions understand the reasons for students dropping out. It helps institutions put in place corrective measures to reduce the dropouts. This chapter serves as a platform for predicting dropouts using a Naïve Bayesian classifier. Institutions can therefore have a clear idea of students who are dropping out, who are likely to get dropped, along with the reasons for this. In the future, prediction of the rates of students dropping out can be made.

References

1. Rai S (n.d.) Student's dropout risk assessment in undergraduate course at residential university, pp 1–69
2. Abu-oda GS, El-halees AM (2015) Data mining in higher education: university student dropout case study, vol 5(1), pp 15–27
3. Yathongchai W, Yathongchai C, Kerdprasop K (n.d.) Factor analysis with data mining technique in higher educational student drop out, pp 111–116
4. Jadri M (2010) Student dropout analysis with application of data mining methods, pp 31–46
5. Bayer J, Bydˇ H, Jan G (n.d.) Predicting drop-out from social behavior of students, (Dm)
6. Abu-Oda GS, El-Halees AM (2015) Data mining in production management and manufacturing. Int J Data Mining Knowl Manage Process (IJDKP) 5(1):97–106
7. Muzenda A (2014) Students perceptions on determinants of dropouts from colleges, vol 5, no 1, pp 114–118

A Survey on Energy-Efficient Techniques to Reduce Energy Consumption in Delay Tolerant Networks

Lalit Kulkarni, Nimish Ukey, Jagdish Bakal and Nekita Chavan

Abstract There are several approaches to reduce the energy consumption. Among them, one approach which allows for reducing the transmission energy consumption is Delay Tolerant Networks (DTNs). Several techniques are adopted to optimize the energy efficiency. Adjusting the resource allocation and decreasing the equipment usages are the basic two categories, which are used to reduce the energy consumption. With the increasing data transmission rate the energy consumption increases, which result in the diminishing of its efficiency. So, it is a necessity to manage the trade-off between energy and performance in Delay Tolerant Networks. The DTN nodes commonly operate on low-power battery resource; hence, there is need of improving the energy efficiency by using appropriate technique to increase the lifetime of the node and also to increase the probability of delivering the bundle. For that purpose, this paper describes the various techniques to achieve energy efficiency in Delay Tolerant Networks.

Keywords Delay Tolerant Networks · Mobility · Opportunistic forwarding · Power management · Store-and-forward · Efficiency

L. Kulkarni (✉) · N. Ukey (✉)
Department of Information Technology, Maharashtra Institute of Technology, Pune, India
e-mail: lvkulkarni@gmail.com

N. Ukey
e-mail: nimishukey@gmail.com

J. Bakal · N. Chavan
Department of Information Technology, GH Raisoni College of Engineering, Nagpur, India
e-mail: bakaljw@gmail.com

N. Chavan
e-mail: nekita.chavan@raisoni.net

H.R. Vishwakarma and S. Akashe (eds.), *Computing and Network Sustainability*, Lecture Notes in Networks and Systems 12, DOI 10.1007/978-981-10-3935-5_9

1 Introduction

The DTN does not have the end-to-end path, so it is very highly challengeable to perform the communication between the source and the destination because of the sparsely connected or intermittently connected network and the mobility of the node. DTN nodes have the very limited resource devices, such as the mobile nodes, sensor nodes, and war tanks which also majorly operated on a battery. Hence, because of less or few energy resources it is necessary to reduce the consumption of energy in such networks. DTN has various applications such as underwater networks [1], Vehicular Ad hoc NETworks (VANETs) [2], military networks [3], and interplanetary networks [4]. The traditional routing protocols/mechanism was developed for end-to-end connections, hence they will not work here. Therefore, store, carry, and forward technique have been used in DTNs.

The end-to-end path establishment was necessary for a traditional routing scheme before transmitting a data. So IRTF (Internet Research Task Force) proposed a new technology called DTN which uses the opportunistic forwarding that uses the "store-carry-forward" manner. IRTF also introduced the concept of "bundle". The bundle is the block of data which is stored in the MT (Mobile Terminal). MT transmits the data to another MT or directly transmits it to the destination. The main problem is to decide when to activate the interface and start the probing for the message transmission. Hence, the continuous searching will result in fast discovery, but with more consumption of the battery. So, it needs to decide when to activate it for transmitting the message. The discovery process of another MT is very energy consuming as it continuously tries to search for MT to transmit the message until and unless it finds it. Hence, there is need of the trade-off between the quicker discovery and less power consumption.

In DTN, the massive energy consumption brings unimaginable issues to the energy sources. Hence, by reducing the energy consumption in DTN will increase the life span of devices or terminals and also it will help to reduce the cost of communication. Decreasing the energy consumption in DTN can not only cut down the cost of communications, but also benefit the long-term development of wireless communications.

In this paper, the main issue of energy efficiency is discussed, which includes optimization and trade-off. Firstly, the optimization of energy efficiency in DTNs is discussed. Then, two methods of energy efficiency optimization are described. Performance and energy are directly proportional to each other in DTNs. So in the case of saving energy, the performance will automatically degrade. Hence, there is a need to trade-off between the energy and performance, which is discussed with an example later on. At last, various techniques are discussed to improve the energy efficiency in delay tolerant networks.

2 Related Work

The energy efficiency metric is explained in the first section of the paper. In the second section, the energy efficiency categorization is discussed which includes the two kinds, which are resource allocation adjustment and equipment usage decrements. In the third section, the trade-off between the energy and performance is discussed.

2.1 Energy Efficiency Metric

In existing energy efficiency metric, the main issue is that they are designed for parts of wireless networks rather than the complete networks. Furthermore, they are not principally planned considering the different elements in diverse situations, e.g., traffic load.

2.2 Energy Efficiency Optimization

Figure 1 illustrates the various schemes of energy efficiency optimization. The illustrated schemes are already existing approaches which reduce the energy consumption by reducing the consumption of the equipment and the allocation of resources.

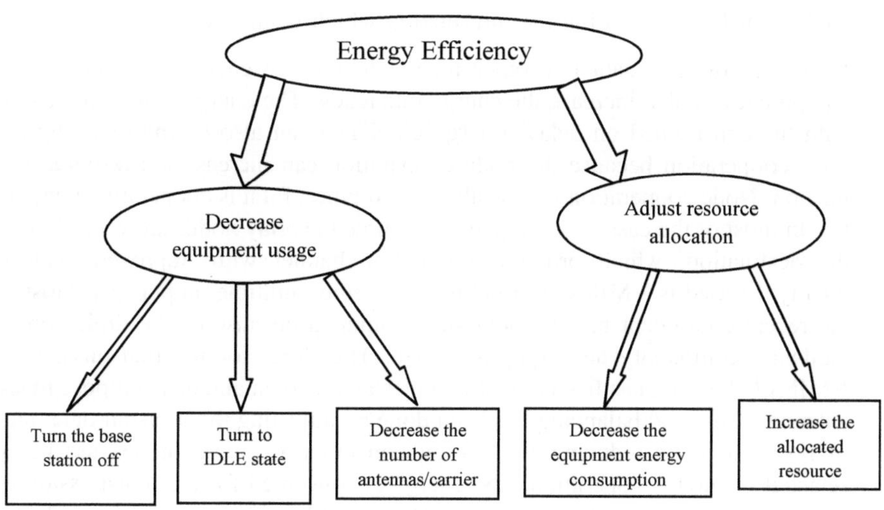

Fig. 1 Categories of energy efficiency optimization

Decrease the equipment usage: To reduce energy consumption, the most common way is to turn off the base station when the traffic load of BS (Base Station) is very low (i.e., changing the BS to IDLE).

- The usage of equipment can be decreased by turning off the BS. When a base station coverage area is overlapped in such a way that it covers the coverage area of another base station, then we can able to turn off the base station according to the traffic distribution to conserve the energy.
- When within the range of BS nobody accesses the BS then BS needs to change the state to IDLE.
- To preserve energy, the number of antennas needs to be adjusted in a frequent manner. When the number of antennae is changed, the information needs to be updated or reconfigure it.

Adjusting the resource allocation: To reduce the consumption of energy, there is a need to adjust the resource allocation. The resource allocation can be adjusted by the two ways. The first way is to decrease the energy consumption of equipment, and another way can be done by incrementing the resources for the allocation purpose.

- If some nodes of the equipment are in sleep mode, then by reducing the energy consumption of equipment we can reduce the energy consumption of nodes. The nodes are very less functional in sleep mode; hence, the energy can be saved. The energy can be saved by energy efficient routing and scheduling. The trade-off between the residual energy and energy consumption is balanced. Another way to conserve the energy is by decreasing the energy consumption of equipment for transmitting the information which is unnecessary.
- Reducing transmit data rate other than reducing the energy consumption of equipment can also increase the energy efficiency. The energy can be improved with the constrained on delay. Energy efficiency can also be improved by the node cooperation because the node cooperation can increase the resource utilization. Node cooperation is basically of two types. First is cooperative relaying [5]. In most of the cases, as compared to source the relay nodes are very close to the destination, which provides the extra channel with improved channel quality. Second is VMIMO (virtual multiple input multiple output) [6]. Most of the recent equipment has the only single antenna because of the limitation of hardware complexity and equipment size. Therefore, for the transmission of VMIMO, the antennas from the other nodes and its own antenna compose many antennas. The load balancing has been done by adjusting the transmit data rates of nodes and for reducing the energy consumption, fewer nodes are being selected for node cooperation. The energy is consumed for the transmission of data and also for the control signal exchange and equipment exchange. These

are the two main things, which consume more energy by the node. The use of all nodes is not a good idea to select it for the transmission purpose because even if the node does not participate in the transmission of a data the energy will get consumed by a node. Hence, the energy cannot be saved by the node.

Trade-off between the energy and performance: With a reduction of performance, the energy consumption is reduced. We can say that the performance and energy are directly proportional to each other. So, it is needed to trade-off between the energy and performance.

In real-time systems, there is a time constraint. We cannot delay the data delivery beyond the time limit. If the transmission data rate is decreased, then the energy consumption will also get reduced. The data rate cannot be very low. It needs to manage in such a way that it can be slow as much as possible with the consideration of time constraints and performance, which is one of the methods for balancing the trade-off. Hence, it is necessary to trade-off between the consumption of energy and transmission performance.

3 Techniques

3.1 Distance-Based Energy-Efficient Opportunistic Forwarding (DEEOF)

The two schemes have been proposed for DEEOF [7] in mobile DTN. It will help to improve the efficiency of energy and also increase the message delivery ratio. The forwarding equivalent energy efficient distance (FEED) algorithm will help for energy efficiency by providing the same energy efficiency for different time intervals. These schemes compare the future energy efficiency with current energy efficiency and regarding that decision is taken. Opportunistic forwarding is an effective and efficient way for achieving a trade-off between the performance of the network and consumption of energy for achieving maximum energy efficiency. There are basically two DEEOF algorithms as follows:

Distance-based energy-efficient opportunistic forwarding: In this algorithm, if there is any relay node existed within the maximum transmission range of the source node and if its value is less than the threshold value, then the source node forwards the packet to that nearest node without copying the data. This algorithm is performed at the source at every contact time.

Algorithm 1: Distance-based energy-efficient opportunistic forwarding

1. Set the threshold frequency
2. Then it checks if there is any relay node within its maximum transmission range. If yes, then the distance of relay node is calculated by without copying a packet. it measures the distance of the relay without packet copy
3. Calculate the forwarding equivalent energy-efficiency distance (FEEDs i.e. to define the relationship between the two nodes and delay for equivalence ene rgy efficiency) and P (The probability that the distance to the source node is smaller by at least onerelay).
4. If P is smaller than the threshold value, then it forwards the copy of the packet to the nearest relay node. Else there is no need to forward the packet copy.
5. Set the threshold frequency

Distance-based energy-efficient opportunistic forwarding: It is based on the probability distribution function (p.d.f.) of forwarding energy efficiency. It can predict how better energy efficiency is achieved with more accurate prediction; hence, the better forwarding decision can be taken.

3.2 *Message-Driven Based Energy Efficient Routing*

The DTN nodes commonly work on low-power battery resource; hence, there is need of improving the energy efficiency by using an appropriate protocol, or scheme to increase the lifetime of the node and also to increase the probability of delivering the packet. The message-driven based energy efficient routing [8] improves the energy efficiency by enabling the forward based on the delivery requirements and individual message lifetime.

ER: In an epidemic routing [9], the node forwards the message to all the nodes which come in contact with that node, i.e., it floods the message. Here, the message delivery rate has been increased. The node which receives the message will carry forward the message to the next node until and unless the destination node receives the message. This process reduces the message delivery delay but increases the cost of delivery.

2HR: In Two Hop Routing [10], whenever a source node comes with a contact of any node, then it forwards the message to it. But the restriction is that the relay node can be able to transmit the data to the destination only, i.e., it cannot able to transmit the data to another relay node. It reduces the delivery cost but increases the delivery delay.

Energy efficient routing algorithm: This algorithm takes the message lifetime and message delivery probability as an input and after that, it returns the optimum number of message copies which are generated.

Here, it calculates the amount of message copies which is forwarded to the nodes which fulfill the constraints of the message lifetime and message delivery lifetime. Hence, the amount of message copies forwarded to a node can be any value from 1 to N, where the message lifetime time with the 1–N copies of the Delay Cumulative Distribution Function (CDF) of the system should be greater than or equal to the message delivery probability. If the set of a number of message copies spread to the node (G) is zero, then it cannot able to find the optimal number of message copies, it then returns the maximum number of possible copies (i.e., same as the number of nodes in the network). If the G is greater than or equal to one element, then it calculates the energy corresponds to each element in G and the elements which utilize the very less energy is then corresponds to optimal values.

3.3 *DTN-Oriented Wireless Activation Mechanism Based on Radio Fluctuations (DWARF)*

As the mobile terminal has a very limited amount of battery resource, it needs to save energy while searching for 'terminal discovery' because this process is very energy consuming. The DWARF [11] mechanism is used for terminal discovery which will consume very less amount of energy as compared to others.

DWARF (energy efficient adaptive interface activation for DTN): The activation of the interface is depending on the situation. The interface is activated after the particular interval of time; if the MT cannot able to discover another MT, then it needs to change the interval of time to reduce the battery consumption, which has been accessed by the MT for continues discovery of the other mobile terminal.

Performance: The energy consumed for the NDREQ (neighbor delivery request) equals to the number of interface activation. Hence, if the number of interface activation has been reduced to half, then the energy can also get reduced by half.

Evaluation: On the basis of the signal fluctuation of the MT, the activation of the interval has been set. Let, the mobile terminal fluctuates at a high rate then it can be set to 10[s] otherwise with the fluctuation gets slow, then the activation of the interval will be set to 40[s].

DWARF enables the MTs to reduce the consumption of energy by reducing the number of NDREQ. Even after reducing the number of NDREQ, the success of discovering the other MT is equal to the fixed-interval scheme. DWARF achieves the NDRSP equals or slightly more than the fixed-interval scheme because of the fewer collisions. The efficiency of the DWARF is not depending on the number of Mobile Terminals.

4 Conclusion

The survey has been done on the various energy-efficient techniques in DTNs. Based on this survey, the various techniques for reducing the energy consumption in DTNs have been studied. The methods of optimizing the energy efficiency are categories into two kinds, i.e., adjusting the resource allocation and decreasing the equipment usage. It is necessary to balance the trade-off between the network performance and energy consumption. The two DEEOF algorithms are used to reduce the energy consumption (i.e., improves the energy efficiency) and to achieve the higher packet delivery ratio. The opportunistic packet forwarding has been done on the basis of the current energy efficiency and the predicted future energy efficiency. The message-driven based energy efficient algorithm thus improves the energy efficiency up to a great extent as compared to the ER and 2HR protocols. This algorithm forwards the message based on the lifetime of the message and the requirements of the delivery for the DTN. The DWARF has the potential for reducing power consumption as well as enhancing efficiency. Thus, all the techniques play a very crucial role to improve the energy efficiency in DTNs. In the future work, the design and implementation of a hybrid algorithm is possible for better performance and energy efficiency and can also be extended with the consideration of heterogeneous (in terms of transmission range) node.

References

1. Fall K (2003) A delay-tolerant network architecture for challenged internets. In: Proceedings of the conference on applications, technologies, architectures, and protocols for computer communications, vol 1, pp 27–34, Aug 2003
2. Farnoud F, Valaee S (2009) Reliable broadcast of safety messages in vehicular ad hoc networks. In: Proceedings of the IEEE international conference on computer communications, vol 1, pp 226–234, Apr 2009
3. Krishnan R et al (2007) The spindle disruption-tolerant networking system. In: Proceedings of the IEEE military communications conference, vol 1, pp 1–7, Oct 2007
4. Burleigh S, Hooke A, Torgerson L, Fall K, Cerf V, Durst B, Scott K, Weiss H (2003) Delay-tolerant networking: an approach to interplanetary internet. IEEE Commun Mag 41 (6):128–136
5. Laneman JN, Tse DNC, Wornell GW (2004) Cooperative diversity in wireless networks: efficient protocols and outage behavior. IEEE Trans Inf Theory 50(12):3062–3080
6. Dohler M, Lefranc E, Aghvami H (2002) Space-time block codes for virtual antenna arrays. In: IEEE international symposium on PIMRC 2002, vol 1, pp 414–417, Sept 2002
7. Lu Y, Wang W, Chen L, Zhang Z, Huang A (2014) Distance-based energy-efficient opportunistic forwarding in mobile delay tolerant networks. In: IEEE international conference on communications (ICC), vol 1, pp 3564–3569
8. Chaithanya Manam K, Gurav G, Siva Ram Murthy C (2013) Performance modeling of message-driven based energy-efficient routing in delay-tolerant networks with individual node selfishness. In: IEEE COMSNETS vol 1, pp 1–6

9. Vahdat A, Becker D (2000) Epidemic routing for partially connected ad hoc networks. Duke University, Technical Report, vol 1, pp 1–14, Apr 2000
10. Groenevelt R, Nain P, Koole G (2005) The message delay in mobile ad hoc networks. Perform Eval 62(1–4):210–228
11. Izumikawa H, Pitkanen M, Ott J, Timm-Giel A, Bormann C (2010) Energy-efficient adaptive interface activation for delay/disruption tolerant networks. In: IEEE ICACT, vol 1, pp 145–150, Feb 2010

Triband Handset Antenna Designing with a High Gain for UMTS/WiMAX/WLAN Applications

Sonam Parekh, Rajeev Mathur and Payal Jain

Abstract This paper presents a novel design of a triband antenna for wireless communications. Presently, many research groups are working on LTE/4G/5G mobile communications technologies and proposed various types of antenna designs. We have proposed an antenna which not only operates at three resonant frequencies but also having high gain. This antenna also fulfills the requirement of low cost, less weight, small size for wireless mobile devices. The design and simulation of proposed triband mobile antenna carried out with Flame Retardant 4 (FR-4) substrate and dimension of antenna substrate is 95 mm × 50 mm. Performance parameters of this antenna is investigated in terms of gain, return loss, VSWR, and radiation pattern. The resonant frequencies of the proposed antenna are 2.45, 5.09, and 7.65 GHz. High Frequency Structure Simulator software's (HFSS's) optometric is used for the proposed antenna for more accuracy, and results are optimized. The proposed antenna provides an operating band, covering the (UMTS) Universal Mobile Telecommunications System bands (2300–2400 MHz), (WLAN) Wireless Local Area Network bands (2400–2497 MHz), and (WiMAX) World Interoperability for Microwave Access system bands (3300–3790 MHz) simultaneously. The simple configuration and low profile attributes of the proposed antenna made it easy for fabrication and suitable for the application in the UMTS/WiMAX/WLAN and satellite communications.

Keywords UMTS · WiMax · WLAN · Triband antenna · 4G

S. Parekh (✉) · P. Jain
Department of Electronics & Communication, Geetanjali Institute of Technical Studies, Udaipur, Rajasthan, India
e-mail: sonam_parekh@ymail.com

P. Jain
e-mail: payaljain248@gmail.com

R. Mathur
Suresh Gyan Vihar University, Jaipur, India
e-mail: rmathur_2000@gmail.com

H.R. Vishwakarma and S. Akashe (eds.), *Computing and Network Sustainability*, Lecture Notes in Networks and Systems 12, DOI 10.1007/978-981-10-3935-5_10

1 Introduction

In recent years, wireless communication system provides a great interest in an antenna with multiband characteristics for research work. Antennas with small size, low-cost fabrication, light weight, conformability, ease of installation and integration with feed networks have many applications over the broad frequency ranges. This system is having data transfer rate of around 10 times faster than the 3G mobile communication. Nowadays due to increasing demand of high spectral efficiency to transfer data in form of video and multimedia, it is required to develop such antenna which operates in wide range of frequencies. As per the recent demand, a triband antenna, for the next generation wireless communication system, is presented in this paper. The proposed antenna provides operating bands, covering the (UMTS) Universal Mobile Telecommunications System bands (2300–2400 MHz), (WLAN) Wireless Local Area Network bands (2400–2497 MHz), and (WiMAX) World Interoperability for Microwave Access system bands (3300–3790 MHz), simultaneously [1, 2].

2 Antenna Structure

Design of triband antenna based on the basic parameters of resonant frequency fr = 2.4 GHz, dielectric constant of FR4 Substrate i.e., 4.4, loss tangent of 0.0002, and height of 1.6 mm. Width and length of a patch antenna is calculated using standard formula for the design of microstrip patch antenna. The design has been simulated by using high frequency structure simulator software (HFSS) which is a full-wave electromagnetic field simulation package with the criterion of return loss S11 less than −10 dB [3, 4]. The total size of the substrate is 95 mm × 50 mm including ground plane with 65 mm × 50 mm and height of 1.6 mm, and the size of the radiated patch is 30 mm × 28 mm is fed by a microstrip line. Rectangular patch has been investigated for the triple band antennas in the wireless communication systems.

Design of the proposed antenna structure is shown in Fig. 1. Ground plane is kept on the same plane of antenna. The antenna is fed by 50 Ω microstrip line [5, 6].

Following are the design specifications for the triband antenna with antenna patch and ground structure:

Patch width (W) = 28 mm
Patch length (L) = 30 mm
Ground plane width (Wg) = 50 mm
Ground plane Length (Lg) = 65 mm
Dielectric constant 4.4
Height of substrate (h) = 1.6 mm

Fig. 1 Geometry of microstrip patch antenna

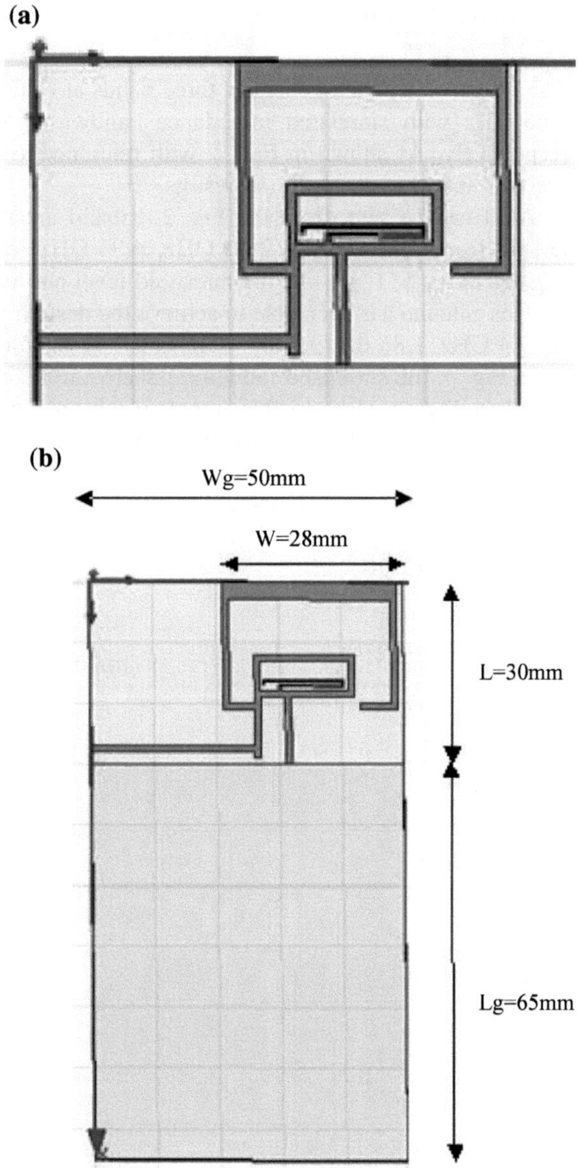

3 Results and Discussion

The proposed antenna generates three bands at center frequency of 2.45, 5.09, and 7.65 GHz with simulated impedance bandwidth of 0.23, 1.86, and 2.62 dBm, respectively. As shown in Fig. 2, with reference to the simulation results, we can see that antenna achieve good results.

As it can be seen from the Fig. 2, triband antenna exhibits wideband characteristic from 2.12 GHz to 2.63 GHz, 4.94 GHz to 5.26 GHz, and 7.47 GHz to 7.79 GHz for S11 ≤ -10 dB threshold level and used for mobile handsets.

This antenna has been able to achieve the desired value of the VSWR of 0.23 dB for 2.4 GHz, 1.86 dB for 5.09 GHz, and 2.62 dB for 7.64 GHz as shown in Fig. 3.

In Fig. 4, the simulated radiation patterns at the center frequency fc = 2.4 GHz are plotted. According to this figure, patch antennas produces a good broadside radiation pattern at 2.4 GHz and the peak gain is obtained to be around 26.85 dBm.

Similarly, in Fig. 5, the simulated radiation patterns at the center frequency, fc = 5.09 GHz is plotted, and the peak gain is obtained to be around 29.50 dBm.

As shown above in Fig. 6, at frequency fc = 7.65 GHz, we have obtained gain is around 31.17 dB.

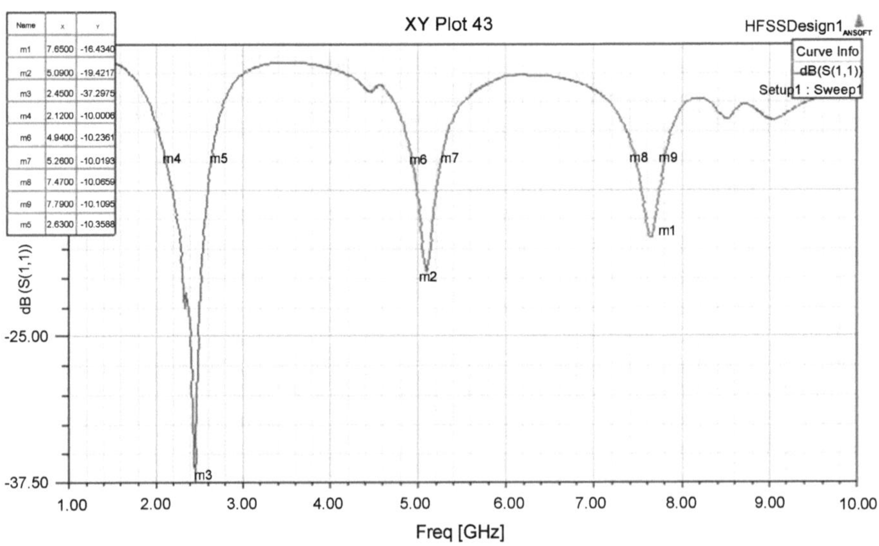

Fig. 2 Frequency response of an antenna

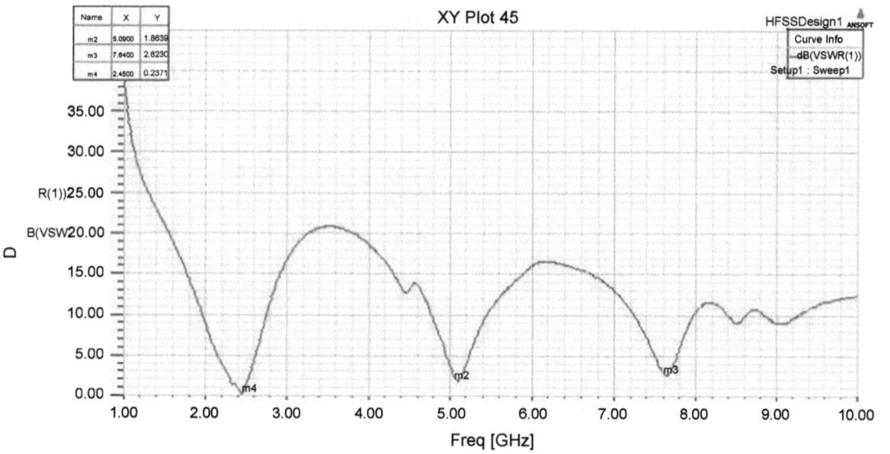

Fig. 3 VSWR response of an antenna

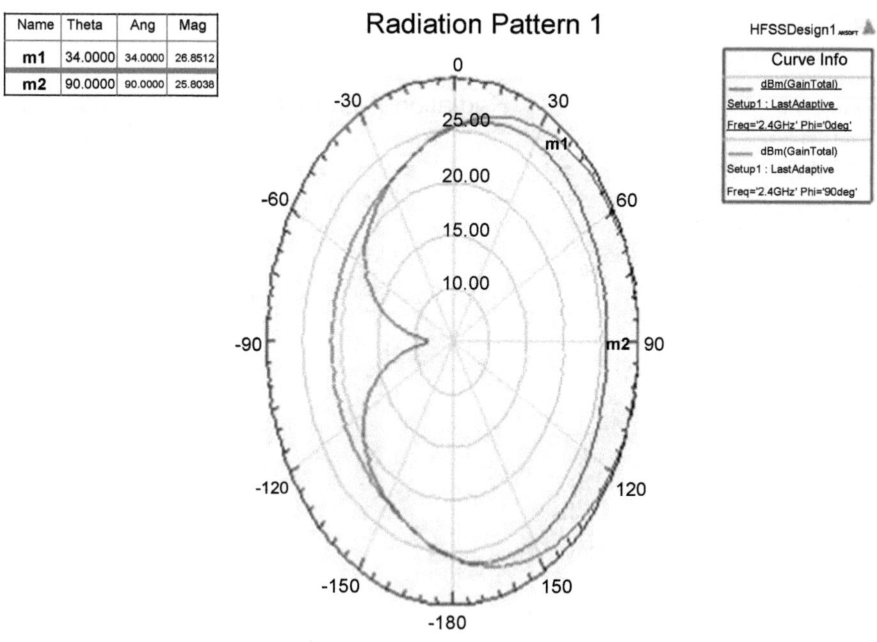

Fig. 4 Radiation pattern of an antenna

Name	Theta	Ang	Mag
m6	50.0000	50.0000	29.5062
m7	44.0000	44.0000	29.4541
m8	54.0000	54.0000	29.2929
m9	38.0000	38.0000	28.9770

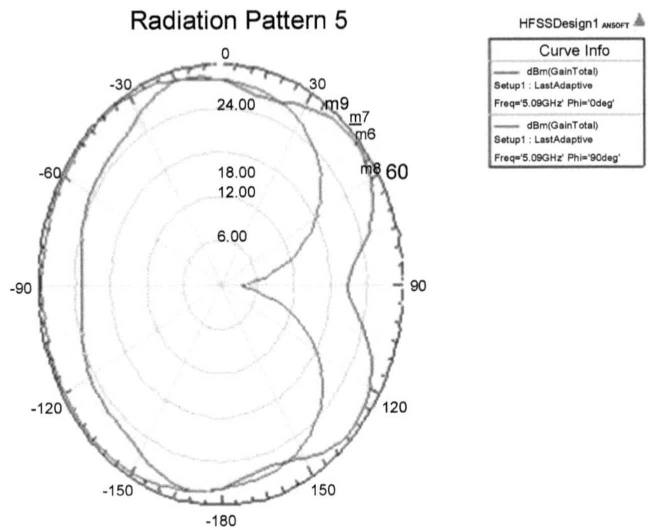

Fig. 5 Radiation pattern of an antenna

Name	Theta	Ang	Mag
m1	60.0000	60.0000	31.1730
m2	64.0000	64.0000	31.0239
m3	54.0000	54.0000	31.0240
m4	68.0000	68.0000	30.7271

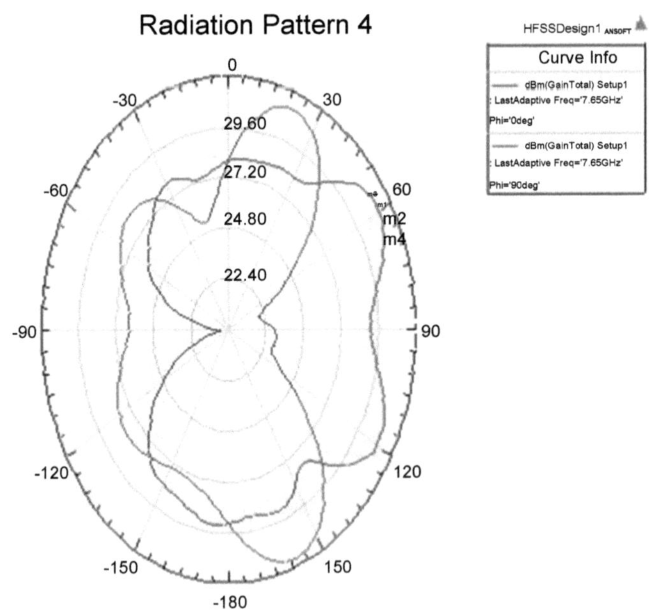

Fig. 6 Radiation pattern of an antenna

4 Conclusion

A triband antenna for wireless communications designed have exhibited a high gain of 26.85 dBm, 29.50 dBm, and 31.17 dBm for the resonant frequencies of 2.4 GHz, 5.09 GHz, and 7.65 GHz, respectively. Besides, size of antenna is kept small considering its use in hand held mobile devices. Likewise, the VSWR value of 0.23, 1.86, and 2.62 dB is obtained.

This antenna thus is a suitable candidate to be used in mobile communication application of (UMTS) Universal Mobile Telecommunications System bands (2300–2400 MHz), (WLAN) Wireless Local Area Network bands (2400–2497 MHz), and (WiMAX) World Interoperability for Microwave Access system bands (3300–3790 MHz) simultaneously.

References

1. Chi YW, Wong KL (2009) Very-small-size printed loop antenna for GSM/DCS/PCS/UMTS operation in the mobile phone. Microw Opt Technol Lett 51(1):184–192
2. Bhatti RA, Yi S, Park SO (2009) Compact antenna array with port decoupling for LTE-standardized mobile phones. IEEE Antennas Wirel Propag Lett 8:1430–1433
3. Wong KL, Huang CH (2008) Compact multiband PIFA with a coupling feed for internal mobile phone antenna. Microw Opt Technol Lett 50(10):2487–2491
4. SEMCAD X by SPEAG, www.speag.com
5. Lizzi L (2011) Dual-band printed fractal monopole antenna for LTE applications. IEEE Antennas Wirel Propag Lett 10(00):760–763
6. Wong KL, Lee GY, Chiou TW (2003) A low-profile planar monopole antenna for multi-band operation of mobile handsets. IEEE Trans Antennas Propag 51(1):121–125
7. Karkkainen MK (2005) Meandered multiband PIFA with coplanar parasitic patches. IEEE Microw Wirel Compon Lett 15(10):630–632
8. Yang D-G, Kim D-O, Kim C-Y (2012) Design of internal multi-band mobile antenna for LTE700/UMTS/WiMAX/WLAN operation

Architectural Outline of Decision Support System for Crop Selection Using GIS and DM Techniques

Preetam Tamsekar, Nilesh Deshmukh, Parag Bhalchandra, Govind Kulkarni, Kailas Hambarde, Pawan Wasnik, Shaikh Husen and Vijendra Kamble

Abstract The crucial task for Indian policy makers and farmers is the decision of crop selection by taking into consideration the various factors, which boosts the precision farming. To overcome this scenario, a decision support system is proposed by using GIS and DM techniques, which helps in deriving a pattern by associating various factors to enhance DSS to suggest potential crop for a region. The system was designed, developed, and implemented across a selected region. This paper narrates the architectural framework of the implemented system.

Keywords GIS · DM · DSS

P. Tamsekar (✉) · N. Deshmukh · P. Bhalchandra · G. Kulkarni · K. Hambarde · P. Wasnik · S. Husen · V. Kamble
School of Computational Sciences, S.R.T.M. University, Nanded 431606, MS, India
e-mail: pritam.tamsekar@gmail.com

N. Deshmukh
e-mail: nileshkd@yahoo.com

P. Bhalchandra
e-mail: srtmun.parag@gmail.com

G. Kulkarni
e-mail: govindcoolkarni@gmail.com

K. Hambarde
e-mail: kailas.srt@gmail.com

P. Wasnik
e-mail: pawan_wasnik@yahoo.com

S. Husen
e-mail: husen09@gmail.com

V. Kamble
e-mail: vijendrakamble5@gmail.com

H.R. Vishwakarma and S. Akashe (eds.), *Computing and Network Sustainability*, Lecture Notes in Networks and Systems 12, DOI 10.1007/978-981-10-3935-5_11

1 Introduction

Geographical information system (GIS) is a computer-based technology which manipulates, explains, stores, and analyses information spatially and helps for planning and decision-making by producing maps and data tables as output. Geographic information system (GIS) has existed for over four decades, the concept of which dates back to the 1960s when computers were used for spatial analysis and quantitative thematic mapping. GIS's present state and its potential trend of future development in the context of mainstream of IT must also be understood by examining the use of geospatial information in various sectors such as industrial sector, business sector, and agriculture sector.

The process of extracting knowledge, pattern, laws, and rules from the spatial database is known as spatial data mining (SDM). SDM helps in improving the accuracy and reliability of decision-making and enables the people to maximize the efficient use of data.

Today's great challenge in the field of agriculture in India is precision farming. In developed countries, policy makers as well as farmers rely heavily on geographic information systems. But, in developing countries, still the predecessor's method for selecting crop for cultivation is used; they rarely consider the scientific facts, which results in higher crop productivity. So in order to make the crop selection process scientific by integrating GIS and DM techniques, will result in better crop productivity. GIS integrated with DM techniques in the sector of agriculture can be useful for making decision in the process of crop selection, yet very less study has been done for crop selection process using GIS integrated with DM techniques, and there is a lot of scope in this field to develop a decision support system by integrating GIS which would be beneficial for farmers in order to do precision farming. GIS technology integrates common database operation such as query and statistical analysis with the unique visualization and geographic analysis benefits offered by map. The major challenges that we are facing in the world are overpopulation, pollution, deforestation, and natural disaster which have geographic dimension. Local problem also has geographic component that can be visualized using GIS technology [1]. Ensuring food security within a changing global climate together with the growing concern in reducing the environmental footprint of farming while increasing the economic viability of agricultural practices has resulted, in the last few decades, in the development of precision agriculture. Research and practice in precision agriculture aim at sustainably optimizing the management of agricultural fields by addressing the spatial variability in plant and environment [2] (Fig. 1).

Frequently, we come across a time to take decision which requires knowledge about the scenario. Sometimes, our problem or scenario is so large and complex; hence, we make decision with incomplete data or information. Yet GIS gives us a simple way to present the complex scenario in a simple manner to understand and take decision [3]. Subsequently, GIS makes the interaction between various factors easy to understand. Whereas precision agriculture is considered as a wide scope

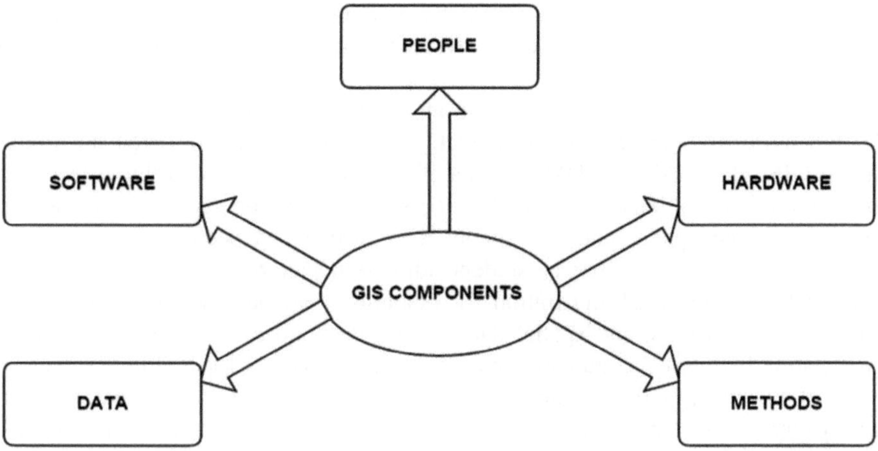

Fig. 1 GIS components

application area which may prosper from novel spatial data mining technique. In accordance with the present theme to develop a decision support system for crop selection, we need to know soil texture, soil biological details, climate, irrigation, etc. The gathered data on its own will not make any sense to take decision. The locations which are spatially referenced are nothing without diagram to understand. With the help of map, one can easily imagine the scenario. The outbreak of cholera in London in the year 1850 is one of the examples depicted by John Snow by using GIS, which helped to take the decision [4].

2 Agriculture Scenarios in India

Agriculture is the main occupation of the people in Maharashtra. Both food crops and cash crops are grown in the state. The food crops sown in Maharashtra are wheat, rice, jowar, bajra, and pulses. Cash crops include cotton, groundnut, turmeric, and tobacco. According to the time of sowing and harvesting, there are two seasons in a year as kharif and rabi.

3 System Outline

It is hypothesized that selection of proper crop by using decision support system can bring improvement in productivity and it could be achieved by using various factors such as soil texture, meteorology, irrigation, and market location in GIS integrated with DM techniques based on decision support system. The study consists of the following objectives: firstly, to develop the thematic cartography of soil, meteorology, irrigation, well inventory, and road network; secondly, to develop a model framework for decision support system for precision farming; and lastly, to validate the approach through an applied case study [5–7].

When we investigated we found that, in [8], author has presented the studies about GIS integrated with DM techniques and identifies the underlying technologies and theoretical background that is necessary to build up such system, and they have integrated spatial and non-spatial data, which has been used to describe, explain, and predict university student admission patterns and thus gives information to university administration for planning strategies for course marketing. GIS, spatial statistics, and spatial data mining techniques are used to explore the association between the students and other various factors for building a decision support system, to enhance the university administration in course marketing. In this paper [9], the author has put in multi-criteria decision-making method. The primary use of the MCE techniques is to look into a number of options in the light of multiple criteria and conflicting objectives. The outcome of this work has indicated that the method employed here was equal to integrate climate, soil, and relief database with different spatial and temporal resolutions in a GIS framework. Using this technique, thematic interpolated map was created, which has taken into account specific characteristics of crops, such as growth cycle or phonological phases. Climate of the MCE of soil and relief environment components proved to be useful to delineate suitable areas for corn and potato crop yield. In [10], the author had developed a decision support system (DSS) for the Guwahati city which bears the principal objective to produce a digital database. For this study, author used the spatial and temporal analysis techniques to develop a decision support system (DSS) which has provided facilities to the planning authorities to take strategic decisions and to set guidelines regarding the new constructions. The use of GIS involves in the integration of spatially referenced data in a problem-solving environment, and the author also explained that GIS-based decision support system (DSS) is an interactive computer-based systems which help decision-makers to utilize the data and models to solve unstructured problems. Author also taught with the aid of this DSS sets a meaningful relationship will be broken which will address zoning and its connection to the existing urban density distribution, the demand of building permits, the rate of urban growth, and the index of saturation. During this study, the author focused some main objective like to aid the decision-makers in laying the groundwork for the development of the Guwahati metropolitan area and to develop a Web-based interactive decision support system (DSS) for quick and ready extraction of plotwise detailed information. For a long-term development plan, this decision support system can be applied in much larger cities with the summation of more data and desired changes. The methodology adopted for spatial decision support system for crop selection is presented in Fig. 2. Several important factors applicable for decision support system such as biological details of the soil, irrigation, seasons, and crop details are taken into consideration in order to keep the model compact and efficient.

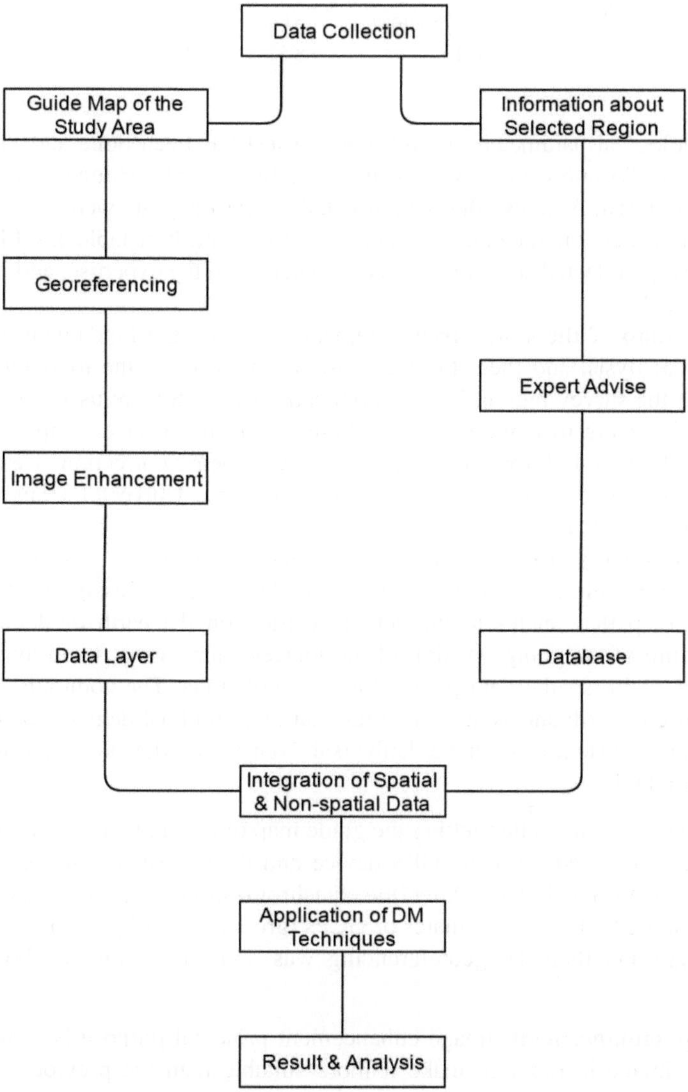

Fig. 2 Proposed DSS for crop selection

4 System Planning and Architectural Components

In order to produce the decision support system for crop selection, the following method has been used.

1. **Data Collection**: After the identification of the necessary fields for the data collection, the next step was to acquire the data. In order to know the chemical

Table 1 pH of collected soil sample

Soil sample	S1	S2	S3	S4
pH	8.51	8.13	8.14	7.82

and biological parameters of soil, the soil test has been done and the pH was acquired. To know the water availability, the records of tube well inventory were acquired. And also the rainfall details of past 5 years were obtained. After the data collection, the data were entered into the attribute table of GIS software. The pH of collected soil samples is calculated from the expertise, and it is shown in Table 1.

2. **Guide map of the study area**: Thematic maps are the fundament of the suitability analysis, and then, the foremost step is to draw the topographic guide map of the survey region. Several applications of GIS were used for overlaying thematic layers to demonstrate databases, and all the layer maps have to be changed into a uniform coordinate system. Geometric corrections were done for the maps of different extractions and were put into Universal Transverse Mercator (UTM) projection.
3. **Georeferencing**: Before processing on map into GIS, the base map must be properly georeferenced. Georeferencing is the process of assigning coordinates to the map that relates to the actual position on the earth or defined as the procedure of assigning position of the images such as maps and aviation photographs with spatial data points, lines, or polygons. The common and worldwide used coordinate system is latitude and longitudinal degree, other than this coordinate system state plane, Universal Transverse Mercator, and many other also available.

In the present study after getting the guide map of the study area, the four control points were collected by using GPS device and then these coordinates were synchronized in GIS software. After the synchronization of coordinates using the georeferencing tool, the coordinates or GCPs (ground control points) were assigned on the map and then the georeferencing was done according to above-defined process.

4. **Image enhancement**: Image enhancement principal purpose is to process the given image in order to make it more suitable than the previous or original image so the result image can be used for further requirement.
5. **Data layers**: Digitization is an important process in GIS; in digitization, the base map is digitized by using the points, lines, polylines, and polygons feature. In the current scenario, the points have been used to show the tube well in the study area, whereas by using polygon, the border map of the study area has been depicted. Also, the irrigated area has been shown by using the polygon.
6. **Information about selected region**: Information about the landholder and land is collected from the authorized government agency. Land information covers the land size, availability of irrigation, and the crops taken by the farmer, and the landowner's personal details were also gathered. By using the irrigation

information, the irrigated area map has been generated and also borehole location map is also generated. Also, this information is considered for crop selection process.

7. **Expert's advice**: To deal with agriculture-related information, expert advice has been taken into consideration. The crops taken into consideration as options for selection and the parameters for selecting the crops that is soil pH, irrigation, and season are done with the advice of the agriculture expert.
8. **Database**: In database creation, the data gathered for this work are organized in order to apply it in the field and for further use in data mining. In present study, data collected about landholder and the owner are entered in GIS and the data gathered from the expert are also preserved in a database.
9. **Integration of spatial and non-spatial data**: After the cartography and data collection, the important step in spatial decision support system is the integration of the spatial data and non-spatial data. Integration of spatial and non-spatial data means joining the database to the respective map.
10. **Application of DM techniques**: By using the clustering algorithm, the clusters of topographic features (elevation, slope, terrain), physical and chemical properties (texture, colour, organic matter content, pH), climatic factors (temperature, rainfall, relative humidity), and distance to market are created. Then, finally by using these clusters, a decision tree is developed which helps in decision-making.
11. **Result and analysis**: After applying the data mining techniques, the results generated are processed in GIS and analysed and then are shown on maps.

5 Conclusion

This paper summarizes how to design and develop GIS-based DSS for crop section in Indian scenario using GIS integrated with DM techniques. The proposed system is spatial decision support system for selecting the crop. It was our experimental finding that the GIS integrated with DM technique has offered broad and easy to use tools for analysis and has been considered as a best practice to be used for decision-making process. This study is also important and relevant to fill the knowledge vacuum of farmers in decision-making process for crop selection by considering the various important parameters for the respective crop.

References

1. Web resource. http://www.westminster.edu/staff/athrock/GIS/GIS.pdf Accessed 30 May 2014
2. Peeters A, Ben-Gal A et al (2012) Developing a GIS-based Spatial Decision Support System for Automated Tree Crop Management to Optimize Irrigation Inputs, iEMSs

3. Seppelt R, Voinov AA, Lange S, Bankamp D (eds) International congress on environmental modelling and software managing resources of a limited planet, sixth Biennial meeting, Leipzig, Germany
4. Web resource. http://www.iemss.org/society/index.php/iemss-2012-proceedings
5. Tamsekar P et al, Implementation of GIS based DSS for crop selection: a case of autumn crops. In: 9-10 ICICT 2015, international conference springer international publications, vol 1. ISBN 978-981-10-0766-8
6. Tamsekar P et al, Architectural outline of GIS based DSS for crop selection. In: 9-10-2015 ICICT 2015, international conference springer international publications, vol 2. ISBN 978-981-10-0754-5
7. Tamsekar P et al, GIS Based decision support system for crop selection In: 17-10-2015 Special edition for NCETCIT- 2015 National Conference IJCRT International Journal, vol 1, Issue-2, Dec 2015, ISSN 2454-7719
8. Tang H, McDonald S (2002) Integrating GIS and spatial data mining techniques for targeting of university courses. In: Symposium on geospatial theory, processing and applications, Ottawa
9. Zhu Z et al (2009) Research on GIS-based agriculture expert system software engineering, 2009. In: WRI world congress on WCSE'09, vol 3. IEEE, pp 252–255. E-ISBN 978-0-7695-3570-8
10. Biswajit S, Reddy D, Venketrao B (2004) GIS based decision support system for seismic risk reduction in urban planning of Guwahati city, India. In: 13th world conference on earthquake engineering, Vancouver, B.C., Canada, 1–6 Aug 2004, Paper No. 2685

Eval Is Evil: Analyzing Performance of Web Applications Based on PHP and JavaScript by Static Analysis

Nilay Shah and Praveen Gubbala

Abstract Transforming text into executable code at runtime with a function Eval() in dynamic languages such as PHP and JavaScript provides the ability to programmers to extend applications at any time. But every extensive power comes with a price, and here, performance security and efficiency are the cost. In prior work, we examine reason behind the performance degradation by Eval() calls. But in PHP, Zend compiler has some limitation, and in JavaScript, browser has few limitations. Though the execution of Eval() remains unchanged, we identified few replacements for the same functionality. As few large-scale PHP frameworks have a common pattern of unnecessary use of this dynamic feature, we targeted moodle (a large-scale PHP framework) to prove performance enhancement by replacing Eval'd code with some other programming features. Our static analysis survey reflects that almost 70% Eval'd code is replaceable.

Keywords Eval() · Static analysis · Meta-programming language

1 Introduction

If we look up on back-hand development language, PHP scripting will definitely be on top. This language provides very high dynamism and its flexible nature engages more programmers with top preference. JavaScript, the dynamic scripting language, enables an easy way to turn text into executable code at runtime. This paper will explain more about runtime execution for both scripting language and performance prediction techniques by static analysis.

Language PHP has started as collection of scripts as to develop personal homepages. But now PHP considered as scripting language which is one of the most popular in the Web world for server-side language. Traditionally, PHP used to develop

N. Shah (✉) · P. Gubbala (✉)
Symbiosis Institute of Technology, Lavale, Pune 411042, Maharashtra, India
e-mail: nilay.shah@sitpune.edu.in

P. Gubbala
e-mail: praveeng@sitpune.edu.in

H.R. Vishwakarma and S. Akashe (eds.), *Computing and Network Sustainability*,
Lecture Notes in Networks and Systems 12, DOI 10.1007/978-981-10-3935-5_12

dynamic Webpage like to retrieve dynamic content from another source but now in modern approach in PHP indicates it as one of the strongest server-side programming language. JavaScript is high-level, dynamic, untyped scripting language. Along with HTML and CSS, JavaScript is one of the third essential elements in Web terminology. The majority of Web sites employ it, and most of the modern browsers support it without any plug-in installation. JavaScript is the most popular and PHP is 4th popular languages in GitHub. We looked up on current stats in GitHub repository. There are 323,938 for JavaScript and 138,771 for PHP active repository till 2015.[1]

PHP has various dynamic features like eval which allows runtime code execution such as string input, special methods (known as magic methods), variable variables, and dynamic include. Eval() is the most usable and performance-affected feature among these. On the another side, JavaScript also allows the similar dynamic execution too. JavaScript provides several dynamic functions, like setTimeout Function, but we discussed Eval() language construct in this paper.

1.1 Eval() in JavaScript

Eval() function call, able to evaluate JavaScript code which is effectuated at runtime.

```
eval(x);
```

Depending upon defined bound of an input string x, the state-of-the-heap-allocated unpredictable value of x and the local variable in scope can be altered as a negative effect of executing input string directly by eval call. While few of data abstraction techniques applied in some languages, JavaScript is restricted in encapsulation mechanism. Hence, the negative impact of an Eval() call can cause to the entire heap. Sometimes this becomes very necessary function call for designers, but in most of the cases, the Eval() presence indicates extremely destructive approach. The Eval() call is known as the most abused feature in scripting languages such as JavaScript.

In above code, if variable x is directly coming from user input or from other Web services, the Web site authority is unknowingly open ups the door for hackers to access JavaScript compiler. Eval() has power to execute anything on every call (like cookie stealing and introducing malware). The presence of Eval() does not point security problem every time and not every Eval() call open up the door for cross-site scripting (XSS) attack. It is just like power comes with the responsibility; you need to know how to use it correctly, that is it [1]. But even if you will not able to use it precisely, the damage potential will rise with every call.

This is all about the security loopholes in eval'd code, but we are here to discuss more on performance impact. Section shows the performance comparison between misused Eval() call and its replacement.

---

[1] https://github.com/.

1.2 Eval() in PHP

Like JavaScript, PHP is also most influenced dynamic scripting language which is mostly used in server-side scripting. Well, too much flexibility and dynamic nature played a big role in its popularity. In addition, PHP includes various object-oriented features such as interface, exception, inheritance, and platform independency as well as flexible dynamic features same as other scripting language. Eval() expression is also included which is able to compile the dynamically built code and able to run arbitrary PHP code. It is correspondent to echo() in the sense it generates output for every input but instead of text, it executes input string as PHP code.

```
<?php
$friendname1 = 'Malik';
$friendname2 = 'Anupam';
$input = 'My friends are $friendname1 and $friendname2';
print_r($input . "<br>");
eval("\$input = \"$input\";");
print_r($input . "<br>");
?>
```

This code generates output as My friends are $friendname1 and $friendname2 when first call with the print_r statement, but next output will be My friends are Malik and Anupam when print_r has been called after running eval() call.

2 Literature Review

Both scripting language allows evaluation of any argument (input as string) where the input string is always mutable. There are plenty of tools available for dynamic analysis of Web application based on input and output requirements. Apart from explained dynamic features in the introduction, these tools are good to derived performance of an application. Firstly, dynamic analysis tools fail to identify the performance of eval'd call because every time input will be unknown. Testing tools may not be able to derived input and output sets for eval'd code every time. Hence, it creates saturation point for dynamic analysis and opens the door to static analysis technique. Secondly, we believe there should be static analysis techniques or tools which can help developer team to judge or predict the performance before it will hand over to testing team. To analyze and predict performance of any Web application, analyzer needs a tool to identify every Eval() call and its nature through static analysis [2]. The difficulty level of static analysis will defiantly increase if there are the existence of Eval() calls with the unpredictable input string. There will not be any specified time limit, no memory bound, and not even termination guarantee. So it is essential to know about various static analysis techniques in these scripting languages, catch each and every Eval() call, and suggest its replacement (if possible) to enhance the performance.

From our survey, the eye-opening factor was the usage pattern of Eval() calls in leading PHP framework [3]. Too much flexibility sometimes becomes a reason for laziness in hard work. The various papers reflect the analysis pattern of unnecessary Eval() calls. If we consider WordPress framework, the static analysis shows that on every release ratio of Eval() call went down. In JavaScript, there are set of code/calls exist nowadays where Eval was the only option to get desired execution. The reason behind popularity and performance of most popular PHP Web site facebook.com is HHVM compiler. HHVM compiler does not support dynamic evaluation by Eval() at all [4].

In this paper, we examine results to show performance degradation by Eval() calls in both PHP and JavaScript and comparison with suggested generic solution. At last, static analysis techniques for both PHP and JavaScript to analyze existing Web system.

3 Proposed System

It is essential to establish static analysis tool for developer-end to enhance the performance of a specific code segment. But it is very difficult to identify each and every occurrence of Eval() call and other performance degradation factors from million lines of code. There are some static analysis techniques which can convert entire coding into meta-programming for analysis purpose and able to identify required code segments. Our proposed idea is to integrate two different analysis techniques (PHP and JavaScript) as a tool to identify code segments which can affect the performance of an application. As prior work, we implemented the mentioned idea for PHP applications. At the initial stage, we used PHP AiR and PHP-analysis to identify Eval() call and output will be in a text file which describes each and every Eval() call in the application and its nature. Secondly, a separate tool is used to generate statics of performance comparison between original code and replaced code. To measure performance in PHP we used mircotime() function, which helps to calculate execution time of input code segment. JavaScript has widely adopted performance benchmark jsperf to measure the performance of JavaScript code.

Figure 1 describes the general description of our system. Initially, PHP AiR framework has been used for analysis purpose. PHP AiR takes the source code of PHP application as input and its meta-programming conversion (Rascal language [5]) will be stored in a binary file. LoadBinary() function used to initiate analysis procedure. The syntax mentioned in next block represents filter for the specific Eval() call. X stands for the argument passed in Eval() (to identify Eval for specific argument). In the last stage of this module, the system will store these results in a text file which is in user understandable format. Next module is concentrated as performance measurement. There is not any defined benchmark to calculate the performance of

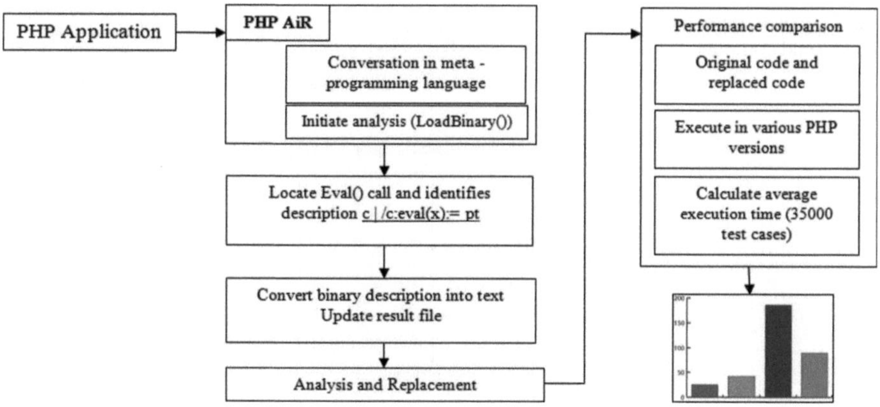

Fig. 1 General description of proposed system

PHP code snippet. Hence, we used microtime() to calculate execution time. The system will execute both input code in various PHP versions (min 35000 test case consider for each code and each version). The average execution time comparison will be displayed by chart graph.

In this paper, we examine results in this explained system to show performance degradation by Eval() calls in both PHP and JavaScript and performance comparison with suggested generic solution.

4 Performance and Eval

In this section, we examine a comparison of performance between Eval() calls and its generic replacement.

4.1 Performance Factor in PHP

Heavy usage of Eval() call in PHP 4.0+ will take performance issue as parser takes the tremendously high time to initialize because few limitations of Zend engine. Zend engine considers input string in Eval() call as new PHP file. The Zend compiler converts the whole PHP code to a binary structure, and then parses converted the binary structure to generate output. But every time where an eval call has been identifying in binary structure, and it has to reactivate the parsing procedure and convert input string in eval into binary format again as shown in Fig. 2.

PHP AiR [2] converts PHP code into meta-programming language. In part of our research work, moodle one of the leading PHP framework targeted most often time. Framework spread across 5500 PHP files and around 14 Lakhs lines of source

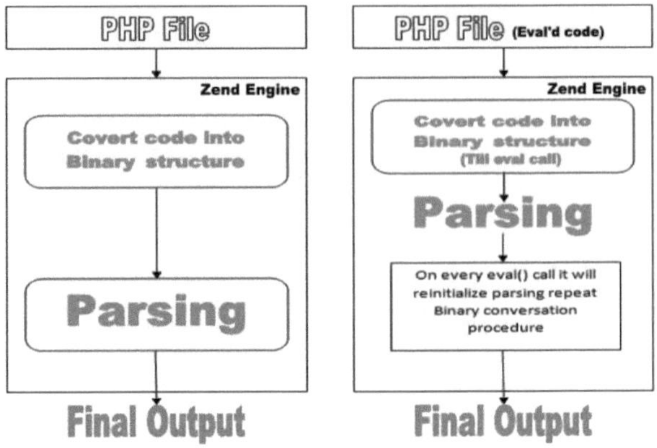

Fig. 2 Execution of PHP file in Zend compiler

code. We converted moodle version 2.9 into meta-programming language and tried to identify nature of Eval() calls in this large-scale PHP framework. PHP AiR is not only designed to convert code into meta-programming, but it actually helps in static analysis too. With this, we classified Eval() calls and its nature [6].

Below code shows actual PHP code of one Eval occurrence. We used Rascal language for static analysis because Rascal is specialized in both meta-transformation and static analysis. In this example, at line 339 in File.php file eval has been called to concate two string variable code and value. We replaced string concatenation operation through Eval() by array_fill operation and execute both codes in different PHP versions. Figure 3 shows execution time comparison of eval'd code and its optimum replacement.

Moodle 2.9 "file.php" line 339

```
Actual PHP code:

  return eval($code."return \$value;\n}\n");
```

This explanation reflects only one Eval() call occurrence and performance improvement. We examine all Eval() call occurrences and it shows 70% of eval() call can be replaced without affecting output. In our work, we tried to identify nature of these Eval() call and suggested an optimal replacement for each call.

4.2 Performance Factor in JavaScript

JavaScript also treats eval call same as PHP does. It also considers Eval() input as new JavaScript file. Firstly, we concentrated on Eval calls in moodle (PHP code). And now, we are dealing with Eval in JavaScript for the same purpose. Here in this section, we showed the performance factor with Eval in JavaScript. For JavaScript,

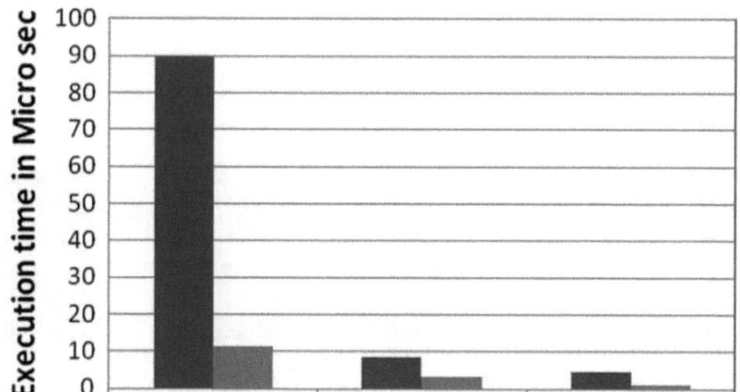

	PHP version 5.5.12	PHP version 5.6.13	PHP version 7.0.6
Eval'd code	89.73	8.48	4.56
Non eval'd code	11.36	3.15	1.03

Fig. 3 Performance comparison of two PHP code segments

we measured performance in operations/sec (higher is better). Execution time for specific functionality as performance measurement is meaningless for JavaScript because it takes just 1ms to execute 1kb of JavaScript code on the mobile device and it varies device to device. Ops/sec is measured with well-known jsperf.com performance benchmark. We used benchmark.js to conduct all test in various browser.

On our primary lookup, we observed that in most of the Web services Eval() call has been used to parse JSON code. Traditionally, there was no such replacement in the browser to parse JSON code. But in a recent update, JSON.Parse is considered as loophole replacement and it is more secure than the previous one.

Sample JavaScript code with JSON

```
var jsontext = '{ "employees" : [' +
'{ "firstName":"Manish" , "lastName":"Mishra" },'
+ '{ "firstName":"Irfan" , "lastName":"Khan"
},' + '{ "firstName":"Sid" , "lastName":
"sonvane" } ]}';

+ var firstobj = JSON.parse(jsontext);
- var secondobj = eval ("(" + jsontext + ")");
```

In above code, variable sample code contains small piece of JSON code, and +p and −p are the function calls to parse-defined JSON code. Figure 4 shows performance comparison between these calls. This result reflects almost 10x performance enhancement by given suggestion.

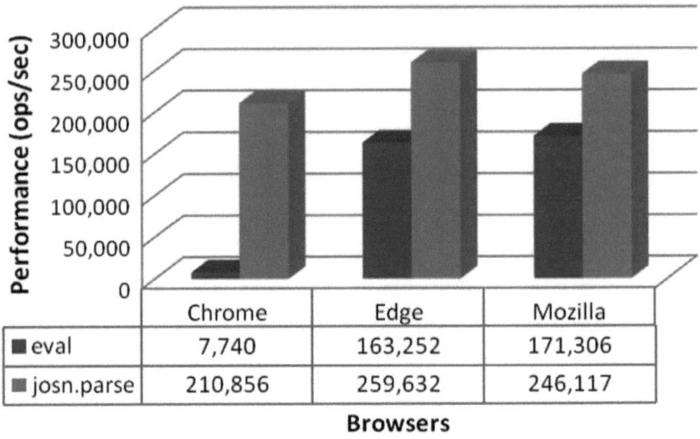

Fig. 4 Performance comparison between JSON.Parse and Eval()

5 Conclusion and Future Work

Dynamic scripting languages such as PHP and JavaScript provide a number of dynamic features to shape language flexibility into language popularity. Eval() expression is one among those flexible features. In PHP and JavaScript, it provides execution of arbitrarily written code at runtime, although the more powerful features make programs harder to understand and as runtime input mutable for every next execution dynamic analysis techniques cannot able to predict the performance of the system. Hence, it knocks door for static analysis. But static analysis cannot able to help in performance prediction unless every potential of the input string of Eval() is known.

As a part of our work, we used PHP AiR, which converts source code into the meta-programming language. We analyzed transformed code to understand nature of every Eval() calls and tried to remove Eval'd code segment to enhance the performance. The results reflect enhancement of performance by 2x. In future work, we are looking for applying the same terminology in JavaScript code.

References

1. Meawad F, Gregor R, Morandat F, Vitek J (2012) Eval begone!: semi-automated removal of eval from javascript programs. ACM SIGPLAN Not 47(10)
2. Hills M, Klint P (2014) PHP AiR: analyzing PHP systems with Rascal. In: 2014 software evolution week-IEEE conference on software maintenance, reengineering and reverse engineering (CSMR-WCRE), pp 454–457
3. Hills M (2015) Evolution of dynamic feature usage in PHP. In: 2015 IEEE 22nd international conference on software analysis, evolution and reengineering (SANER), pp 525–529

4. Zhao H, Proctor I, Yang M, Qi X, Williams M, Gao Q, Ottoni G et al (2012) The HipHop compiler for PHP. ACM SIGPLAN Not 47(10):575–586
5. Klint P, Van Der Storm T, Vinju J (2009) Rascal: a domain specific language for source code analysis and manipulation. In: Ninth IEEE international working conference on source code analysis and manipulation, 2009. SCAM'09, pp 168–177
6. Hills M, Klint P, Vinju J (2013) An empirical study of PHP feature usage: a static analysis perspective. In: Proceedings of the 2013 international symposium on software testing and analysis, pp 325–335
7. Morandat F, Hill B, Osvald L, Vitek J (2012) Evaluating the design of the R language. In: ECOOP 2012 object-oriented programming. Springer, Berlin, pp 104–131

A Study on IDS (Intrusion Detection System) and Introduction of IFS (Intrusion Filtration System)

Rita Dewanjee and Ranjana Vyas

Abstract Network security in organizations is not limited to tangible systems but beyond the physical existence, its focusing on security of non-tangible data flowing in network inside and outside of organization while communicating through Internet. In this paper, we will discuss about different types of intrusion detection system (IDS) available and comparison of their various aspects. Finally, I propose my research work as intrusion filtration system (IFS), which will be a new methodology for network security.

Keywords IDS (intrusion detection system) · ID (intrusion detection) · FWN · Network security · IFS (intrusion filtration system)

1 Introduction

Network users are getting more dependent on online transaction. The dependency of customers is increasing day by day for doing all their routine works through online. The maximum mass of white collar people are now dependent on online services because of their hectic schedule. All day-to-day services of a normal user now depend on e-commerce only. Not only for paying bill of any traditional routine work but also for food, clothing, health, insurance, travelling, banking, etc.

R. Dewanjee (✉)
MATS School of Information Technology, MATS University, Raipur, India
e-mail: rita.dewanjee1@gmail.com

R. Vyas
IIIT Allahabad, Allahabad, Uttar Pradesh, India
e-mail: ranjanavyas@gmail.com

H.R. Vishwakarma and S. Akashe (eds.), *Computing and Network Sustainability*, Lecture Notes in Networks and Systems 12, DOI 10.1007/978-981-10-3935-5_13

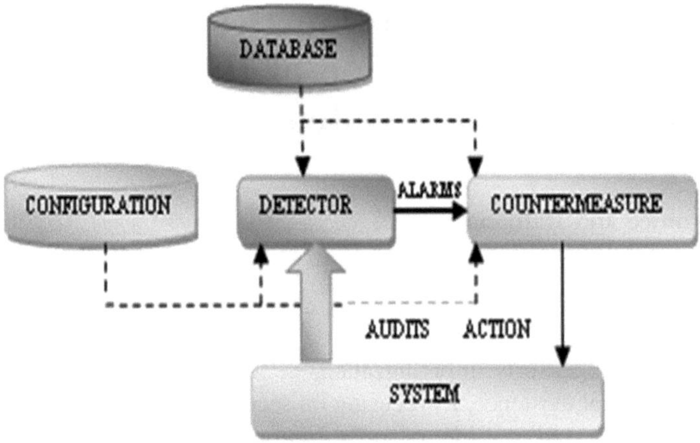

Fig. 1 Simple intrusion detection system [12]

Hence, the chances are also increasing for attackers to hack the data and make their session vulnerable. Hackers are also targeting mass of network users to exploit the authenticity of confidential communication. Intrusion detection systems are the tools that help us to secure the network as well as to find out the reasons for failure.

After Denning's work in 1981, many prototypes of the IDS were developed [1]. For infrastructural security, IDS is an important protection mechanism. IDS needs to be accurate, adaptive and extensible. Looking to these parameters and the magnitude of complexities of today's network environments, there is a need for more organized and automated IDS. The development of improved automated IDS is need of situation [2].

Detecting the error or unusual happenings in any network system done intentionally or unintentionally is the objective of intrusion detection system. IDS is used in many ways. IDS is used to monitor the systems from attacks and alarms the user to prevent it (Fig. 1).

2 Intrusion Detection System

In this section, we will discuss the available IDS in the market. The types of IDS available are as follows:

- Anomaly-/Heuristic-/Behavioural-based
- Signature-/Knowledge-/Pattern-based
- Host-based
- Packet-/Network-based

2.1 Anomaly/Heuristic/Behavioural based (AIDS)

Anomaly-based IDS detects the abnormal behaviour in data traffic. The deviation from the normal behaviour is considered as attack [3]. Anomaly detection-based IDS uses the program profiles which show significant deviation from normal activities. Such IDS detects abnormal behaviour of any system and also finds intrusion which may be a new attack. It is not necessary that IDS is familiar with all abnormal behaviour of a system.

Detection of an intrusion is a process of detecting abnormalities in passing on data inside or outside the network. Anomaly-based intrusion detection model works on two major findings: low false positive rate and a high true positive rate [4]. A variety of anomaly detection techniques were suggested, and the comparison of their strengths and weaknesses is quite difficult. This technique helps to find high detection rate and low false alarm rate [5].

2.2 Signature-/Knowledge-/Pattern-Based (SIDS)

The SIDS is based on the available signature or patterns of intrusions. SIDS can detect only such attacks which are previously defined in database. Signature-based IDS matches the signatures of already-known attacks that are available with the database of known attack signatures to detect the computer attacks [3].

The advantage is that signatures are very easy to develop and understand, if we know what network behaviour we are trying to identify [5]. Network traffic is monitored by signature-based intrusion prevention system to match the signatures. Once a match is found, the IPS takes the necessary and required action [3].

2.3 Host-Based Intrusion Detection System (HIDS)

HIDS basically monitors the functionality and felicitation of host to filter the intrusive activities. HIDS focuses to concentrate on the available information at host and operations performed on host. These IDS are less prone because HIDS detects the intrusion of the host application, and HIDS is aware about the application's normal status.

It is directly monitoring the host application and processes. With HIDS, it is easier to identify the intrusions from process, generating abnormal activities in system [6]. HIDS can work proactively and can sniff the traffic of network of host and alert the user in real time [3].

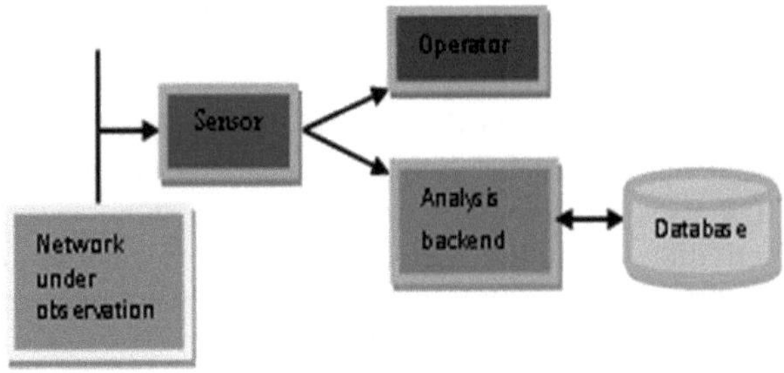

Fig. 2 Common architecture of NIDS [9]

2.4 Packet/Network-Based Intrusion Detection System (NIDS)

NIDS can be compared using the parameters to find out various categories of intrusions with development community and to enhance its functionality in new versions [7]. Nessus, Satan and Ballista are used for state-based, and NetRanger or Real Secure is used for transition-based tools in NIDS, as detection paradigm tools [8].

A network-based IDS collects data in the network level on a segment or from subnet transparently. NIDS sensors may be located anywhere in the network and monitors the traffic of network [9] (Fig. 2).

NIDS captures packets of data from network media and finds matchings from stored attack signatures' database. If any data packet is matched with an intrusion signature, an alert is generated and packet is logged into a file or database. Snort is one of the majorly used NIDS [3].

NIDS is an important tool for protecting critical data and infrastructure. The quality of any NIDS depends on the percentage of true attacks detected and combined with the generated total number of false alerts [10] (Table 1).

3 Performance Discussion of IDS

Anomaly-based systems work with minimum FPR. Frequent improvement in all the database is required for better performance of IDS, but anomaly-based systems are giving better performance in comparison with others in typical situations. These IDSs use logs of computer immunology as an information resource [8].

These IDS also have IPS to protect the systems from novel intrusions. Knowledge-based IDS or SIDS depends on the available pattern and signatures in the database. SIDS can only filter intrusions which are found in IDS database. These IDS' network usage differs, depending on the product IDS database. These IDS have high false alarm rate.

The performance of HIDS depends on the OS in which they are used. HIDS uses various logs of host systems as information source [8]. Host-based IDS uses network tools such as state and transition as detection paradigm tools. USTAT is used

Table 1 Comparative table of IDS

Parameter	Anomaly-/Heuristic-/ Behavioural-based	Signature-/ Knowledge-/ Pattern-based	Host-based	Packet-/ Network-based
Example	Firestorm, Bro, Dragon	Suricata, Prelude	Dragon Squire, Real Secure	Network Flight Recorder, Snort, Cisco secure ID, Tripwire
Information sources from computers	Statistics, Expert systems, Neural Networks, Compute Immunology	Expert systems, Signature analysis	Accounting, Syslog, C2 security audit	SNMP information, Network packets
Installation and deployment	Typical	Mixed	Intermediate	Easy
Throughput	Max.	Moderate	Max.	Moderate
Network usage	Less	Very Less	Less	Medium
IPS capability	Yes	No	No	Yes
User friendliness	Less	Yes	Less	Yes
Performance	OS independent	OS dependent	OS dependent	OS independent
Identification of unknown attack	Yes	No	No	No
Maintenance required	Comparatively less	Needs frequent updating	Comparatively less	Comparatively less

for transition-based HIDS, and COPS and Tiger are used for state-based. NIDS can be compared using different factors such as performance, scalability, accuracy and ability to find out various categories of intrusions. The improvement of NIDS depends on the features included in its database, manuals issued to users, help documentation and its interaction with development community, etc. [7, 8].

4 IFS (Intrusion Filtration System)

We have gone through all the available types of IDS in this paper. All have their distinct advantages and methods of protection from unwanted security issues. I propose a new intrusion filtration system. Rather than detection, the system will use filtration process (Fig. 3).

The concept is to mark the filtered files, and only these filtered and marked files will be transferred and used for data. The methodology will be very simple to mark

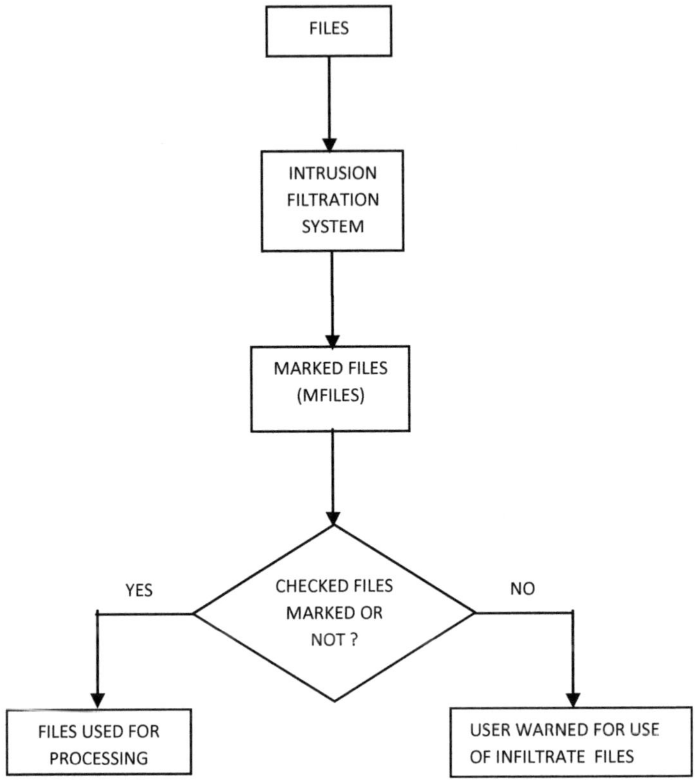

Fig. 3 Proposed intrusion filtration system (IFS)

a file as a safe file. The system can be used and implemented in both client and server machines. The financial implication will be lesser than the cost used to implement the IDS.

Although the proposed IFS system is under study of research process still, we are giving the following few comparisons that are found as per the study—(Table 2).

5 Conclusion

The different categories of IDS are explained here with all their advantages. After deploying firewall technology in network, the IDS is also becoming next logical step for many organizations at the network perimeter [11]. IDS is capable of offering protection from attackers, whether internal or external. IDS even can be used into those areas where traffic does not pass the firewall or uses it the least.

Table 2 Comparative table of IDS and IFS

Parameter	Intrusion Detection System (IDS)	Intrusion Filtration System (IFS)
Information sources from computers	Statistics, Expert systems, Neural Networks, Compute Immunology, Signature analysis, Accounting, Syslog, C2 security audit, SNMP information, Network packets	IFS will check all the files available in system in any format coming to the System and moving out of the system through LAN or Internet
Installation and deployment	Typical	IFS default utility program for system
Throughput	Moderate	MAX
Network usage	Less or high dependent on the IDS implemented	System in-built facility, work off-line also
IPS capability	Less or good dependent on the IDS implemented	YES
User friendliness	Less or good dependent on the IDS implemented	YES
Performance	OS independent	Antivirus of IDS dependent
Identification of unknown attack	Dependent on the feature of IDS implemented	IFS will mark the scan files so depends on antivirus or IDS
Maintenance required	Comparatively less	NO, updating on IDS or antivirus will suffice the purpose

The IFS will be a system which will work internally as a utility program and mark the files as filtered or unfiltered. The security of IFS will be ensured through cryptography technique. By proposing IFS, we are trying to avoid the use of corrupted files and subsequently their distribution in the network.

References

1. Denning Dorothy (1987) An intrusion-detection model. IEEE Trans Softw Eng 13(2):222–232
2. Rathore JS (2012) Survey on intrusion detection and prevention system and proposed cost effective solution using software agent 1(3)
3. Kaur T, Kaur S (2013) Comparative analysis of anomaly based and signature based intrusion detection systems using PHAD and Snort. In: Proceeding of security and privacy symposium, Feb 28 to Mar 2
4. Ghassemian M (2011) Analysis of an anomaly-based intrusion detection system for wireless sensor networks. Int J Comput Appl (0975 – 8887) 28(7)
5. Jyothsna V, Rangampet A, Rama Prasad VV (2011) A review of anomaly based intrusion detection systems. Int J Comput Appl 28(7):26–35
6. Patil S, Meshram BB (2012) Network intrusion detection and prevention techniques for DoS attacks. In J Sci Res Publ 2(7)
7. Debar H (2000) An introduction to intrusion-detection systems. In: Proceedings of connect

8. Schaelicke L, Slabach T, Moore BJ, Freeland C (2003) Characterizing the performance of network intrusion detection sensors. In: RAID. Lecture notes in computer science, vol 2820. Springer, pp 155–172
9. Vigna G, Kruegel C (2005) Host based intrusion detection. In: Handbook of information security. Wiley, Dec 2005
10. Albag H, Network & agent based intrusion detection systems. Istanbul Technical University, TU Munich
11. Intrusion detection—systems for today and tomorrow. In: SANS institute 2001
12. Whitea JS, Fitzsimmonsb TT, Matthewsca JN, Coulter WH (2013) Quantitative analysis of intrusion detection systems: Snort and Suricata. In: Proceedings of the SPIE, vol 8757, id. 875704, 12 pp
13. Red Hat Enterprise Linux 4, Security Guide. http://www.centos.org
14. Evaluating-intrusion-detection-systems-and-comparison-of-intrusion-detection-techniques-in-detecting. http://www.intechopen.com
15. Jacob Víctor G, Rao Meda S, Venkaiah VCH (2010) False positives in intrusion detection systems. http://www.academia.edu
16. Dreger H, Kreibich C, Paxson V, Sommer R (2005) Enhancing the accuracy of network-based intrusion detection with host-based context. In: Proceedings of the second international conference on detection of intrusions and malware, and vulnerability assessment. Springer, pp 206–221
17. Debar H, Dacier M, Wespi A Towards a taxonomy of intrusion-detection systems. Int J Comput Telecommun Networking—Spec Issue Comput Netw Secur Arch 31(9):805–822 (1999) (Elsevier)
18. Intrusion detection systems: definition, need and challenges. In: SANS institute 2001
19. Karthikeyan KR, Indra A (2010) Intrusion detection tools and techniques—a survey 2(6)
20. Petersen C (2003) An introduction to network and host based intrusion detection
21. Sandhu UA, Haider S, Naseer S, Ateeb OU (2011) A survey of intrusion detection & prevention techniques. In: IPCSIT, vol 16. IACSIT Press, Singapore 2011

Simulation and Performance Analysis of Modified Energy Efficient DSR Protocol in MANETs

Siddalingappagouda C. Biradar and Prahlad Kulkarni

Abstract In wireless communication, Mobile Ad hoc networks (MANETs) offer multi-hop communication for mobile nodes which are bounded within limited transmission range. Thus, to transmit data through multiple hops, improve the network lifetime and utilization of power, routing of packets plays a vital role. There are various routing protocols such as reactive, proactive, and hybrid. In this paper, we have developed a new routing protocol called as Modified Dynamic Source Routing Protocol (MDSR) which builds multi-path routes and selects the optimum path among several paths to destination based on threshold level of energy and distance, and it also improves network lifetime. MDSR gives improved performance result, such as a residual energy, active routing path energy ratio, packet loss ratio, energy consumption, and overhead, under different pause time. Compared to existing conventional DSR routing protocol, the proposed MDSR is implemented with certain simulation parameters using Network Simulator (NS-2) tool.

Keywords MANETs · DSR · MDSR · Performance metrics

1 Introduction

This paper describes about newly developed routing algorithm, which provides better performance to that of DSR routing protocol, which has dissimilar concepts compared to the conventional routing protocols in wireless networks [1]. In the

S.C. Biradar (✉)
Department of Electronics and Communication Engineering,
Don Bosco Institute of Technology, Bangalore 560060, Karnataka, India
e-mail: siddubiradarr@gmail.com

P. Kulkarni
Department of Electronics and Telecommunication Engineering,
Pune Institute of Computer Technology, Pune 411043, Maharashtra, India
e-mail: ptkull1@gmail.com

H.R. Vishwakarma and S. Akashe (eds.), *Computing and Network Sustainability*,
Lecture Notes in Networks and Systems 12, DOI 10.1007/978-981-10-3935-5_14

MDSR, Hopping paths may help to balance the energy consumption and distribute incoming traffic load across the network. Also, distribution of data packets helps to decrease overall power consumption [2]. We especially centered on distinctive execution measurements, for example, network residual energy, active path energy ratio, packet loss ratio, energy consumption, and overhead [3]. The flow of our work in this paper is explained as: In Sect. 1, it explains about the general concept of MANETs. In Sect. 2, we discuss a routing procedure performed by present DSR routing protocol. In Sect. 3, it tells about proposed MDSR routing protocol. In Sect. 4, our state about the performance parameters that are used during simulation. We formulate later in Sects. 5 and 6 the simulation and performance analysis of existing DSR and modified MDSR routing protocol by network simulator tool NS2, considering set of performance metrics and by simulating several scenarios and finally the conclusion and future work.

2 DSR Routing Protocol

DSR is used in MANETs; it does not maintain any type of information initially to transmit the packets. So they start route discovery process in which node discovering route when it is necessary because of which they call DSR as reactive routing protocol [4]. From the Fig. 1a, initially source node asks about route to destination node by sending control packets (RREQ and RREP) [5]. Source node N1 broadcast Route Request (RREQ) packet to its nearest neighbor node (N2 and N3), RREQ packet contains Unique ID, intermediate node detail, source address and destination address, after receiving RREQ by node N2, it adds its own address to RREQ packet and broadcast to its nearest nodes N5 and N1, Node N1 rejects the RREQ packet sent by N2 because of same packet ID. Even node N3 broadcast the RREQ packets to node N1 and N4, Node N1 reject the packet, Node N4 in turn broadcast RREQ packet to node N3, N5, N6, and N7, node N5 floods the RREQ packets to node N8, N2, and N4, Finally, RREQ packet reaches (by node N5 and N7) to destination node N8. Destination node N8 uses the same route path for sending RREP packet that RREQ packet followed [6]. Route Replay (RREP) packet contains route record, which follow the route path as N1, N2, N5, and N8 as show in Fig. 1b. Now whatever the data packets to be sent by source node to destination node, it will include this route in the header of data packet, hence with the help of RREQ and RREP packets, we determined route path. This route path will be stored in route cache which will be helpful for sending packet whenever source node needs to send [7]. As MANETs are infrastructure less, there is a chance of link breakage, so in such case, route maintenance technique is used. Whenever any link is broken, that information will broadcast in the network by intermediate nodes through Route Error (RERR) control packet. When RERR packet reaches source, then it will remove the broken route path from route cache and update with new route path [8].

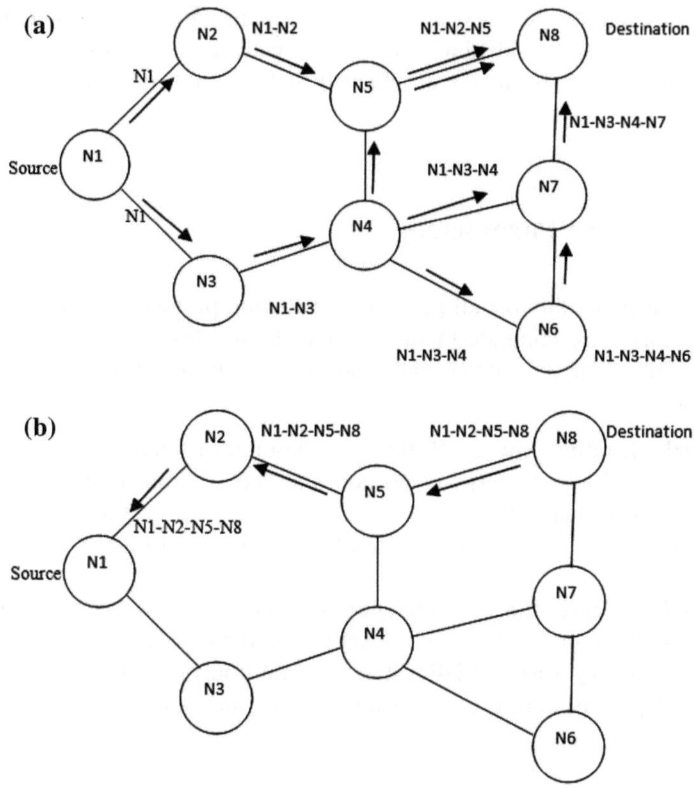

Fig. 1 **a** Building the route updates while route discovery. **b** Transmission of the route reply packets with the route information

3 Proposed MDSR Routing Protocol

In this part, we are going to represent a complete overview and purpose of MDSR routing protocol.

As conventional DSR routing protocol will not consider energy expenditure by individual nodes and will rely on optimum path without information of residual energy and distance of the neighbor node, Proposed MDSR will consider remaining energy parameter since from the first phase of route (route request phase), The delay for ACK packet is made to depend on residual energy of node in that path. Therefore, overall delay that exists between sending of RREQ control packets by source node and reception node due to DSR will be minimum as possible. The first RREQ control packet that will be forwarded to node ID is what which will be submitted via best route from point of view of energy, and this algorithm has a provision of changing the next alternative optimum path for destination by comparing residual energy with threshold energy with the other nodes in alternative path, by which energy balancing is achieved such that network lifetime is relatively prolonged.

4 Performance Parameters

In order to calculate the overall performance routing protocol for MDSR and DSR routing protocol, we compare them with set of execution measurements such as, network residual energy, active path energy ratio, packet loss ratio, and energy consumption.

- **Network Residual Energy Ratio**: It is the ratio of total residual energy of all nodes by total initial energy of all nodes at end of simulation time.
- **Active Path Energy Ratio**: It is the ratio of total nodes residual energy in route path from source and destination node by total number of nodes initial energy in route path.
- **Packet Loss Rate**: It is the difference between total numbers of packet sent by source node to total number of packet received by destination node.
- **Packet Delivery Ratio (PDR)**: It is a proportion of total packets received to total packets sent during certain simulation period, it is given by

$$PDR = PR * 100\% / PS$$

where, PR is sum of packet received by destination node, PS is sum of packet sent by source node.

- **End to End Delay**: It is defined as average time taken by packets to transmit from source to destination across MANETs. It includes all types of delays caused by buffering during routing discovery and route maintenance

  End to End Delay = Σ (packet Arrive Time – Packet Send Time)

- **Throughput**: It is defined as average transform rate or bandwidth of route, it is given by

$$TP = PR * SZ / SE$$

where, SZ is Packet Size, SE is Simulation End Time.

5 Result and Discussion

The performance analysis of existing DSR and MDSR routing protocols in MANETs is carried on NS 2.35 simulator which is used under Linux/Windows platform [5]. The performance analysis is done by both protocols through the simulation parameters. Table 1 shows the simulation parameters that are used for MANETs.

Figure 2 shows deployment of 11 nodes in a network, in which node '0' is the source and a node '10' is D destination, the red color node indicates a node having less energy than the threshold level. Green node indicates nodes residual energy more than threshold energy level. Yellow color node indicates node reaching threshold energy level. The shortest path for data packets between source and destination in DSR routing protocol, the route path is 0-1-7-10 till the completion of simulation time, In MDSR routing protocol, the route path is 0-1-7-10 and 0-2-7-10 till the completion of simulation time.

In Fig. 3, it shows total energy remaining in the MANETs at different simulation time (pause time). The energy remained at simulation with respect to MDSR routing protocol is more when compared to DSR routing protocol. The reason is MDSR routing protocol transmits data packets based on the remaining energy within the nodes, as the energy in the node becomes less than threshold level (0.2 J) then route path is changed to a node which contains the more energy, but this will not happen in existing DSR routing protocol.

In Fig. 4, It shows active path residual energy ratio with respect to pause time. The active path during simulation of DSR is 0-1-7-10 and during MDSR, it is 0-1-7-10, 0-2-7-10. In MDSR routing protocol, route path is 0-1-7-10 stays till 0 to 20 s, later route path 0-2-7-10 is carried out based on residual energy in communicating path till end of simulation.

In Fig. 5, we can observe nodes in MDSR routing protocol takes more time to die (less than threshold level) because of new routing procedure which is based on

Table 1 Simulation environment

Simulator	NS-2.35
Routing protocol	DSR, MDSR
Simulation period	0–60 s
Simulation area	800 m × 800 m
Number of nodes	11 nodes
Queue size	50
Transmission range	250
Interference range	550
Packet size	1500 bytes/packet
Application type	FTP
Agent type	TCP
Initial energy	10 J
Threshold energy	20% of initial energy

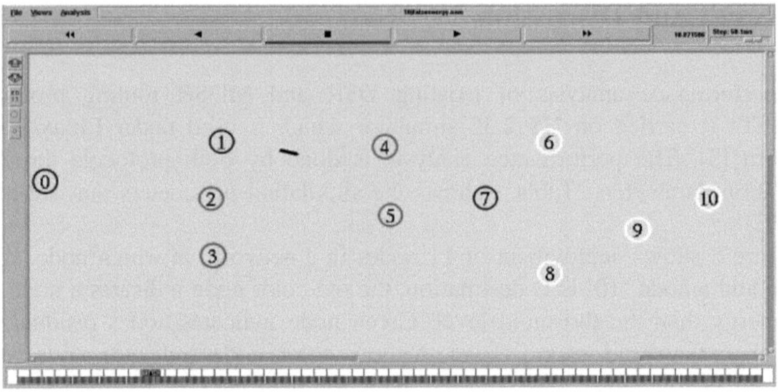

Fig. 2 11 nodes MANETs topology

Fig. 3 Total energy remaining in MANETs versus pause time

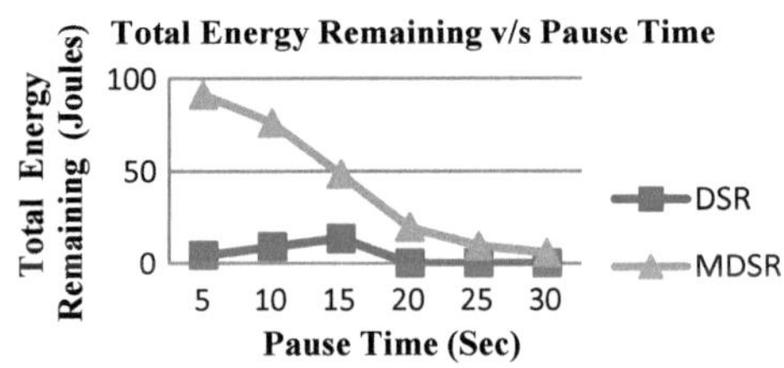

Fig. 4 Active path residual energy ratio versus pause time

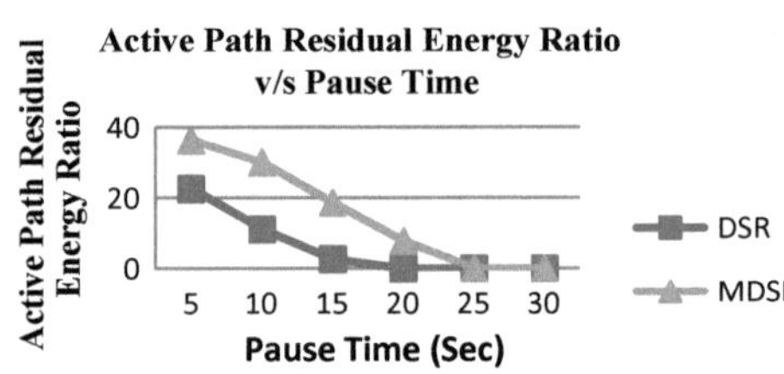

residual energy as residual energy is compared periodically with threshold level and next alternative path is selected, hence nodes death can be prolonged and improve network lifetime.

In Fig. 6, it shows packet loss rate with respect to pause time. Packet loss rate is more in DSR when compared to MDSR routing protocol. During DSR routing

Fig. 5 Node dead versus pause time

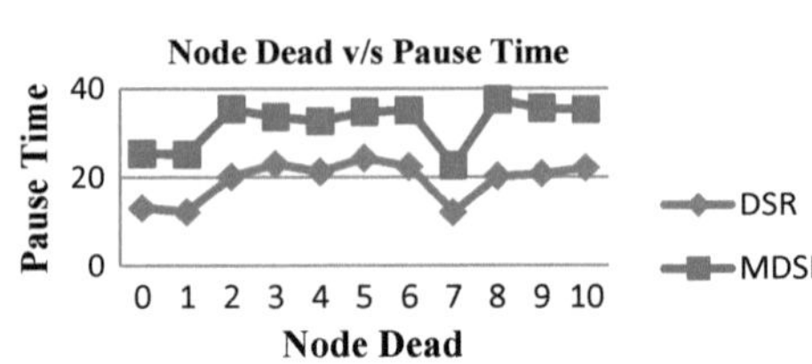

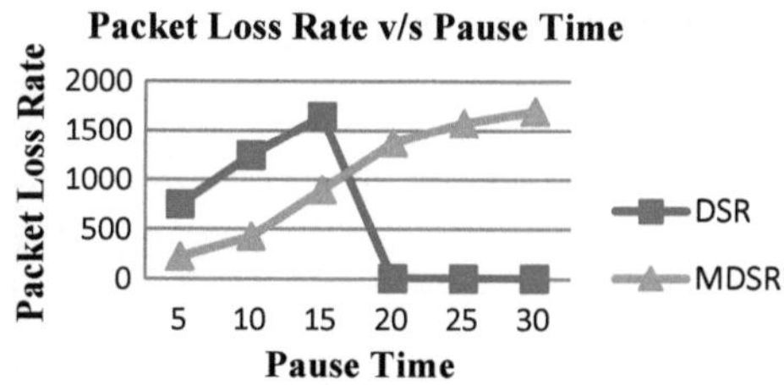

Fig. 6 Packet loss rate versus pause time

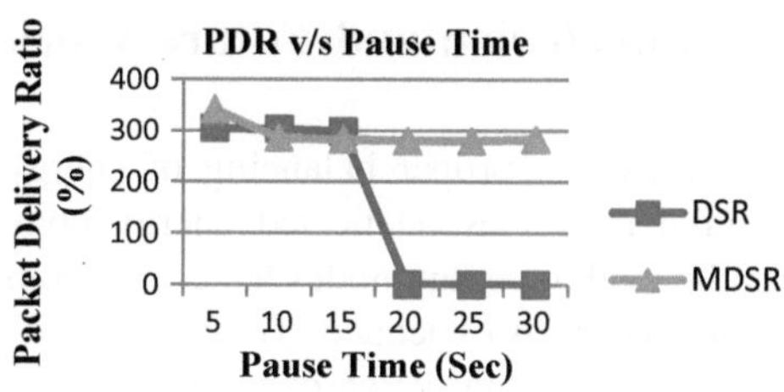

Fig. 7 PDR versus pause time

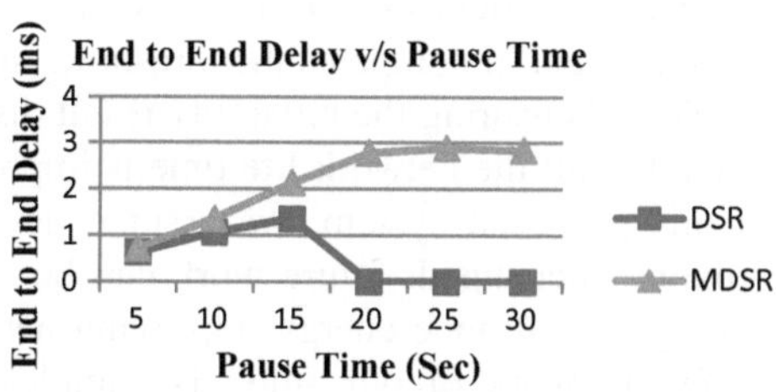

Fig. 8 End to end delay versus pause time

protocol, data packets transmission takes place till 0 to 15 s through routing path 0-1-7-10; later there will be no any routing path for communication because of loss of energy within the node. But whereas MDSR routing protocol consist of two route path 0-1-7-10 and 0-2-7-10 to transfer the data packets to destination node, this is based on energy present within the node.

In Fig. 7, it shows that MDSR has maximum PDR than DSR. That is number of packets delivered by MDSR is more when compared to DSR. The reason is in MDSR, the route path depends on energy present in node if node has less energy it switch to next route path but this does not happen in DSR.

In Fig. 8, it shows that MDSR has maximum end to end delay than DSR. The reason is, in MDSR, it changes the route path based on energy present at each node; as a result, every time there will be a delay in determining the routing path to destination.

In Fig. 9, it shows NLR at different pause time. DSR has less number of control packets overload when compared to MDSR protocol. The reason behind this is MDSR will switch to different route paths based on the energy present in nodes as a result they require more control packet (RREP and RREQ) to confirm correct route path.

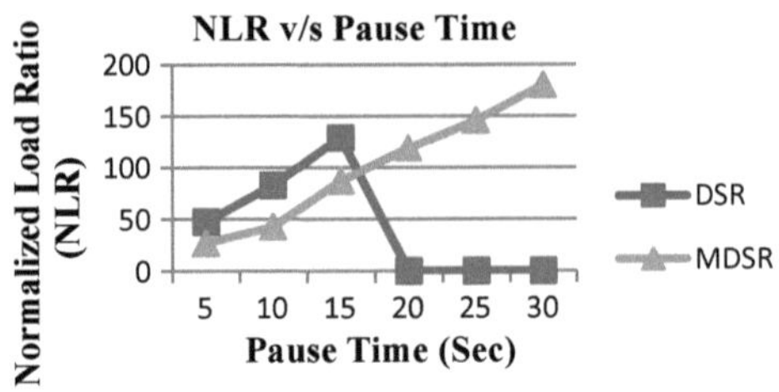

Fig. 9 NLR versus pause time

6 Conclusion and Future Work

In MANETs, proper balancing of energy consumption and traffic load are main significant issues. Deflected energy consumption amongst nodes and congested traffic load to certain nodes leads to serious destruction to routing process for which more energy is expended. To avoid these problems, we proposed a new concept in MDSR, where in which existing DSR routing protocol selection routing path was based on shortest path with cost as metric and energy of individual node was not taken into account our proposed work will select a route taking energy into consideration and it compares the threshold energy value based on that alternative path is selected ensuring the reliability result discussed above we observe that in MDSR overall in all the network life time is improved and is better compared to DSR. As in this proposed system, we assume node to be immediate for which topology remains unchanged, future work can be focused on dynamic change in topology saving overall node energy improving network life span.

Future work of our study is completely focused on improving the network lifetime by modifying different routing protocol such as DSDV, TORA, OLSR, and AODV in ad hoc network. We also concentrate on implementing it on large scale wireless area network.

References

1. Baisakh (2013) A review of energy efficient dynamic source routing protocol for mobile ad hoc networks. Int J Comput Appl 68:6–15
2. Rishiwal V, Yadav M, Verma S, Bajapai SK (2009) Power aware routing in ad hoc wireless networks. J Comput Sci Technol 9:101–109
3. Alilou M, Dehghan M (2007) Upgrading performance of DSR routing protocol in mobile ad hoc networks. Int J Electr Comput Energ Electron Commun Eng 1:762–764
4. Biradar SR, Sharma HHD, Shrama K, Sarkar SK (2009) Performance comparison of reactive routing protocols of MANETs using group mobility model. IEEE Int Conf Signal Process Syst 8:192–195
5. Sultana S, Begum S, Tara N, Chowdhury AR (2010) Enhanced-DSR: a new approach to improve performance of DSR algorithm. Int J Comput Sci Inf Technol 2:113–123
6. Kumar S, Chaba Y (2010) Quality of service aware—modified DSR protocol in mobile ad-hoc network. Int J Adv Res Comput Sci Softw Eng 4:876–882

7. Biradar SR, Sharma HHD, Shrama K, Sarkar SK (2009) Performance comparison of reactive routing protocols of MANETs using group mobility model. IEEE Int Conf Signal Process Syst 3:192–195
8. Almobaideen W, Hushaidan K, Sleit A, Qatawneh M (2011) A cluster based approach for supporting QoS in mobile ad hoc networks. Int J Digital Content Technol Appl 5:112–117

Emotion Recognition on the Basis of Eye and Mouth

Arnima Chaurasia, Shailendra Pratap Singh and Divya Kumar

Abstract It is very interesting to identify human facial expression in the domain of image processing. Since last decade, a lot of research has been focused on this area. The role of this field in human–computer interaction has increased its importance multifold. In this paper, we have used manual database of eyes and mouth which is further classified on the basis of emotion. The expressed emotions included anger, sadness, happiness, and neutral. This classified database is then used to retrieve corresponding facial emotion. We have used Haar cascade technique for detecting the eyes and mouth and nearest neighbor for classification of emotion. The accuracy of this technique is measured on 400+ sample images taken arbitrarily. From the result, we conclude that presented algorithm is able to precisely classify emotion in frontal images on the basis of eyes and mouth only.

Keywords Emotion · Facial feature · Haar cascade · Nearest neighbor · Image retrieval

1 Introduction

Emotion is involved in our daily life. According to Cabanac [1], an emotion is a complicated psychological state that involves three different components: subjective experience, physiological response, and a behavioral response. It is composed of two words, E-Motion that means Energy in Motion. It includes anger, joy, panic, grief, and many more feelings. There is no universal accepted definition of emotion

A. Chaurasia (✉)
Computer Science Department, Banasthali Vidyapith, Jaipur, India
e-mail: arnimachaurasia7@gmail.com

S.P. Singh · D. Kumar
Computer Science and Engineering Department, Motilal Nehru National Institute of Technology Allahabad, Allahabad, India
e-mail: shailendra.cs09@gmail.com

D. Kumar
e-mail: divyak@mnnit.ac.in

H.R. Vishwakarma and S. Akashe (eds.), *Computing and Network Sustainability*,
Lecture Notes in Networks and Systems 12, DOI 10.1007/978-981-10-3935-5_15

in the literature. Emotions are considered as complex phenomena and have combination of subjective feeling, physiological response (body), and expressive behavior [2]. Subjective feeling cannot be measured; however, physiological response of emotion can be measured from pounding heart, facial expression, and blood rushing to the face. Physiological behavior, voice, and face are modalities which are used to recognize the emotion. Each of which has its own weakness and strength. Facial expression is an important aspect in human–machine intersection. Human–computer interaction provides ability to the machines to recognize human emotion which help in perception, decision making, learning, and prediction which influence the rational thinking. Mehrabina [3] has shown that human conveys messages 55% times by facial expression and only 5% times by language. Later, Ekman and Friesen [4] have reported that facial expression is universal tool to express emotions in human race. Kobayashi et al. [5] studied a machine recognition method using neural network for static image facial expression. They have developed a dynamic recognition system of human facial expression by taking images of six basic facial expressions and sequentially changing these images to detect facial expression. Metallinou et al. [6] have used facial as well as vocal modalities to achieve improved emotion recognition system. They have used Gaussian Matrix Models (GMM) to model each modalities. They have also used Bayesian classifier weight scheme and support vector machines to combine multiple modalities. Dy et al. [7] developed a multimodal emotion recognition system that was trained using a spontaneous Filipino emotion database. Proposed system could extract voice and facial feature and then use support vector machine to classify correct emotion label.

In this paper, we have discussed a multiple attribute image retrieval system on emotion, in which we create a manual database of emotion and use this database to detect exact facial expression. We have described a general approach to retrieve images on the basis of facial expression which are detected on the basis of eyes and mouth feature. To achieve this objective, flow of the paper goes in this way: In Sect. 2, we have outlined basic functioning of Haar cascade for facial feature detection, in Sect. 3, we have described our proposed algorithm, and result of experiments is discussed in Sect. 4.

2 Feature Detection Using Haar Cascades

In this section, we will discuss feature detection using Haar cascade [8]. Face detection is the first step in face recognition system. Purpose of the face detection is to localize and extract the facial feature, leaving out the background information. Paul Viola and Michael Jones proposed an effective object recognition method using Haar feature-based cascade classifiers [9]. Haar cascade is a feature detection method for visual object with higher detection rate. They have introduced a new image representation (known as integral image), which allows computation of feature very quickly. They have also proposed AdaBoost-based learning algorithm [10], which uses a small number of essential visual features from a large set and produces efficient classifier. Using Haar cascade allows to detect faces at the rate up to 15 frames per second. It does not require any auxiliary information (such as pixel color in image or image difference in video sequence) to be stored. It works only with information present in grayscale image. This technique does not work directly with image intensities [11]. Integral image is computed using basic operation on every pixel and allows fast feature evaluation on image. Paul Viola and Michael Jones [9] have also suggested a method to construct a classifier with small number of important features using AdaBoost. This technique reduces large extent of computation used in classification. To make classification efficient and fast, only a small number of features are selected in learning process. Cascade [12] function is trained using a large number of positive and negative images. This classifier is then used to detect object in the sample image. For more details on Haar feature, we redirect reader to [13]. Every feature is a single value which is calculated by subtracting sum of pixels under black area from sum of pixels under white area from the applied feature on image. There are a large set of pixels in a image, and a number of features are much more than the number of pixels. To calculate this large number of features, concept of integral image comes as a handy tool. To calculate sum of pixels using integral image requires just four pixels. Then, we need to select fewer features out of these large number of features using AdaBoost algorithm. Initially, we apply all the features on the images and find out the best threshold to categorize positive and negative faces. At starting stage, all images have identical weight but after every classification, miss-classified-image weight is increased. We repeat above process until we get required certainty with required number of features. Finally, classifier is a weighted sum of all selected features. We do not apply all the features on a window, instead we segregate feature in different stages and apply next-stage feature only when it passes the previous stage. The above-mentioned technique is called cascading of classifier.

Steps to Create a Haar-like Classifier

1. Set of positive and negative training images.
2. Mark confident images by using objectmarker.exe tools.
3. Create a .vec (vector) file which is based on positive-marked images using createsamples.exe.
4. Guide the classifier using haartraining.exe.
5. Run the classifier using cvHaarDetectObjects().

3 Proposed Work

This section details out the algorithm used to classify the frontal images on the basis of only eyes and mouth. Feature extraction is one of the most significant steps to successfully analyze and recognize facial expressions automatically. The system can be broadly categorized into different stages: query processing, image classification, and image retrieval stage. Haar cascade and nearest neighbor method are used for feature detection and image classification. Eyes and mouth are identified as the critical features, and these features are used to classify the emotion. These features are detected with the help of Haar cascade method. After feature detection, nearest neighbor [14] approach is used to retrieve the emotions contained within the face. In this work, the system has efficiently recognized the four universal emotions from face images. First of all, we have created a manual database using Algorithm 1. For this purpose, we have taken 91 images with happy faces, 81 images with sad faces, 77 images with angry faces, and 78 images with neutral faces. Then, we have used Algorithm 2 to recognize emotion present in a given image. Calculated average Euclidean distance is depicted in the Table 1, where ed_i represents calculated average Euclidean distance for ith image and ED is minimum Euclidean distance for all values of i.

Algorithm 1 Database creation using Haar cascade

Generate database D composed of D_i
  Where $i \in \{Happy, Sad, Neutral, Angry\}$ and D_i contains two set
  $D_iE = Set\ of\ eyes$ and $D_iM = Set\ of\ mouths$
  $D_iE = \{D_iE_1, D_iE_2, \ldots, D_iE_n\}$ and
  $D_iM = \{D_iM_1, D_iM_2, \ldots, D_iM_n\}$

  for each facial feature i
    Generate Sub database D_iE and D_iM as
    Select Image with feature i
    Apply hear-cascade to detect facial feature (eyes and mouth)
    Copy detected feature in D_iE and D_iM
  end for

Algorithm 2 Multiattribute image retrieval on emotion

Step 1 **for** each experiment with facial feature k
    Where $k \in \{Happy, Sad, Neutral, Angry\}$
    Take I_k as input image
    Applying $haar - cascade$ to I_k and generate template I_kE and I_kM

    for $\forall i \in \{Happy, Sad, Neutral, Angry\}$

$$ed_i = \sum_{j=1}^{n}(ed\ of\ I_kM\ with\ D_iM_j + ed\ of\ I_kE\ with\ D_iE_j)$$

    end for

  Calculate Average value of ed_i
  end for

Step 2 Find minimum $ED = MIN(ed_{happy}, ed_{sad}, \ldots)$
Step 3 return label of I_k as ED

---

4 Experimental Result

To demonstrate our algorithm, we have performed two experiments. In Experiment 1, we have taken a single image of each category and then performed Algorithm 1 to extract facial feature and then used Algorithm 2 to calculate average Euclidean distance. It is clearly observed that for happy face image, average Euclidean distance is 2.1281 from happy templates, 2.5292 from sad templates, 2.7361 from angry templates, and 2.6957 from neutral templates. Due to minimum distance from happy template, assigned label for this image is 'Happy.' Smiler calculation is done for sad-, angry-, and neutral-faced images, and results are depicted in Table 1. For the vague images, average calculated distance is much more than threshold value, and hence, we have not assigned any label to this type of image. Experiment 2 is performed by taking a set of 400 images of each category, and we try to label all the image base on facial expression. When actual label is matched against predicted label, the accuracy for the happy faces image is 91.8%, for sad face image is 85.7%, for angry face image is 70.8%, and for neutral face image is 70.9%. Result for experiment-2 is shown in Fig. 1, where H represents Happy, S represents Sad, A represents Angry, and N represents Neutral. Accuracy of retrieved images is calculated as follows:

$$Accuracy = \frac{Total\ number\ of\ relevant\ images}{Total\ number\ of\ retrieved\ images} \tag{1}$$

Table 1 Calculated average Euclidean distance

Face to be labeled			Euclidean distance from templates			
			Happy	Sad	Angry	Neutral
Exp for	Happy	Eyes	1.0901	1.0170	1.4210	1.2734
		Mouth	1.0380	1.5122	1.3421	1.4223
		Total	**2.1281**	2.5292	2.7631	2.6957
	Sad	Eyes	0.2031	0.1612	1.8025	0.4010
		Mouth	1.3161	0.2232	0.8165	1.2160
		Total	1.5192	**0.3844**	2.6190	1.6170
	Angry	Eyes	1.0471	0.8042	0.3262	0.6240
		Mouth	1.4321	0.8012	0.2334	1.1440
		Total	2.4792	1.6054	**0.5596**	1.7680
	Neutral	Eyes	0.5107	1.0610	1.0556	0.6132
		Mouth	1.4001	1.0131	1.0542	0.5012
		Total	1.9108	2.0741	2.1098	**1.1144**
	Vague		47.2840	51.8419	38.0792	43.6952

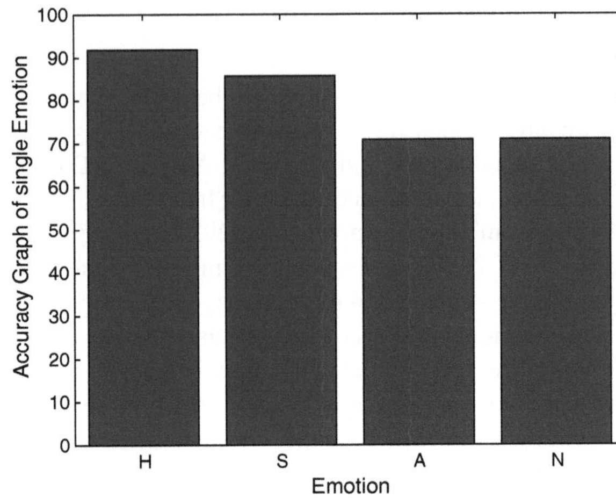

Fig. 1 Accuracy graph of single emotion

5 Conclusion

Many facial features are being used to classify emotion in the image which are cheeks, eyes, mouth, lips, and nose. We have demonstrated that using only eyes and mouth can give accuracy upto 91%. We have used k-nearest neighbors and Haar cascades to detect facial feature and classification of emotion. In this system, classification of images can be obtained on the basis of only two facial features which are eyes and mouth. Result shows that the system was able to identify the facial expressions accurately from the images.

References

1. Cabanac M (2002) What is emotion? Behav Process 60(2):69–83
2. Kleinginna PR, Kleinginna AM (1981) A categorized list of emotion definitions, with suggestions for a consensual definition. Motiv Emotion 5(4):345–379
3. Mehrabian A (1972) Nonverbal communication. Transaction Publishers
4. Ekman P, Friesen WV (2003) Unmasking the face: a guide to recognizing emotions from facial clues. Ishk
5. Kobayashi H, Hara F, Ikeda S, Yamada H (1993) A basic study of dynamic recognition of human facial expressions. In: Proceedings of 2nd IEEE international workshop on robot and human communication, 1993. IEEE, pp 271–275
6. Metallinou A, Lee S, Narayanan S (2008) Audio-visual emotion recognition using gaussian mixture models for face and voice. In: Tenth IEEE international symposium on multimedia, ISM 2008. IEEE, pp 250–257
7. Dy MLIC, Espinosa IVL, Go PPV, Mendez CMM, Cu JW (2010) Multimodal emotion recognition using a spontaneous filipino emotion database. In: 2010 3rd international conference on human-centric computing (HumanCom). IEEE, pp 1–5
8. Wilson PI, Fernandez J (2006) Facial feature detection using haar classifiers. J Comput Sci Coll 21(4):127–133

9. Lienhart R, Maydt J (2002) An extended set of haar-like features for rapid object detection. In: Proceedings of 2002 international conference on image processing, 2002, vol 1. IEEE, pp I–900
10. Wu Y, Ai X-Y (2008) Face detection in color images using adaboost algorithm based on skin color information. In: First international workshop on knowledge discovery and data mining, WKDD 2008. IEEE, 339–342
11. Papageorgiou CP, Oren M, Poggio T (1998) A general framework for object detection. In: Sixth international conference on computer vision, 1998. IEEE, pp 555–562
12. Lim C, Bearden JN, Smith JC (2006) Sequential search with multiattribute options. Decis Anal 3(1):3–15
13. Viola P, Jones M (2001) Rapid object detection using a boosted cascade of simple features. In: Proceedings of the 2001 IEEE computer society conference on computer vision and pattern recognition, CVPR 2001, vol 1. IEEE, pp I–511
14. Zhang H, Berg AC, Maire M, Malik J (2006) Svm-knn: discriminative nearest neighbor classification for visual category recognition. In: 2006 IEEE computer society conference on computer vision and pattern recognition, vol 2. IEEE, pp 2126–2136

Exploration of Machine Learning Techniques for Defect Classification

B.V. Ajay Prakash, D.V. Ashoka and V.N. Manjunath Aradya

Abstract To develop good quality software product, there is a need of continuous defect identification and classification in each module before delivering a software product to the customer. Developing software needs proper managing of the available software resources. To deliver a software product on time, developing quality software products, Information Technology (IT) industries normally use software tools for defect detection. Based on severity, defects are detected and classified. This can be automated to reduce the development time and cost. Nowadays, machine learning algorithms have been applied by many researchers to accurately classify the defects. In this paper, a novel software defect detection and classification method is proposed and neural network models such as Probabilistic Neural Network model (PNN) and Generalized Regression Neural Network (GRNN) are integrated to identify, classify the defects from large software repository. Defects are classified into three layers based on the severity in the proposed method abstraction layer, core layer, and application layer. The performance accuracy of the proposed model is compared with MLP and J48 classifiers.

Keywords Software defect · Neural network models · Software quality · Defect tracking · Defect classification

B.V. Ajay Prakash (✉)
Department of Computer Science and Engineering, SJBIT, Bangalore, India
e-mail: ajayprakas@gmail.com

D.V. Ashoka
Department of Computer Science and Engineering, JSSATE, Bangalore, India

V.N. Manjunath Aradya
Department of Master of Computer Application, SJCE, Mysore, India

H.R. Vishwakarma and S. Akashe (eds.), *Computing and Network Sustainability*, Lecture Notes in Networks and Systems 12, DOI 10.1007/978-981-10-3935-5_16

1 Introduction

Software defect detection and classification are important and expensive activities in software development. According to Shull et al. [1], in IT industries, manual software reviews and software testing activities can detect only 60% of software defects. A study by [2] found that the probability of defect prediction is 71% from software defect predicators. Software defects or bugs can be defined as "state in a software product that unable to meet the requirement specification in software." Defects in software program can be logical error which makes to behave differently from actual functionality. Defect management is the essential phase in software development life cycle (SDLC). Defect management consists of defect identification, defect prevention, defect classification, defect prediction, and defect reporting.

Defect prevention action exercises are to discover blunders in programming necessities and outline archives, to survey the calculation executions, imperfections logging, documentation and root cause analysis. Defect identification is to check the code standard infringement as the code is produced, changed, and adjusted. Defect reporting is to obviously depict the issue connected with specific module in programming item so designer can fix it effectively. Defect classification is the process of classifying the defects based on severity to show unexpected behavior of the program output impact on quality of software. The main objective of defect prediction is to predict how many defects or bugs in the product developed, before the deployment of software product, also to estimate the likely delivered quality and maintenance effort.

The organization of this paper is as follows: Sect. 2 presents comprehensive related review of different neural network model explored on different aspects of the defect management activities while Sect. 3 proposed method to identify and classify the defects. Section 4 presents the application of GRNN and PNN model and results obtained. Finally, in conclusion section, work has been concluded and in future work section, goals for future research are presented.

2 Related Work

The problem of software defects detection and defects classification has become a major research topic in the field of software development due its need in the software industries. Researchers have been working in applying machine leaning classification algorithms to optimize the result. Classification algorithms have been explored for software defect prediction and classification [3], neural network [4], naive bayes [2], and decision trees (Taghi et al. 2005). Neural network (NN) has ability to process of nonlinear dynamic software defect data. In practice, it is difficult to select suitable parameters in neural network model which includes number of learning rate, hidden layers, weights, and training cycles [5]. NN architecture parameters setting are obtained by used trial and error methods or rule of thumb. Best possible parameters setting is difficult to obtain in NN architecture

[6]. Honar and Jahromi [7] proposed a novel framework for analyzing program codes such as classes and methods using call graph. Kim et al. [8] proposed new techniques for classifying the change requests as buggy or not buggy. According to Antoniol et al. [9], all bug reports are not associated to software problems, may be from changes request from bug reports. Fluri et al. [10] used semi-automated approach called agglomerative hierarchical clustering to discover hidden pattern from program code changes. Jalbert and Weimer [11] identified bug report duplication automatically from bug software repositories. Cotroneo et al. [12] made malfunction investigation in Java Virtual Machine (JVM). Guo et al. [13] identified many factors which may affect in fixing the bugs for windows family systems. According to Kalinowski [14], if defect rates are reduced by 50%, rework can be reduced. Used Defect Causal Analysis (DCA) improves the quality of the program. Then, DCA is enhanced and named it as Defect Prevention-Based Process Improvement (DPPI) to conduct and measure. Davor [15] has proposed an approach for automatic bug tracking using text categorization.

3 Proposed Method for Defect Identification and Classification

The proposed method basically classified into four major steps: defects identification phase, applying data mining techniques and defects classification based on severity measures. Figure 1 shows the block diagram of the proposed method.

Defect data sets are retrieved and stored in the file system. For pre-processing, the defect data attributes are parsed and some attributes are selected for measurements using various software metrics. Proposed method classifies the bug into three major layers, namely abstraction layer, application layer, core layer. Abstraction layer defects are related to major functionality failures which lead to data loss in the system. Application layer defects are associated with minor defects. Application layer may be graphical user interface functionality behave differently from expected. Core layer defects severity is high which can cause software failures. Defects which fall into core layer needs be investigated properly in order to reduce software failures. In order to classify defects into abstract, core and application layer neural network models are used. In our proposed method, GRNN and PNN are applied to classify defects in software source code.

3.1 Generalized Regression Neural Networks

One of the supervised learning neural network models is Generalized Regression Neural Network. GRNN may be used time series predictions, classification, and regression. The architecture of GRNN is as shown in Fig. 2.

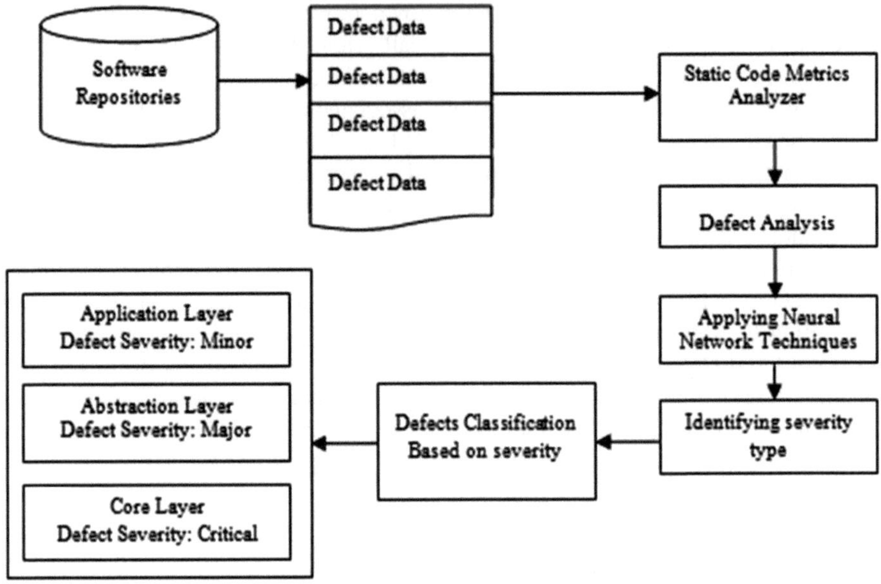

Fig. 1 Steps in defects identification and classification

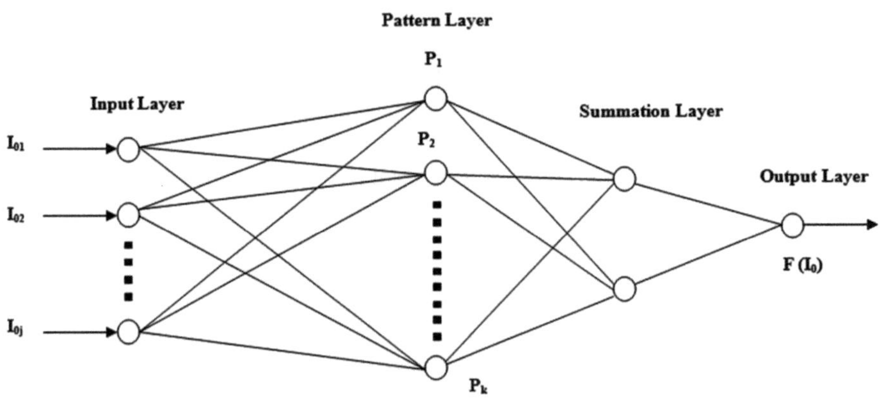

Fig. 2 Architecture of GRNN model

GRNN has four layers: input layer, summation layer, pattern layer, and output layer.

The input for the input layer depends on the number of attributes used in the defect classification. It contains attributes which are dependent to each other i.e., an input vector 'I' (feature matrix F_i). Patter layer contains neurons which need training, and the output of patter layer is given as input to the summarization layer. Summarization layer performs normalization of the output result. The weight vector is calculated using the following equations:

$$W_i = e\left[\frac{||I - I_t||^2}{2h^2}\right]$$

$$F(I) = \frac{\sum_{i=1}^{n} T_i W_i}{\sum_{i=1}^{n} W_i}$$

Where the output F (I) is weighted average of the target values T_i of training cases I_i close to a given input case I.

3.2 Probabilistic Neural Network (PNN)

The PNN classifier is based on Bayes–Parzen classifier [16]. The foundation of the approach is well known decades ago (1960s). PNN model follows bayesian classifier and reduces the risk of misclassification errors. Due to the absence of data about the class, PNN make use of nonparametric techniques. The advantages of PNN are better convergence and generalization properties. The architecture of PNN is as shown in Fig. 3. Training the PNN model is faster than backpropagation. The operations in PNN are organized into multilayered feed forward network with four nodes, namely input, hidden, decision, and class nodes.

Input layer has different input features which are dependent to each other. The difference between GRNN and PNN is PNN, which does not carry weights in the hidden node.

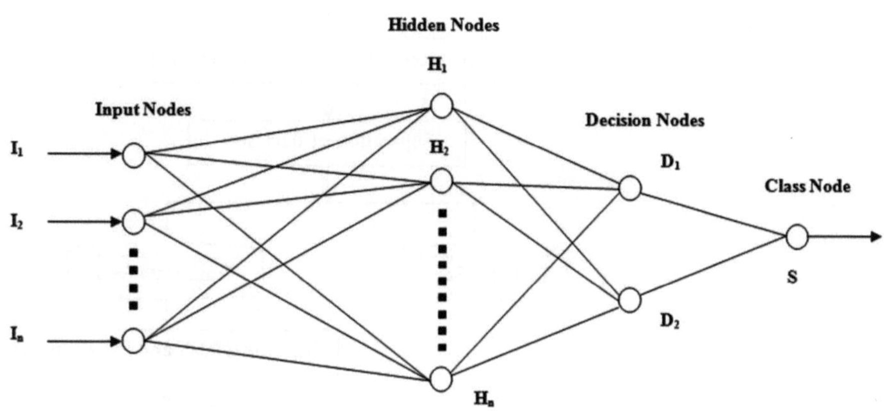

Fig. 3 Architecture of PNN model

Every hidden node acts as weights for sample vector. The hidden node creation is defined as the product of sample vector '*E*' and input feature vector '*F*' given as $h_i = E_i \times F$. The following equation is used for class activation process.

$$S_j = \frac{\sum_{i=1}^{n} e^{\frac{(h_i - 1)}{\varphi^2}}}{N}$$

Where '*N*' is example vectors belonging to class 'S', 'h_i' is hidden node activation, and 'φ' is smoothing factor.

4 Applying GRNN and PNN Model for Defect Classification

Openly available promise repository NASA and MDP software defect data sets are collected. GRNN, PNN, J48, and Multilayer Perceptron (MLP) classification algorithms are applied on various data sets such as MC1, MC2, MW1, CM1, JM1, KC1, PC1, PC2, PC3, and KC3. MATLAB 11 is used for implementing the neural network models. The evaluating procedure of different neural network model is shown in Fig. 4.

In each data set, some attributes were selected based on more impact. Features were extracted based on the dependencies. Datasets have been separated into

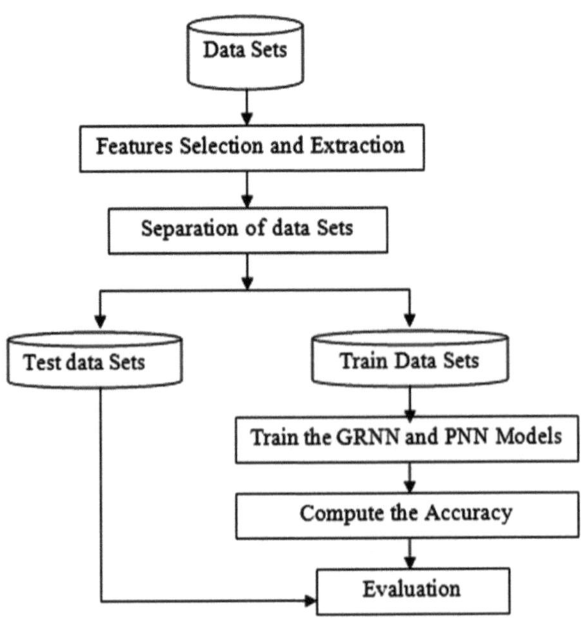

Fig. 4 Procedure for evaluating GRNN and PNN Architecture

Table 1 Accuracy result of different classifiers on 11 datasets

	Classifiers			
Data sets	GRNN	PNN	MLP	J48
KC3	0.874	0.889	0.72	0.776
MC1	0.978	0.968	0.952	0.915
MC2	0.892	0.814	0.669	0.778
MW1	0.912	0.928	0.852	0.838
PC1	0.918	0.911	0.907	0.88
PC2	0.968	0.944	0.94	0.95
PC3	0.901	0.891	0.823	0.831
PC4	0.889	0.886	0.814	0.82
CM1	0.824	0.811	0.797	0.801
JM1	0.882	0.872	0.75	0.725
KC1	0.813	0.882	0.733	0.721

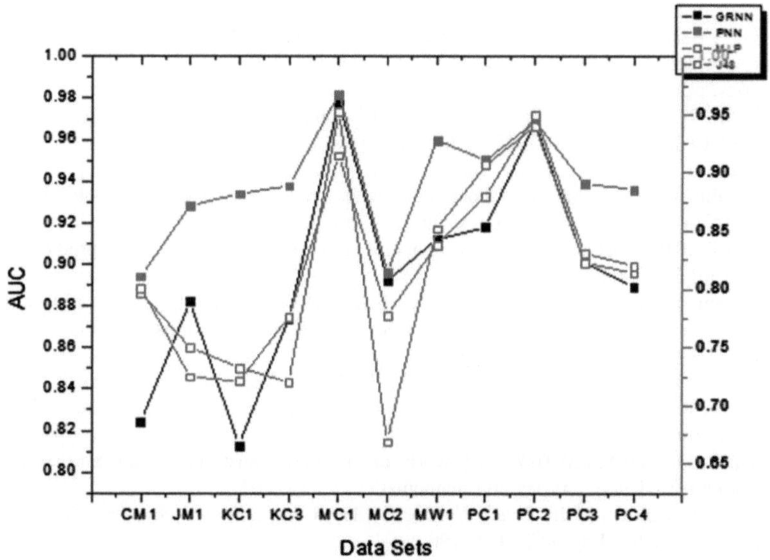

Fig. 5 Accuracy results comparison of GRNN, PNN, MLP, and J48 classifiers on 11 data sets

training and testing datasets with 20% and 80%, respectively. For accuracy indicator, area under curve (AUC) is used to evaluate various classifiers in our experiments. According to Lessmann et al. [17], use of AUC improves cross study comparability. Figure 4 shows the procedure followed in applying GRNN and PNN model. Table 1 shows the accuracy result of different classifiers on 11 datasets. A comparison result of various classifiers is shown in Fig. 5.

5 Conclusion

Software industries are facing challenges in improving the quality of software and reducing the defects in the software. To overcome these problems, there is a need of better algorithms and software tools. Manually finding and classifying software defects takes much time. Several authors have applied various machine learning algorithms for defect classification. To automate the defects identification and classification, a novel method is proposed. In our proposed method, GRNN and PNN models are integrated in software defects classification process. Openly available MDP and NASA data sets are used for evaluation of the proposed neural network models. Evaluation results shows better accuracy compared to MLP and J48 classifiers. Further, these work will enhanced to apply on real-time data sets.

References

1. Shull F, Basili V, Boehm B, Brown AW, Costa P, Lindvall M, Zelkowitz M (2002) What we have learned about fighting defects. In: Proceedings eighth IEEE symposium on software metrics, pp 249–258
2. Menzies T, Greenwald J, Frank A (2007) Data mining static code attributes to learn defect predictors. IEEE Trans Softw Eng 33(1):2–13
3. Denaro G (2000) Estimating software fault-proneness for tuning testing activities. In: Proceedings of the 22nd international conference on software engineering—ICSE '00. ACM Press, New York, New York, USA, pp 704–706
4. Zheng J (2010) Cost-sensitive boosting neural networks for software defect prediction. Expert Syst Appl 37(6):4537–4543
5. Lessmann S, Baesens B, Mues C, Pietsch S (2008) Benchmarking classification models for software defect prediction: a proposed framework and novel findings. IEEE Trans Softw Eng 34(4):485–496
6. Lin S-W, Chen S-C, Wu W-J, Chen C-H (2009) Parameter determination and feature selection for back-propagation network by particle swarm optimization. Knowl Inf Syst 21 (2):249–266
7. Honar E, Jahromi M (2010) A framework for call graph construction. Student thesis at school of computer science, physics and mathematics
8. Kim S, Whitehead EJ Jr, Zhang Y (2008) Classifying software changes: clean or buggy? IEEE Trans Softw Eng 34(2):181–196
9. Antoniol G, Ayari K, Penta MD, Khomh F, Guéhéneuc YG Is it a bug or an enhancement? a text-based approach to classify change requests. In: Proceedings of the 2008 conference of the center for advanced studies on collaborative research, New York, pp 304–318
10. Fluri B, Giger E, Gall HC (2008) Discovering patterns of change types. In: Proceedings of the 23rd international conference on automated software engineering (ASE), L'Aquila, 15–19, pp 463–466
11. Jalbert N, Weimer W (2008) Automated duplicate detection for bug tracking systems. In: IEEE International conference on dependable systems & networks, Anchorage, 24–27, pp 52–61
12. Cotroneo D, Orlando S, Russo S (2006) Failure classification and analysis of the java virtual machine. In: Proceedings of the 26th IEEE international conference on distributed computing systems, Lisboa, 4–7, pp 1–10

13. Guo PJ, Zimmermann T, Nagappan N, Murphy B (2010) Characterizing and predicting which bugs get fixed: an empirical study of microsoft windows. In: ACM international conference on software engineering, Cape Town, 1–8, pp 495–504
14. Kalinowski M (2010) Applying DPPI: A defect causal analysis approach using bayesian networks. In: Ali Babar M (ed) Product-focused software process improvement, vol 6156. Springer, Berlin, pp 92–106
15. Čubranić D (2004) Automatic bug triage using text categorization. In: SEKE 2004: proceedings of the sixteenth international conference on software engineering & knowledge engineering
16. Masters T (1995) Advanced algorithms for neural networks. Wiley, New York
17. Lessmann S, Baesens B, Mues C, Pietsch S (2008) Benchmarking classification models for software defect prediction: a proposed framework and novel findings. IEEE Trans Softw Eng 34(4):485–496
18. Masters T (1995) Advanced algorithms for neural networks. Wiley, New York
19. Ajay Prakash BV, Ashoka DV, Manjunath Aradhya VN (2014) Application of data mining techniques for defect detection and classification. In: Proceedings of the 3rd international conference on frontiers of intelligent computing: theory and applications, vol-1. Springer, pp 387–395

Design and Development of a Real-Time, Low-Cost IMU Based Human Motion Capture System

P. Raghavendra, M. Sachin, P.S. Srinivas and Viswanath Talasila

Abstract This paper presents the design of a portable low-cost wireless wearable embedded system to provide motion capture in real-time. The system consists of multiple wireless nodes strapped on to the subject. Each node contains a Wi-Fi module, battery, inertial sensors, magnetometer and a microcontroller sealed inside a 3D printed enclosure. The microcontroller runs an attitude estimate algorithm and streams the data to a Blender game engine. Data from all the nodes is collected using round-robin algorithm and given to an avatar model which mimics the human gait.

Keywords Gait · Blender game engine · Complementary filter · Quaternion · IMU · Data streaming · Motion-capture · Attitude estimation · Real-time

1 Introduction

The measurement of human motion—gait measurement—is crucial in many disciplines ranging from medicine (movement disorders with a neurological basis [1] (e.g. cerebral palsy) or with an orthopedic basis [2] (e.g. osteo-arthritis, limb injuries)) to the animation industry. At a basic level, gait measurement is simply the measurement of joint angles (velocities and accelerations as well) in 3D. In

P. Raghavendra (✉) · M. Sachin · P.S. Srinivas · V. Talasila
Department of Telecommunication Engineering, M.S. Ramaiah Institute of Technology, Bangalore, India
e-mail: raghavendra.karnad@gmail.com

M. Sachin
e-mail: sachin.gkf@gmail.com

P.S. Srinivas
e-mail: sumukh110@gmail.com

V. Talasila
e-mail: viswanath.talasila@msrit.edu

H.R. Vishwakarma and S. Akashe (eds.), *Computing and Network Sustainability*, Lecture Notes in Networks and Systems 12, DOI 10.1007/978-981-10-3935-5_17

medicine, often more complex gait measurements are required, such as the shear forces induced during specific walking motion, or the load distribution across various joints in a limb, or the energy consumed in each gait cycle [3]. In this paper we are concerned specifically with only joint angles (velocities and accelerations); and thus we shall discuss the different tools of measurement for these parameters.

A typical gait analysis system is equipped with multiple cameras and a patient who wears markers on various reference locations on the body, (Gavrila 1996; Karaulovaa 2002; Cappozzo 2005; Goffredo 2008; Chiari 2005). With this, when the patient walks the cameras record the trajectory of each body marker, and an underlying model then gives the overall 3D gait of the person. This gives a complete breakdown of the movement of each joint. The cameras can be of the normal visible range types or in some cases include infra-red cameras as well. The movement of each joint is captured in the form of Euler angles in 3D; and this information is sent to a stick figure—which is then animated to mimic the motion of the human with the measured joint angles. The motion of the stick-figure defines the gait of the person 5. The camera based technology is reasonably mature; but it suffers from important drawbacks. A specific environment/room needs to be created in order to use the cameras for analyzing the gait; for e.g. the lighting conditions need to be appropriate for the cameras to function effectively and a careful camera calibration is required. Performing a camera re-calibration is a difficult procedure and often needs a computer vision expert to be present. Furthermore whenever the system has to be deployed in another location, e.g. a rural hospital or at an outdoor location for an animation movie capture, a complete re-calibration is required as well as ensuring the lighting conditions are optimal, which is usually not possible. Further, such systems require a dedicated laboratory which makes the system more expensive, requires maintenance and takes more setup time [4].

This has given rise to a new generation gait analysis systems, developed using the concept of wearable sensors, essentially inertial sensors strapped onto a patient's body, (Tao Weijun 2012; Bonato 2003; Engin 2005). The use of inertial sensors for motion tracking of individual limbs and joints is a recent trend and is currently of great interest. For example in Zhou (2007), an inertial motion system has been developed to track the motion of the upper limbs, using a Kalman filter and standard kinematic models of limb motion. The University of Kent and East Kent Hospitals, in the UK, have performed clinical studies of the application of inertial sensors to monitor the rehabilitation of patients with neurological disorders; they used a standard Xsens IMU module with in-built tracking algorithms and performed various clinical tests (such as drinking water and performing more complex manipulation tasks) and the studies concluded that the use of inertial sensors for specific rehabilitation is clinically feasible, (Lu Baia 2012).

Inertial sensing however suffers from a perennial problem of drift 6. In this paper we present a detail mathematical technique to cancelling out the drift—through the construction of a complementary filter—and using the drift-free accelerometer angular information to correct the drift prone gyroscope data. This is presented in Sect. 4.1. In Sect. 3 we present the entire experiment set up for the data capture.

Finally, in Sect. 4.2 we transfer the measured angular positions to an animation engine, called Blender, to visualize the actual motion.

2 Existing Methods of Data Capture

Number of technologies have been proposed for recognizing and analyzing human gait. All the methods have many advantages and disadvantages. Disadvantages in real time analysis will be accuracy, number of sensors, cost, range and scan rate. The existing methods can be divided into two categories they are wearable sensors and non-wearable sensors. Wearable sensors include GRF plates [5], pressure sensor [6], EMG [7], Ultra sound [8], UWB [9] and goniometer [10]. Non wearable sensors include single camera with image processing [11], stereoscopic vision [12], time of flight camera [13], structured light and IR thermography. The IMU based motion capture system are relatively less expensive and easy to setup as they do not require are light controller studios with high resolution cameras. IMU based mo-capture systems can give accuracy usually up to 1° while the camera based mocap system can easily give subdegree accuracy. Therefore, IMU based motion capture system is gaining popularity in applications which doesn't require subdegree accuracy.

3 Experimental Setup

We aim to place sensor nodes at each joints of the body. These nodes consists of a Wi-Fi SoC (ESP8622-12E) which communicates with the accelerometer (ADXL345), gyroscope (L3G4200D) and the magnetometer (HMC5883L) using I2C protocol. The angular rate (wx; wy; wz), from the gyroscope is integrated to get the angular position while the data from the accelerometer and magnetometer is used to correct the drift caused due to the gyroscope bias and noise. The detailed working of the orientation estimation algorithm is explained in Sect. 5. This data is sent to a UDP server running in the Blender game engine to obtain a real-time animation of the avatar. The attitude estimation algorithm runs at around 600 Hz and the animation is displayed at 20FPS. Block diagram of the system is as shown in Fig. 1.

Figure 2 shows multiple sensor nodes strapped on to the body, each of them placed precisely on the predefined position. Every node runs a quaternion based complementary filter to compute the orientation/attitude with respect to the inertial frame and thereby capturing the entire body dynamics.

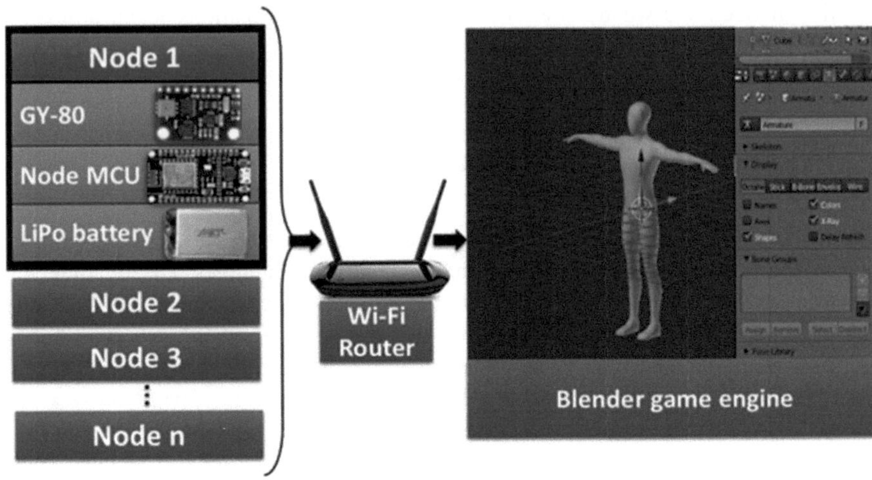

Fig. 1 Block diagram of IMU based gait analysis system

Fig. 2 Illustration of sensor nodes placed on the biceps, fore arm and terminal part of the hand

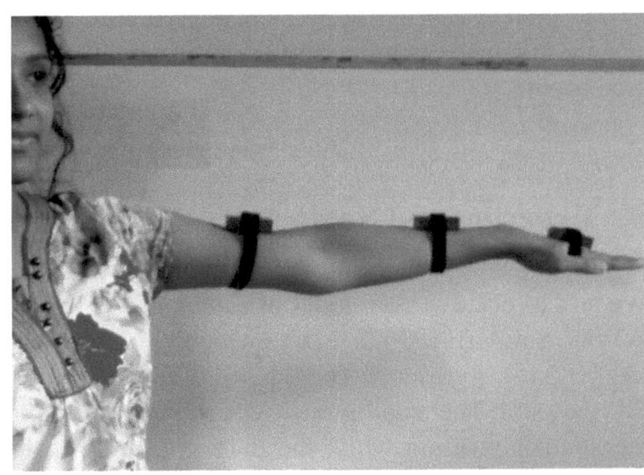

4 Design and Analysis

4.1 Quaternion Based Complementary Filter

Introduction the MEMS gyroscope has the ability to respond to high frequency dynamics and resist noise which makes it the primary sensor. The gyroscope data drifts with time due to numerical integration. Hence, the low frequency component of the accelerometer and magnetometer is used to correct the gyroscope drift as it does not include any numerical integration which means there is no drift but has bounded errors. The overview of the filter is as shown in Fig. 3. Accelerometer outputs (ax; ay; az); denote (ax(0); ay(0); az(0)) as the initial accelerometer

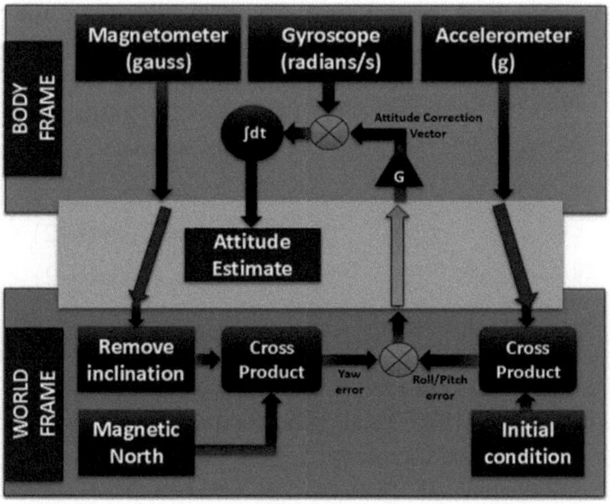

Fig. 3 Block diagram of attitude estimation algorithm

values 7. Similarly gyroscope outputs (wx; wy; wz); and denote (wx(0); wy(0); wz (0)) as the initial gyroscope values. Finally, magnetometer outputs (mx; my; mz); and denote (mx(0); my(0); mz(0)) as the initial magnetometer values. We define the reference 3 axis vector of accelerations (in inertial frame) to be

$$\alpha^{i}_{ref} = [0\,0\,1] \tag{1}$$

α^{i}_{ref} is basically a reference initial condition which corresponds to the accelerometer perfectly levelled and only z axis having nonzero reading[8]. Note that

$$\alpha^{i}_{ref} - \alpha^{i}_{meas} \neq 0 \tag{2}$$

In other words, the measured initial accelerations may *not* be the true initial condition: due to measurement/mounting errors. In a similar way, we define a reference magnetic measurement, assuming the magnetometer is pointing perfectly to magnetic north, as

$$\mathbf{m}^{i}_{ref} = [1\,0\,0] \tag{3}$$

here, m is the normalized version of the actual magnetic field measurements when the sensor is pointing exactly at magnetic north. Note that

$$\mathrm{m}^{i}_{ref} - \mathrm{m}^{i}_{meas} \neq 0 \tag{4}$$

In other words, the measured initial measured magnetic fields may *not* be the true initial condition; due to measurement/mounting errors.

Equations (2) and (4) imply that the sensor frame is not be level. In Sects. 4.1 and 4.2 we present a method to level the sensor platform.

Incremental quaternion form gyroscope data let us define an incremental quaternion $\Delta q = q_w + q_x i + q_y j + q_z k$. Converting the axial rotation from the gyroscope to quaternion form, we have:

$$\begin{aligned} q_w &= \cos(\theta/2) \\ q_x &= U_x \sin(\theta/2) \\ q_y &= U_y \sin(\theta/2) \\ q_z &= U_z \sin(\theta/2) \end{aligned}$$

where, $\vec{U}$ is a unit vector obtained by normalizing the gyroscope values, $\vec{\boldsymbol{\omega}}$ and $\boldsymbol{\theta}$ is the angle of rotation around the unit vector obtained from the gyroscope data. The quaternion which relates the inertial frame and the world frame is the cumulative sum of all the incremental quaternions.

Accelerometer based leveling Let Q_b^i denote the quaternion that relates body frame to inertial frame. Given $ax^b(t)$, $ay^b(t)$, $az^b(t)$ be the accelerometer values in body coordinates, let $ax^i(t)$, $ay^i(\mathrm{t})$, $az^i(t)$ be the representation in inertial coordinates, with

$$Q_b^i\colon \{ax^b(t),\, ay^b(t),\, az^b(t)\} \mapsto \{ax^i(t),\, ay^i(t),\, az^i(t)\}$$

Denote $\mathbf{a}$ for the vector of accelerations $\{ax(t), ay(t), az(t)\}$, so we have $Q_b^i\colon \mathbf{a}^b \mapsto \mathbf{a}^i$. Since the sensor frame is not level at the start of the experiment, in this section we present a technique to level the frame by computing the deviation of the measured vector from the reference vector. Since the measurements, $\mathbf{a}^i_{meas}$ and the reference, $\mathbf{a}^i_{ref}$ are basically vectors, their cross product will result in the error deviation, i.e.

$$\varepsilon^i_{acc} = \mathbf{a}^i_{ref} \times \mathbf{a}^i_{meas} = \begin{bmatrix} \varepsilon^i_{ax} \, \varepsilon^i_{ay} \, 0 \end{bmatrix} \tag{5}$$

Then we obtain the error vector in body coordinates as follows

$$\varepsilon^b_{acc} = Q_i^b \varepsilon^b_{acc} \tag{6}$$

where Q_i^b is the inverse of Q_b^i.

Magnetometer based leveling Given $mx^b(t)$, $my^b(t)$, $mz^b(t)$ in body coordinates, let $mx^i(t)$, $my^i(t)$, $mz^i(t)$ be the representation in inertial coordinates, with

$$Q_b^i\colon \{mx^b(t),\, my^b(t),\, mz^b(t)\} \mapsto \{mx^i(t),\, my^i(t),\, mz^i(t)\}$$

Denote m for the vector of magnetometer measurements $\{mx(t), my(t), mz(t)\}$ so we have

$$Q_b^i\colon \mathbf{m}^b \mapsto \mathbf{m}^i$$

Denote tn for the vector of magnetic fields, m = $\{mx(t), my(t), mz(t)\}$ as measured by the 3axis magnetometer, so we have

$$Q_b^i = \mathbf{m}^b \mapsto \mathbf{m}^i$$

We are assuming that the initial position is with the frame leveled which need not be true. So we aim to level the frame by computing the deviation of the measured vector from the reference vector. Since both the measurements are basically vectors, their cross product will result in the error deviation. In other words we obtain

$$\varepsilon_{mag}^i = \mathbf{m}_{ref}^i \times \mathbf{m}_{meas}^i = \left[0\,0\,\varepsilon_{mz}^i\right] \tag{7}$$

Then we can obtain the error vector in body coordinates as follows

$$\varepsilon_{mag}^b = Q_i^b \varepsilon_{mag}^i \tag{8}$$

The two error vectors, ε_{acc}^b and ε_{mag}^b, are normalized and we obtain the attitude correction factor for platform leveling as $\left[\varepsilon_{ax}^i\,\varepsilon_{ay}^i\,\varepsilon_{az}^i\right]$. Thus the obtained unit error vector is scaled by a factor G to get the attitude correction vector. This is subtracted from the gyroscope data before forming the incremental quaternion,

$$\vec{\omega}^{corrected} = \vec{\omega} - \left[\varepsilon_{ax}^i\,\varepsilon_{ay}^i\,\varepsilon_{az}^i\right] \tag{9}$$

Since we are feeding this result back into the gyroscope integration operation, it will be divided by the rate at which the loop is running. If loop is running at 500 Hz, the value of the filter coefficient (α) is 0.002 * G. The value of G is the tradeoff between amount of noise let in from the sensors and the rate at which it corrects the error. In order to find the value of G consider a traditional complementary filter which follows the equation

$$y_n = \alpha^*(y_{n-1} + \omega^* dt) + (1-\alpha)^* x_n \tag{10}$$

where,

$$\alpha = \tau/(\tau + dt) \tag{11}$$

Here, y_n is the filtered roll/pitch/yaw, y_{n-1} is the previously estimated angle, ω is the angular rate from the gyroscope, x_n is the angles from the accelerometer and magnetometer and τ is the time constant. Therefore, the lilter gain G is given by

$$G = \alpha/dt. \tag{12}$$

Note that this is the filter response to error due to drift and not movement. During fast orientation changes, the response of the filter is directly related to response of the gyroscope. The filter coefficient is manually varied to get the optimum/desired filter output.

4.2 Blender Game Engine

Blender is an open source game engine our aim is to create a character and control it using a wireless node, there are three engines blender render, game engine and cycle render. Both blender render and cycle render is used for creating a character while blender game engine is used to animate the character. Blender render has variety of geometric primitives including polygon meshes, fast subdivision surface modelling and armatures. Armatures have bones which can be scaled and positioned to our character requirements. These bones are then connected to the character such that any movement in bones create movement in the character. The effect of the bone on particular part of the body can be varied by using weight paint as shown in Fig. 4.

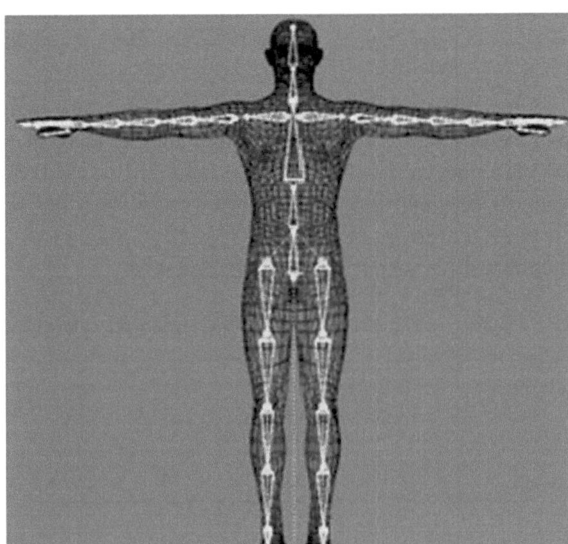

Fig. 4 Character with bones connected to it

Initially there is a bone appropriately placed on the character and from that bone a new bone is extruded and scaled. All the extruded bones will have parent relation with the adjacent bone it is removed to make it an individual/independent bone. If this relation is not removed then movement of one bone will affect the movement of other bone. This bones are controlled using blender python scripting in game engine. Game engine has sensor, controller and actuator. These are blocks used to control the movements of the bone in the character. Once we access the controller we can access all of the sensors and actuators connected to it, and even access information about the object that owns the controller. Before accessing the controller we receive the data from wireless node which are in quaternions form. The communication happens over user datagram protocol (UDP). For UDP communication to happen host and port are defined and bound together to receive the data in string format. So received string is type casted into float data type. Further float data is stored in a buffer for further use. Then we take control of the sensor as stated above we can control the movements of the character by taking control of sensor. All the bones in the character are accessed individually and rotated using the values stored in the buffer. The above step will repeat until all the bones are rotated in the character. The blender python code which enables all the automation must repeat until wireless node sends data hence we include an always block.

5 Results

A nine DOF IMU sensor consisting of three axes gyroscope, accelerometer and magnetometer is used for the experimental validation. The simulation results from Fig. 5 depicts the bias observed from the gyroscope during an oscillatory motion, this constant bias eventually grows linearly with time. Once the bias is known it can be subtracted from subsequent measurements to minimise the effect of bias.

Similarly the jitter from accelerometer can be seen over time. The complementary filter is used to eliminate the bias from the gyroscope and the jitter from the accelerometer. The data were sampled at around 250 Hz and the passed to the complementary filter. The amount of trust on the gyroscope and the accelerometer

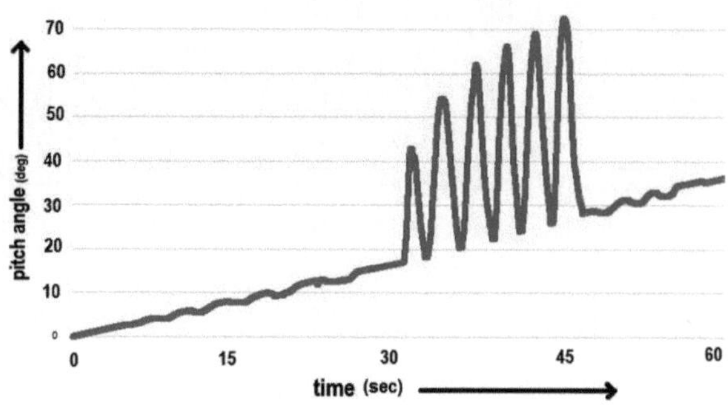

Fig. 5 Pitch angle from the gyroscope

determines the smoothness of the filtered output. In Fig. 6 the red plot depicts the raw gyro roll with the drift, the blue plot depicts the accelerometer roll with jitter, and the green is the complementary filtered output eliminating both drift and jitter.

The filtered output then passed to the blender using UDP protocol from the node to blender is animated using this software. It is used to determine the angle between two different bones, relative velocities and accelerations, position etc. can be analysed depending on the parameters of interest. Figure 7 shows the angle at the elbow joint. The nodes are placed between the joints of the limbs and the motion is captured.

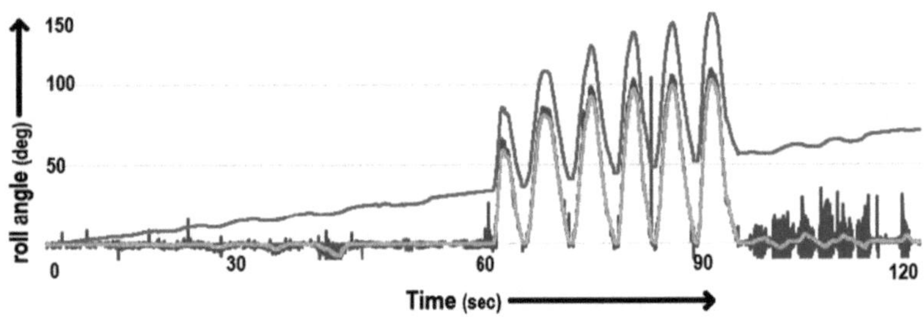

Fig. 6 Roll angles from the gyroscope (*red*), accelerometer (*blue*) and the filter output (*green*)

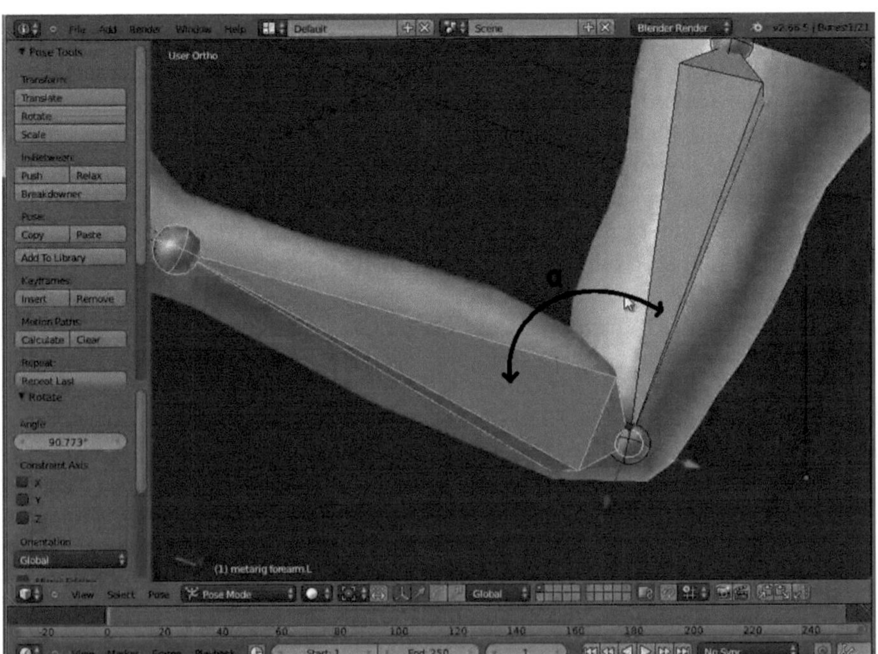

Fig. 7 Measuring the relative position of the radius bone w.r.t the humerus bone

Acknowledgements The first and last author are indebted to Dr. Ramesh Debur, a biomechanics expert from the Center of Rehabilitation at the MS Ramaiah Hospitals, for many illuminating.

References

1. Damiano DL, Abel MF (1996) Relation of gait analysis to gross motor function in cerebral palsy. Dev Med Child Neurol 38:389–396
2. Kay RM (2000) The Effect of preoperative gait analysis on orthopaedic decision making. Clin Orthop Relat Res 372:217–222
3. Baker Richard (2006) Gait analysis methods in rehabilitation. J NeuroEng Rehabil. doi:10.1186/1743-0003-3-4
4. Bamberg SJM (2008) Gait analysis using a shoe-integrated wireless sensor system. IEEE Trans Inf Technol Biomed 12(40):413–423. doi:10.1109/TITB.2007.899493
5. Bokovia R Analysis and interpretation of ground reaction forces in normal gait TEA MARASOVI, MOJMIL CECI, VLASTA ZANCHI laboratory. Mechanical Engineering and Naval Architecture University of Split
6. Tao W, Liu T, Zheng R, Feng H (2012) Gait analysis using wearable sensors. Sensors 12:2255–2283
7. Frigo C, Crenna P (2009) Multichannel SEMG in clinical gait analysis: a review and state-ofthe-art. Clin Biomech 24:236–245
8. Qi Y, Soh CB, Gunawan E, Low KS (2013) Maskooki, using wearable UWB radios to measure foot clearance during walking. In: 35th annual international conference of the IEEE engineering in medicine and biology society (EMBC), Osaka, Japan, 37 July 2013; pp 5199–5202
9. Bamberg S, Benbasat AY, Scarborough DM, Krebs DE, Paradiso JA (2008) Gait analysis using a shoe-integrated wireless sensor system. Trans Inf Tech Biomed 12:413–423
10. Wahab Y, Bakar NA (2011) Gait analysis measurement for sport application based on ultrasonic system. In: 15th international symposium on consumer electronics (ISCE), Singapore, 1417 June 2011, pp 20–24
11. Pratheepan Y, Condell JV, Prasad G (2009) The use of dynamic and static characteristics of gait for individual identification. In: 13th international machine vision and image processing conference, Dublin, Ireland, 24 September 2009, pp 111–116
12. Derawi MO, Ali H, Cheikh FA (2014) Gait recognition using time-of-flight sensor. http://subs.emis.de/LNI/Proceedings/Proceedings191/187.pdf. Accessed 17 Feb 2014
13. Gomatam ANM, Sasi S (2004) Multimodal gait recognition based on stereo vision and 3D template matching. In: CISST, pp 405–410
14. Whittle MW (2007) Gait analysis: an introduction, 4th edn. Edinburgh, Elselvier
15. Tao W, Liu T, Zheng R, Feng H (2012) Gait analysis using wearable sensors, 3rd edn, vol 12, no 12, pp 2255–2283
16. Sahasrabudhe Sameer, Iyer Sridhar (2009) Creating 3D animations of laboratory experiments using open source tools. Proceedings of ICEL, Toronto, Canada

Design and Fabrication of Level Sensor for Remote Monitoring of Liquid Levels

Fouzan Javeed, Uha Durbha, Fakruddin Baig and Khan Samida

Abstract This paper presents the design of a level sensor for the purpose of liquid level sensing and monitoring using capacitive sensing techniques. Conventional techniques have many drawbacks with respect to accuracy and they cannot be used reliably for remote sensing. The design proposed here uses three capacitive sensors and a capacitance to digital converter IC integrated onto a sensor Printed Circuit Board. The sensor PCB is attached to the walls of the container and this allows the measurements to be made without any contact with the liquid. The sensing mechanism is independent of the liquid or the environment in which it is placed. The levels are uploaded onto a database located at a local server, and this data can be easily accessed by the user as per his requirement. Further, an android application has been developed for monitoring the liquid levels conveniently.

Keywords Liquid level monitoring · Capacitive sensing · Sensor design · Fringing capacitance · Database · Android application

1 Introduction

Level sensing of liquids finds an important application in various process industries where the liquid used is to be monitored carefully. The use of capacitive sensing for this purpose is preferred because it is a more accurate technique that can support a

F. Javeed (✉) · U. Durbha · F. Baig · K. Samida
Department of Telecommunication Engineering, M.S. Ramaiah Institute of Technology, Bangalore, India
e-mail: hunt4fouzan@gmail.com

U. Durbha
e-mail: uhadhurba@gmail.com

F. Baig
e-mail: mfakruddinbaig@gmail.com

K. Samida
e-mail: superbsamkhan@gmail.com

H.R. Vishwakarma and S. Akashe (eds.), *Computing and Network Sustainability*, Lecture Notes in Networks and Systems 12, DOI 10.1007/978-981-10-3935-5_18

wide range of applications. The sensors have no moving parts and can be made insensitive to environmental variations. They make no contact with the liquid to be measured, and can be used for remote sensing applications. Conventional capacitive sensing techniques have severe limitations. External interferences due to human presence or other electronic devices cause significant parasitic capacitance that causes drastic variations in the capacitance values, thereby producing erroneous results.

The method proposed here introduces a new approach to overcome the limitations of the conventional methods. This approach relies heavily on the sensor layout which needs to be symmetric. Our sensor design comprises of shields and three parallel plate capacitors made of copper electrodes which act as three different sensors namely, the minimum reference sensor, level height sensor, and the environment sensor. The dimensions of the minimum reference and environment sensor are exactly the same. The width of the level height sensor is same as the other two sensors, but the height is much larger. The shields are used to focus the sensing toward the liquid and minimize the effects of external interferences. The sensor works on the principle of measuring the fringing capacitance between the level height electrode and a ground electrode. The fringing capacitance will be proportional to the variation in dielectric between air and the liquid. The height of the liquid column can be computed from the difference in capacitance values of the level height electrode and reference electrode (Fig. 1).

The data from the sensor is converted to digital format using a capacitance to digital converter—the FDC1004. It has 4 channels, each with a range of 15 pF. A major advantage of the FDC1004 is that it has shield drivers for sensor shields, which can reduce EMI interference and help focus the sensing direction of the sensor. In the approach proposed here, eliminating the human body capacitance effects from the measurements is given importance. This is achieved by maintaining symmetry of the channel and shield electrodes. If there is any mismatch, there will

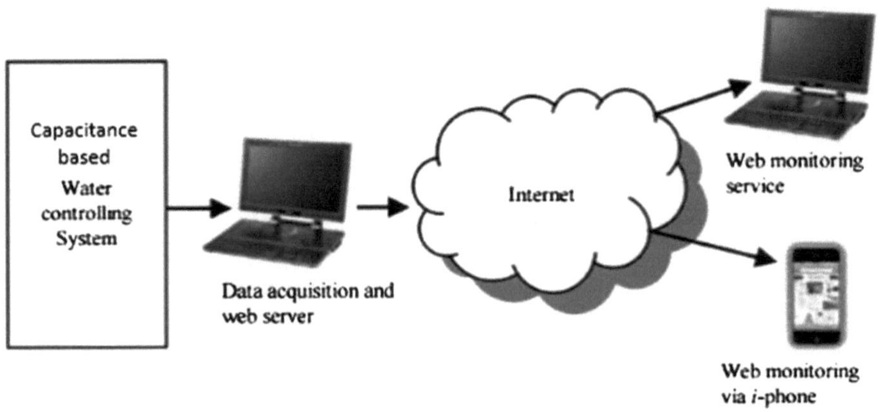

Fig. 1 Functional overview of proposed system

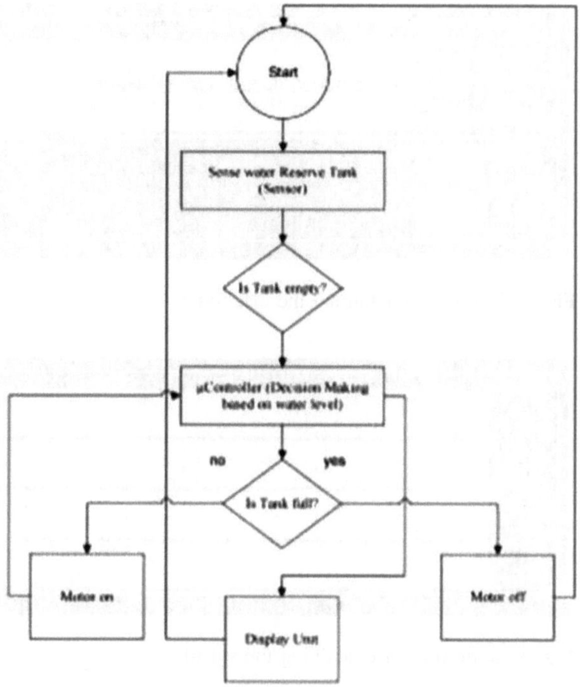

Fig. 2 Flowchart of the proposed system

be significant drifts in the capacitance values. The FDC1004 can be configured for differential mode of operation wherein two of the input channels of the FDC1004 are 180° out of phase with each other. Most other capacitive to digital converters cannot be configured this way. This feature of the FDC1004 enables us to minimize the effects of proximity interference. The capacitances from the sensors as measured by the FDC are fed into the micro controller to perform liquid level computations.

In this paper, first, the sensor layout is described and the test setup is briefed. Then, an arrangement using FDC1004 and the micro controller that can measure the level is presented. The results are discussed and the conclusions are presented in the last section (Fig. 2).

2 Sensor Layout and Fabrication

The design of the sensor plays an important role in determining the reliability and accuracy of the proposed method. Symmetry of the electrodes that constitute the sensor is very important. The sensor designed for use in this project is implemented with a two-layer PCB. On the top layer, which faces the tank, there are the 3 electrodes (reference environment, minimum reference, and level height) with a ground plane surrounding the electrodes. The dimensions of the reference

Fig. 3 Top layer containing the electrodes

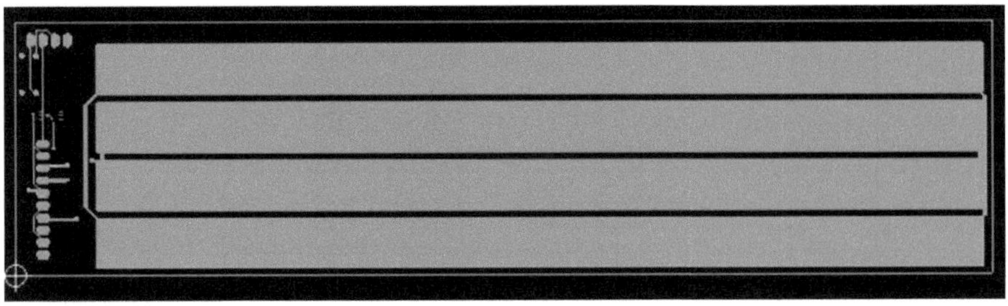

Fig. 4 Bottom layer containing the shields

environment and minimum reference electrode are the same. The distance between the two metal plates is same for all three electrodes. The bottom layer consists of shields for mitigating the effects of external interference. The same set of shields are used for the minimum reference and environment sensors. This shared shield is of the same height as the shield for level height sensor. Shielding the side of the sensor which does not face the container focuses the sensing direction toward the liquid target and provides a barrier from any interference affecting the measurements from the backside. The FDC1004 capacitance to digital converter is built into the PCB. This is done to reduce the effect of parasitic capacitance that builds up when the FDC is to be connected externally using wires. Including the FDC in the PCB gives more accurate results, and the errors are reduced. The sensor was designed using EAGLE software (Figs. 3, 4, 5, and 6).

Fig. 5 X-ray plot of the PCB

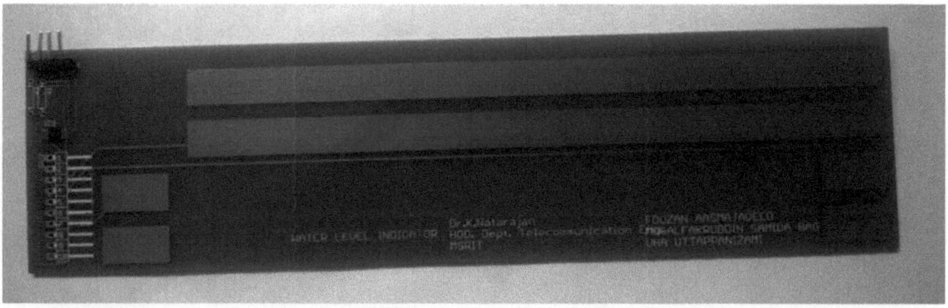

Fig. 6 Fabricated capacitance liquid level sensor

3 Experimental Setup

The sensor PCB is attached to the walls of a container using hot glue ensuring no air gap is present between the side walls and the PCB. The reference sensor is placed at the bottom of the container for liquid dielectric reference. The environment sensor is placed at the top of the container for air dielectric reference. Water was used as the liquid for the test data. Other liquids can also be used in place of water. Liquids that are viscous and that leave a lm or residue when dried will not have consistent measurements since the remnants of the liquid on the sides of the container will affect the capacitance seen by the sensors. The connections are made between the FDC1004, which is situated on the PCB itself and the Wi-Fi enabled micro controller. Sensing that the liquid levels are below the minimum threshold, the pump is automatically turned on. As the liquid levels rise in the tank, the capacitance values change correspondingly due to changes in the dielectric. At any point, the user can log into the database and view the level of the liquid. The water level is also updated in real time on the android app. When the liquid level reaches the maximum threshold, the pump is automatically turned off. The minimum and maximum thresholds can be changed by the user as per their requirements (Fig. 7).

3.1 Calculations

Capacitance Calculations Parallel plate capacitance (measured in Farads) is calculated by:

$$C = \frac{\varepsilon_r \varepsilon_0 A}{d} \tag{1}$$

where ‘A’ is the area of the two plates (in meters). ‘ε_r’ is the dielectric constant of the material between the plates. ‘ε_0’ is the permittivity of the free space. ‘d’ is the separation between the plates (in meters).

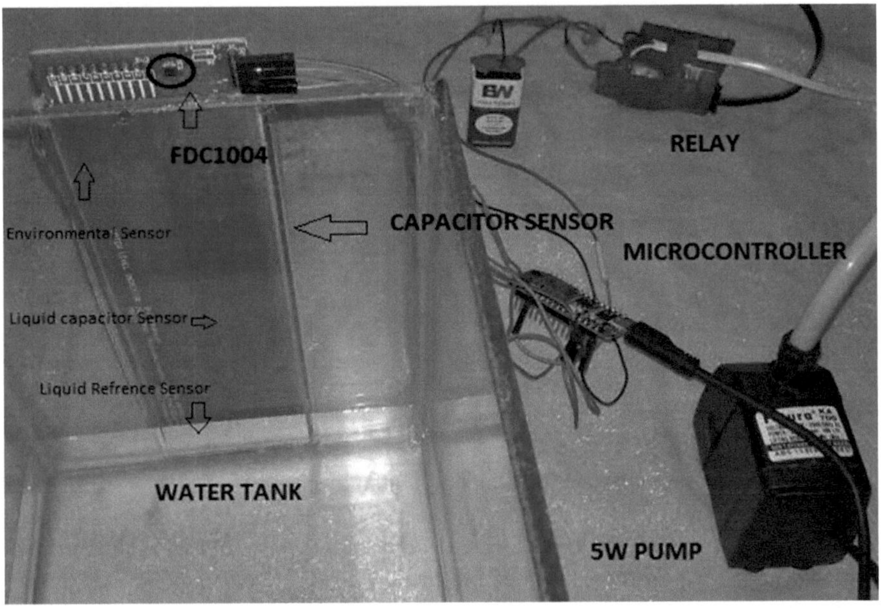

Fig. 7 Test setup

In liquid sensing, fringing capacitance is measured which is the function of dielectric variation and proportional to the liquid height and is calculated by:

$$C_{meas} = h_w e_w + (h_L - h_w) e_a \tag{2}$$

where 'h_L' is maximum liquid height. 'h_w' is liquid height. 'e_w' is liquid dielectrics. 'e_a' is air dielectrics (Fig. 8).

Level Calculations: The liquid level at any interval height is given by

$$\text{Level} = h_{RL} \frac{C_{level} - C_{level}(0)}{C_{RL} - C_{RE}} \tag{3}$$

where 'h_{RL}' is the unit height of reference liquid sensor. 'C_{level}' is the capacitance of level sensor. '$C_{level}(0)$' is the capacitance of level sensor when no liquid present. 'C_{RL}' is the capacitance of reference liquid sensor. 'C_{RE}' is the capacitance of environmental sensor.

In order to avoid the variations in the desired level to actual level, obtained gain and o set is added. First order linear correction algorithm is applied to compensate the variations.

$$\text{level} = \text{Level Gain} + \text{Offset} \tag{4}$$

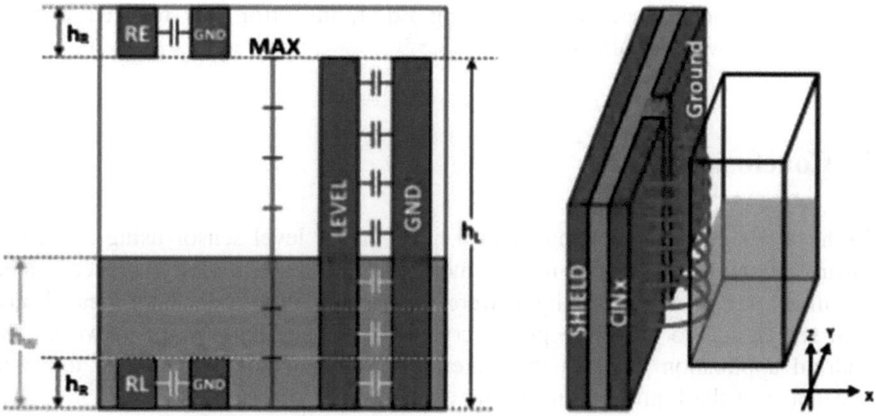

Fig. 8 Ratiometric liquid level measurement

4 Results

From g.9 it can be seen that the two capacitance obtained from environmental sensor and liquid reference sensor are nearly constant where as the liquid level sensor capacitance changes with the quantity of the liquid, linearly when plotted using Eq. 3. The error increased with increase in the level and became worse. The offset (−0.05) and gain (0.9) were measured by minimizing the overall error between the obtained level and corrected level which is obtained by applying first

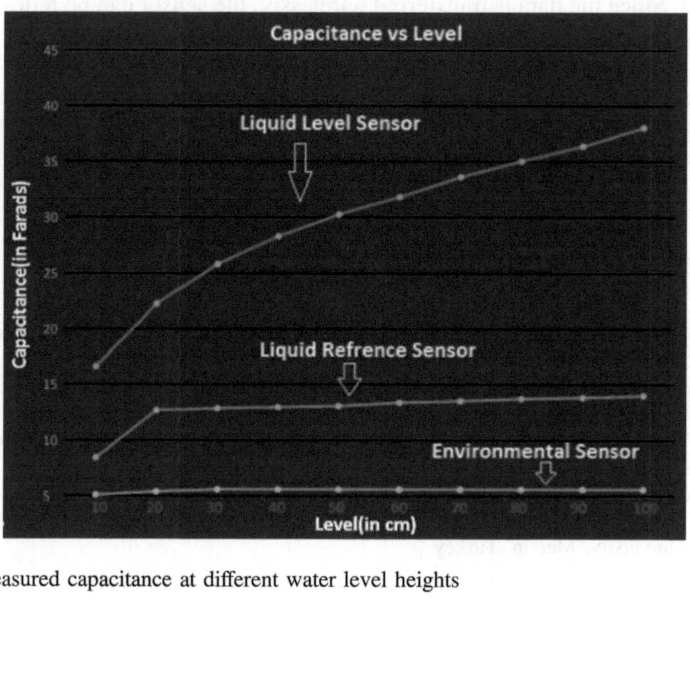

Fig. 9 Measured capacitance at different water level heights

order linear correction algorithm. Using Eq. 4, the error was reduced and the desired levels were obtained (Fig. 9).

5 Conclusion

We have designed and implemented a novel liquid level sensor using capacitive sensing technique that can be used to monitor liquid levels remotely in a container. The monitored levels of the liquid were wirelessly sent to a database located on a local server and to a mobile phone connected to the same Wi-Fi network. The Android application displays the water level on the mobile phone in terms of percentage of the liquid present in the container.

Conventional level sensors should be calibrated for each type of liquid. Furthermore, they are affected from humidity, temperature, and dust. We have designed a smart sensor to overcome the limitations of conventional methods of level sensing. The designed sensor compensates variation of different physical parameters such as temperature, liquid type, humid air gap, and dust. The sensor design allows for accurate results to be displayed. The result is displayed to the user through a mobile phone using an android application and also a database. Since the FDC1004 (the capacitance to digital convertor) IC has also been built into the sensor PCB design, the effects of parasitic capacitances are reduced and the drifts in the capacitance values are largely eliminated. The monitoring of liquid levels plays a vital role in many process industries to control the liquid levels up to a great precision. The sensor designed can be used in these applications since it can be used for remote monitoring and does not require direct contact with the liquid being monitored. Since the data is transferred wirelessly, the sensor can be redesigned as a standalone unit.

Acknowledgements The authors would like to thank Raghavendra Padhmanabha for his indispensable contribution and Dr K Natarajan, Professor and Head, Department of telecommunication engineering, MSRIT for his guidance and support.

References

1. Baxter LK (1997) Capacitive sensors. Design and applications, 3rd edn. New York Press
2. Babu CSS, Somesh DH (2006) Design of self-compensated non-contact capacitive sensors and proficient signal conditioning circuit for multi threshold liquid level control a novel approach. In: IEEE international conference on industrial technology, 2006. ICIT 2006
3. Bande V, Ciascai I, Pitica D (1997) Lowcost capacitive sensor for wells level measurement, 3rd edn. New York Press
4. Canbolat H (2009) A novel level measurement technique using three capacitive sensors for liquids. IEEE Trans Instrum Measur 58(10). Department of Electrical Electronics Engineering, Mersin University, Mersin, Turkey

5. Narayana KVL, Kumar VN (2013) A Bhujanga Rao improved linearized network based liquid level transmitter. In: 2013 international conference on control, automation, robotics and embedded systems (CARE)
6. Reverter F (2010) Interfacing differential capacitive sensors to microcontrollers. Trans Instrum Measur 59(10)
7. Wang D (2006) Capacitive sensing: out-of-phase liquid level technique, SNOA925, Texas Instruments

A Novel Outlier Detection Scheme (ODS) in Wireless Sensor Networks

Shantala Devi Patil and B.P. Vijayakumar

Abstract Outlier is referred to as deviated behavior from normal. Outlier detection is crucial for normal functioning of the network. Suitable actions are needed to thwart such behavior. Event is an expected network behavior, caused by change in the state of the networks, whereas attacks are unexpected behavior in network that cause difficult situations to the network. Event detection and attack detection are many times mistook as same. Event detection is followed by suitable actions to handle the event, whereas attack detection should be followed by countermeasures to subdue the situation. In this paper, we propose an outlier detection scheme (ODS) for detecting events and attacks in the wireless sensor network. We evaluate the efficiency of scheme through simulations.

Keywords Outliers • Wireless sensor networks • Events • Node compromise • Malicious activity

1 Introduction

Sensor devices [1] are used to sense and monitor the physical environment in remote hostile locations. These devices monitor environmental parameters and transmit this data across the network for further processing. If the behavior of network and its elements differs from the expected behavior, it is termed as an outlier or anomaly [2]. The outlier can be caused by noise, events, and attacks. The outlier detection aids in secure functioning of the network by thwarting the

S.D. Patil (✉)
Computer Science and Engineering, REVA Institute of Technology and Management, Bangalore, India
e-mail: shantaladevipatil@gmail.com

B.P. Vijayakumar
Information Science and Engineering, M.S. Ramaiah Institute of Technology, Bangalore, India
e-mail: vijaykbp@yahoo.co.in

H.R. Vishwakarma and S. Akashe (eds.), *Computing and Network Sustainability*, Lecture Notes in Networks and Systems 12, DOI 10.1007/978-981-10-3935-5_19

erroneous behavior. Other advantages of outlier detection in wireless sensor networks (WSN) are as follows: It ensures reliability of the data collected by sensors, reports events occurring in the network, and triggers alarm. The outlier detection schemes [3] for WSN can be applied to *environmental monitoring* for detecting an event and raising an alarm to the base station. In *health monitoring* based on the data monitored by the body sensors, outlier detection helps in providing timely medical attention by identifying the potential risks to the patient. For *habitat monitoring*, endangered species and their behavior are analyzed with outlier detection. In *target tracking*, outlier detection scheme eliminates faulty data and increases tracking efficiency. In *surveillance,* unauthorized access to critical areas is restricted by outlier detection. In *industries,* outlier detection schemes can help identify faulty machineries and processes. Outlier detection scheme in *structural monitoring* helps identify obstructions in the bridges, tunnels, and other structures.

In this paper, we propose a novel outlier detection scheme for fast detection of events and attacks in the network. The rest of the paper is organized as follows. Section 2 discusses the related work. Section 3 gives the prerequisites for proposing our scheme for event detection with notification and attack detection with defense. Section 4 proposes a novel outlier detection scheme (ODS). Section 5 analyzes the ODS scheme based on evaluation metric and compared with the existing schemes such as IADS [4] and ODCS [5]. In Sect. 6, we conclude the discussion.

2 Related Work

Events are predefined based on the sensor node readings and underlying applications. All the nodes are aware of events. Triggers are set for each of these events for notifying their occurrence. The base station periodically queries the network for event occurrence. In contrast to events, attacks are not predefined. They can be detected by comparing with the behavior of nodes. Whenever the behavior of node is abnormal or suspicious, alarms are raised in the network. The base station identifies the attack type and takes appropriate countermeasures to thwart the attack [6]. Event detection must not mix the abnormal readings from sensors to detect and report an event. Similarly, malicious attack detection must prevent mixing of normal data to confirm occurrence of an attack.

2.1 Classification of Outliers Detection Schemes

The outlier detection schemes previously published can be classified as follows. *Based on processing,* the schemes can be classified as centralized and distributed. In *centralized outlier detection schemes*, data is processed by a central authority and a decision is taken about existence of an outlier in the system. Such schemes incur more communication overhead. In *distributed outlier detection schemes,* the processing of data can be done by the nodes themselves and a decision can be taken accordingly about the presence of an outlier. These schemes are more efficient than centralized schemes with no bottleneck, but the nodes exhaust their energy fast, due to extra processing incurred for outlier detection. The decentralized schemes can be improvised by taking advantage of clustered arrangement with the detection of outlier performed by cluster head. This will reduce the processing overhead on the nodes.

From the published work, we observe that existing outlier detection techniques have lot of shortcomings. Hardly can we find any scheme that considers the mobility of nodes. Though mobility induces energy overhead, still it is advantageous for the network in terms of handling holes [7]. Many of the schemes are proposed for flat topology and do not consider the clustered structure of WSN. As [8] shows that clustering is more advantageous in terms of energy conservation. Schemes assume that all the nodes have same capability, i.e., homogeneous, but recent development in WSN has shown that using heterogeneous nodes can increase the network lifetime multifold [6]. Little work has been done on distinguishing between errors, events, and attacks. These shortcomings in turn raise a need to develop an outlier detection scheme to effectively detect events and attacks in the network.

3 Preliminaries

Our proposed scheme takes advantage of clustered WSN architecture [8] as shown in Fig. 1. Each cluster (C) is assigned with a cluster head (CH) capable of communicating with the members of cluster (CM), peer cluster heads, and also to the base station (BS). The base station manages operations of WSN.

The detection of events is performed in distributed manner by the CHs, with cluster being the minimum unit for detecting events. When an event or attack is detected, the event report is generated by the node detecting the event and CH is triggered or alarms are raised by the CHs reporting attacks.

3.1 Assumptions

Nodes are assumed to be stationary until the setup phase and then move in random directions using mobility model [7]. Nodes cannot be added to the network post-deployment. Nodes are heterogeneous with different energy levels and mobilities. Base station is assumed to be having very high computation and communication capability. Nodes are aware of their location.

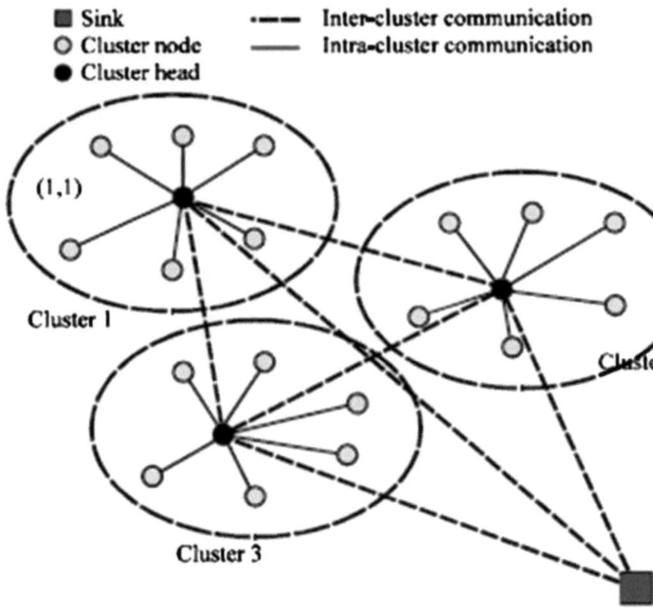

Fig. 1 Clustered WSN architecture

3.2 Adversary Model

For our work, we assume that the adversary can compromise nodes randomly. The goal of adversary is to disguise themselves and remain undetected. In our scheme, the nodes maintain a neighbor node list with relevant attributes when they initially exchange messages with their neighbors. The attributes include node ID, RSSI, location, and message arrival times. An attack is confirmed if the values received in subsequent messages are not consistent [9–12]. We assume that attacks are introduced into the network after some random time.

Defintion Event E: A change in the real-world state, watched by a set of sensors D and minimum K sensors need to conform to it, where D >= K.

Defintion Region of Event RE: The area in the network where an event has occurred.

Defintion Event Report ER: The occurrence of an event is notified to the CH or BS as a report. ER = [NID, TE, CID, LOC(NID)]

Defintion Event Detection Time TE: Average time taken by the nodes to report an event detected to CH or BS.

3.3 Notations

Frequently used notations in the network are enlisted below:

(Xmin, Ymin), (Xmax, Ymax)	Starting and ending of network area coordinates
CHID, CID, NID	Cluster Head ID{1..n}, Cluster ID{1..m}, Node ID {1..p}
LOC()	Location of node, given in terms of X, Y coordinates
EVENT_DETECT	Detect an event (sent by CH to CM)
OUT_NODE	Node moved out of vicinity of cluster
EVENT_REPORT	ER = [NID, TE, CID, LOC(NID)]
EVENT_CONFIRM	Confirmation of event occurrence (Sent by CH to CM)
NODE_INFO	CH requests the node for its details
NODE_INFO_DETAILS	The node replies back with relevant details to CH
t	Delay time the CH waits before confirming an event
attributeValue	Sensor parameter readings
EventCounter	Number of nodes reporting the event
ThresholdE	Threshold value for an event
Message_arrival_time	Time when the receiver receives a message
Message_transmission_time	Time when the sender sent a message
RSSI Buffer	Stores the recent RSSI value
Arrival_time buffer	Stores the max and min message arrival times of a node
Transmission_time buffer	Stores the max and min message transmission times of a node

4 Outlier Detection Scheme

The proposed outlier detection scheme consists of two phases—1: Event Detection Phase and 2: Attack Detection Phase.

Event Detection Phase: In the event detection phase, the events are detected and notified to CH. The CH identifies the region where event has occurred and creates an alert region. The cluster head broadcasts an EVENT_DETECT message to all the members of the cluster. The cluster members upon receiving the message from the CH, checks to see whether it is still within the boundary of the cluster (CXmin, CYmin), (CXmax, CYmax). For the CM within in the boundary limits of the cluster, the sensed attributes are checked with the event thresholds. If the attribute values exceed the threshold values, the CM replies to the CH with the event report. The CH upon receiving the event report from the CM adds the CM to the reporting nodes, updates the event counter by 1, and checks to see whether the counter is equal to k to confirm the event. When the number of nodes reporting the event (event counter) is equal to number of nodes required to confirm the event (k), then the CH broadcast to all CM the occurrence of event through event confirm message. For the nodes that have moved out of the boundaries of cluster, they are just updated in the out node list at the CH.

```
Procedure: EventDetectionPhase
Step 1: CH broadcasts EVENT_DETECT to CM and wait't'.
Step 2: The CM receive EVENT_DETECT from the CH,
          If  Loc(CM)in(CXmin,CYmin),(CXmax, CYmax)
          Go to step 3
          Else
           Reply to CH with OUT_NODE
Step 3: If (AttibuteValue >ThresholdE)
          Reply to CH with EVENT_REPORT
Step 4: CH receives message from CM
          If message received is EVENT_REPORT
          Add NID to REPORT_NODE list
          Update EventCounter by 1
          Go to step 5
          Else If message received if OUT_NODE
          Update the OUT_NODE list
Step 5: Confirm occurrence of event
          If EventCounter >= K
          Broadcast EVENT_CONFIRM to CM
          Call (FindEventRegion)
          Else Repeat step 4
Step 6: End.
```

In the procedure find event region, the CH considers all the nodes in the REPORT_NODE list. From the location of all the nodes, it finds least x value, least y value and maximum x value, maximum y value. Set BXmin = minimum(x) and BYmin = minimum(y); set BXmax = maximum(x); BYmax = maximum(y); using these values constructs a region with (BXmin, BYmin) and (BXmax, BYmax) which cover the event nodes.

```
Procedure: FindEventRegion
Step 1: For all the nodes in the REPORT_NODE list
          Set BXmin=minimum(x); //lowest x value
          Set BYmin=minimum(y); //lowest y value
          Set BXmax=maximum(x); //highest x value
          Set BYmax=maximum(y); //highest y value
Step 2: Set event region (BXmin , BYmin)(BXmax , BYmax)
Step 3: End
```

Attack Detection Phase: In the attack detection phase, when the attacks are detected, they are notified to the CH. The CH initiates the defensive measure by creating an alert and revoking the node from further interactions in the network.

```
Procedure: Detect Attack
Step1: CM node receives message from other neighbor CM
nodes
Step 2: For all messages received from CM nodes
          If (RSSI > RSSI_Buffer ||
          Message_Arrival_Time > Arrival_Buffer ||
          LOC  (CMi)=(!proportional  to  mobility  model
          movement))
          Add node to COMP_NODE List
          Go to Step 4
          Else
          Update VALID_NODE List
Step 3: CM node detecting compromised node sends
          ALERT_MSG to CH
Step 4: If CH receives ALERT_MSG
          Add node to PROBABLE_COMP_NODE List
          Call (Validate_Alert)
Step 5: End
```

In the detect_attack procedure, the occurrence of an attack is detected by checking the inconsistencies. The node under scrutiny is validated by the CH in validate_alert procedure. Here, the CH seeks node details from the nodes under scrutiny and then checks for anomaly; if there exists any, then this node is labeled as compromised node and an alert is passed to all the cluster members to refrain from further communications with such node. If CH does not find any anomalies, then the node is set as valid and a message is sent to the node that made a request to CH for validating.

```
Procedure: Validate_Alert
Step 1: For each node in PROBABLE_COMP_NODE list,
          CH requests NODE_INFO message
Step 2:  when a node receives NODE_INFO message from CH
          Reply NODE_INFO_DETAILS message to CH
Step 3: CH receives NODE_INFO_DETAILS MESSAGE
          If ((RSSI <RSSI_Buffer ||
    Message_Transmission_Time < Message_Arrival_Time ||
    LOC(CMi)=(!proportional     to     mobility     model
    movement)) && consistent )
          Remove this node from PROBABLE_COMP_NODE
          Set as valid node and
          Reply to requested node as valid
          Else
          Broadcast ALERT(NID) to CM
Step 4: CM on receiving ALERT(NID)
          Update IDN to COMP_NODE List
Step 5: End
```

5 Security Analysis of Proposed (ODS) Scheme

We have used MATLAB to simulate the proposed outlier detection scheme. We deployed 100 sensor nodes randomly in a 100 m × 100 m area, having a communication range of 15 m, with initial energy of nodes ranging from 50 to 20 J. We measure the network lifetime, event notification time by taking an average of results obtained from 20 simulation runs. The time taken to detect a compromised node is proportional to the number of nodes in the network. More the number of nodes, more the time taken due to traffic and communication overhead.

We assumed that 1/10 of the nodes in the network are compromised. We need to calculate the number of packets these nodes have injected into the network even before detection of compromised node. The total number of packets for a 100 node network is around 9623, and average number of packets from compromised nodes is around 7.5. The average compromised node traffic out of whole network traffic is round 0.08333%. Also, due to validation from the CH of the nodes compromised, a valid node was never falsely labeled as a compromised node.

Figure 2 shows the efficiency of proposed ODS scheme with IADS and OCDS schemes. In IADS [4], compromised nodes are termed as outliers, and these nodes are used to introduce attacks in the network either actively by cloning nodes or passively by eavesdropping the network communications. In ODCS [5], it has exceptional message supervision mechanism to detect outliers in network and attacks are confirmed by correlating communication and computation workloads of the neighbor nodes. When the number of outliers in the network is around 5, all

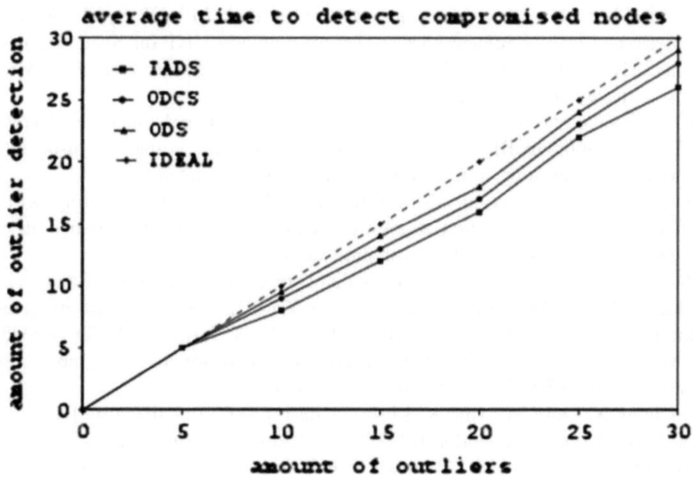

Fig. 2 Comparison of IADS, ODCS, and proposed scheme in detecting outliers

three schemes are able to detect all the anomalies. But as the number of outliers in the network increases, the performance of IADS reduces by 10%, OCDS reduces by 5%, and our proposed scheme performance reduces by 3%. Hence, our scheme has better outlier detection than the previewed schemes.

6 Conclusion

In this paper, we propose a novel outlier detection scheme to detect events and attacks in the network. In the first phase, the CH detects an event, and notifies and creates an event region. And in second phase, the scheme helps detect node attacks and alerts cluster member the presence of such an attack. Simulations are performed to demonstrate the behavior of the nodes against various performance metrics. We observe that the scheme is more efficient in terms of time taken to detect the compromised nodes.

References

1. Bojkovic ZS, Bakmaz BM, Bakmz MR (2008) Security Issues in WSN. Int J Commun 2
2. Zhang Y, Meratnia N, Havinga P (2010) Outlier detection techniques for WSN: a survey. IEEE Commun Surv Tutorials 12(2)
3. Sheng B, Li Q, Mao W, Jin W (2007) Outlier detection in sensor networks. In: Proceedings of MobiHoc
4. Liu F, Cheng XZ, Chen DC (2007) Insider attacker detection in wireless sensor networks. In: INFOCOM, pp 1937–1945

5. Zhang YY, Chao HC, Chen M, Shu L, Park CH, Park MS (2010) Outlier detection and countermeasures for hierarchical wireless sensor networks. IET Inf Secur 4:361–373
6. Halawani S, Khan AW (2010) Sensors lifetime enhancement techniques in wireless sensor networks: a survey. J Comput 2
7. Patil SD, Vijayakumar BP (2016) Mobility coordination model. Int J Adv Netw Appl
8. Patil SD, Vijayakumar BP (2016) Clustering in mobile wireless sensor networks. In: International conference on innovations in computing and networking
9. Bekara C, Laurent-Maknavicius M (2007) A new protocol for securing wireless sensor network against nodes replication attacks. In: IEEE international conference on wireless and mobile computing, networking and communications
10. Ho J-W (2016) Distributed detection of node capture attacks in WSN. Smart Sens Netw
11. Khan WZ, Aalsalem MY, Saad MNBN, Xiang Y (2013) Detection and mitigation of node replication attacks in WSN: a survey. Int J Distrib Sens Netw Article ID 149023
12. Plastoi M, Banias O, Curiac DI (2009) Integrated system for malicious node discovery and self-destruction in WSN. Int J Adv Netw Serv 2(4)
13. Wu W, Cheng X, Ding M, Xing K, Liu F, Deng P (2007) Localized outlying and boundary data detection in sensor networks. IEEE Trans Knowl Data Eng 19:1145–1157
14. Papadimitriou S, Kitagawa H, Gibbons PB, Faloutsos C (2003) LOCI: fast outlier detection using the local correlation integral. In: International conference on data engineering, pp 315–326
15. Branch J, Szymanski B, Giannella C, Wolff R (2006) In-network outlier detection in wireless sensor networks. In: Proceedings of IEEE ICDCS
16. Shi ZJS, Gao H, Li J (2007) Unsupervised outlier detection in sensor networks using aggregation tree. In: Proceedings of ADMA
17. Rajasegarar S, Leckie C, Palaniswami M, Bezdek JC (2006) Distributed anomaly detection in wireless sensor networks. In: Proceedings of IEEE ICCS

Energetic Routing Protocol Design for Real-time Transmission in Mobile Ad hoc Network

Mamata Rath, Binod Kumar Pattanayak and Bibudhendu Pati

Abstract With the emergent natural disasters and numerous emergency situations, the world is looking for interchangeable reliable medium of communication. Mobile ad hoc network (MANET) is the feasible solution toward maintaining connectivity in a pure infrastructure-less scenario. Providing appropriate quality of services (QoSs) in MANET for applications such as voice, video, and data is a challenging task. Due to dynamic network topology changes and also for the delay-sensitive nature, services like voice or video demand specialized treatment compared to their counterpart data service. In this paper, we present real-time AODV (RT-AODV), a novel AODV-based routing mechanism that improves the quality of service for real-time packets in a MANET. In order to build RT-AODV, first we developed another routing protocol, power and delay-optimized AODV (PDO-AODV), which introduces the concept of load balancing over MANET in a best-effort manner. Simulation results reveal that the proposed RT-AODV accomplishes enhanced performance than best effort PDO-AODV routing protocol in terms of delay, data dropped and network throughput .

Keywords MANET · AODV · PDR · Throughput · QoS

M. Rath (✉)
C. V. Raman Computer Academy, Bhubaneswar, Odisha, India
e-mail: mamata.rath200@gmail.com

B.K. Pattanayak
Department of Computer Science and Engineering, Siksha 'O' Anusandhan University, Bhubaneswar, Odisha, India
e-mail: binodpattanayak@soauniversity.ac.in

B. Pati
Department of Computer Science and Engineering, C. V. Raman College of Engineering, Bhubaneswar, Odisha, India
e-mail: patibibudhendu@gmail.com

H.R. Vishwakarma and S. Akashe (eds.), *Computing and Network Sustainability*, Lecture Notes in Networks and Systems 12, DOI 10.1007/978-981-10-3935-5_20

1 Introduction

The great technological innovations of wireless communication have brought a boon in the form of mobile ad hoc networks (MANETs), for today's global communication market. These types of network are capable of operating without any fixed foundation unlike cellular system. Because of their autonomous, distributed, and multihop relaying potential, MANETs are on huge demand for disaster recovery, vehicle tracking, and battlefield management. The network topology changes dynamically due to unpredictable node movements. This creates a serious level of difficulties to maintain network connectivity with neighboring entities. At the end of the day, there is no guarantee of reliable data delivery with acceptable quality of services. Routing protocols play a vital role in establishing end-to-end path, as nodes are not always directly reachable to each other. The power hungry entities over such bandwidth-limited network demand robust and efficient routing strategies for network longevity. So, routing mechanism needs to be designed judiciously to optimize network operation. Conventionally, there are two main types of routing techniques which are table-driven and on-demand basis. Table-based approaches like OLSR needlessly flood the network even without the presence of actual traffic and thus generate huge overhead. The MANET having inadequate bandwidth cannot always afford the luxury of such additional payload. On the contrary, any on-demand protocol such as AODV starts occupying the network resources whenever there is a need from the users to convey any info to their peers. AODV protocol keeps silent until such appeal comes from any network entity. Once the existence of aforesaid task is sensed, AODV starts route discovery procedure by broadcasting a route request (RREQ) packet. This route request message travels through the network and reaches to actual destination and forms a reverse route table. The recipient then responds back by generating a route reply (RREP) packet which is carried by the reverse route created earlier. This process helps to create the forward route for data packets. Here, the route discovery scheme induces delay and also some intermediate nodes get overloaded [1] due to carrying out the data handling frequently. So, traditional routing methods cannot be utilized for mobile ad hoc networks straight away because of their inherent impediments. In this paper, we concentrated on tweaking the primary AODV routing protocol to cater the aggressive QoS demand of real-time network services in a MANET. The rest of the paper is organized as follows: All the previous works related to quality of services in ad hoc network are presented in Sect. 2, and Sect. 3 depicts the encouragement behind this research. Then our main contribution of the proposed approach and design building blocks are illustrated over Sect. 4. Thereafter, results out of extensive simulations and comparison with PDO-AODV protocol are shown in Sect. 5, and finally, Sect. 6 concludes the research.

2 Related Works

Several researches have been conducted in the past on MANET routing protocols. Agbaria et al. [2] devised extrapolation-based technique that considered dynamic scheduling, resource management, velocity, and multipath search to provide real-time and QoS need of a MANET. Sivakumar and Duraiswamy [3] presented efficient algorithm to support quality of service (QoS), by the use of load distributing and congestion avoidance routing method. Their proposed algorithm computes the cost metric based on link loads. The links having lighter loads were preferred for sending traffic to avoid congestion. Srivastava et al. [4] advised an energy-efficient routing to improve the link utilization by equalizing the energy consumption between already exploited and underutilized entities. Their protocol deals with few key factors such as residual energy, bandwidth, load, and hop count for route discovery. In [5], Ze Li and Haiying Shen introduced a QoS-oriented distributed routing protocol for a hybrid network having infrastructure and ad hoc MANET. They analyzed routing by linking it with resource scheduling problem. Their algorithm adaptively adjusts segment size based on node mobility and minimizes the transmission time. Maleki et al. [6] recommended a load balancing algorithm based on DSR that can manage QoS for real-time information. They speculated a node's neighbor count as centrality metric for route selection. They considered link cost among set of nodes, to forward packets through load-optimized path. Tardioli et al. [7] proposed a real-time protocol for MANET with the help of cross-layer design. They combined MAC and routing layer in their protocol and tested with small robot systems, which were exchanging kinematics or laser data. Real-time protocols are discussed and their comparative analysis has been carried with network parameters to validate the network lifetime [8]. Energy- and delay-based routing protocols have been designed in [9] named as PDO-AODV that uses a new approach of selecting the next suitable node during routing in MANET. Detailed survey on MANET routing protocols and real-time applications is presented in [10] by Rath et al. To prevent the MANET, security measures are important. Keeping this in view, an IDS (intrusion detection system) proposal has been presented in [11] based on mobile agent. Delay- and power-based routing protocols in MANET are studied in order to classify them according to their techniques of power efficiency mechanism [12]. A cross-layer approach of protocol design has been offered and implemented in [13]. In [14], Shanrma and Dimri suggested an improved AODV protocol that enhances the packet delivery ratio for mobile ad hoc network. However, they were transmitting more routing packets for real-time scenario, which is adding overhead to the resource-limited network.

3 Motivation

With our detailed study from the literature and analysis of existing AODV-based protocols, it is observed that such protocols are not effective to tackle heavy network payload. This is mainly because shortest hop count-based routing is prioritized, without considering network load and nodes' energy, while selecting the suitable path among nodes. Moreover, the nodes, which take part in forwarding the traffic over same route, get exhausted losing their precious energy. In addition to that, we felt, there needs to be acceptable QoS while users are moving independently. When the idea of QoS was just knocking into our mind, we further realized that there are different requirements of QoS for data and real-time services. The data packets over a wireless link can be retransmitted with some permissible delay, if they are lost. Few deferred data packets will not jeopardize much to the network users, but same thing is not applicable to voice and video packets. The real-time multimedia services are extremely delay-sensitive and are useless, if they fail to meet the delivery deadline by missing series of frames or packets. All the above-mentioned bottlenecks inspired us to ponder in depth, and to design a decisive routing scheme for real-time MANET network, such that (i) It divides the network load on equal fashion among all possible entities, (ii) It minimizes irrelevant transmission saving significant network bandwidth and node energy, and (iii) It caters the need of real-time multimedia services by providing acceptable QoS. Therefore, we came up designing a PDO-AODV routing mechanism, which would balance the uneven load distribution over a burdened MANET. This approach is accepting neighboring nodes' power and delay as routing metrics for route selection. However, PDO-AODV was not having any intelligence for QoS scheduling and works on a best-effort mode. This was not fulfilling our earlier requirement to address the real-time packets. Hence, we further brainstormed and developed RT-AODV, on the top of PDO-AODV framework. We verified through simulation that this proposed approach can help real-time traffic to meet the legitimate timeline and improve the overall network performance.

4 Contribution

We modified the standardized AODV protocols in incremental way of development phases. At first, we aimed to provide load balancing over a dense MANET by implementing PDO-AODV routing method, which was assigning equal priorities among various network services, i.e., they were scheduled on a first come first serve basis. This was conceptualized by taking nodes' power and delay as important metrics for optimized route selection. When this scheme was deployed over the network nodes, the packet loss and congestion issues were improved significantly. This achievement set up the foundation stone for us to venture into the second phase of development, which grants quality of services toward real-time packets.

Each of these phases is elaborated in the following. NS2 simulation environment was used for development and simulation in our experiment.

4.1 PDO-AODV Implementation

Our routing engine core modules were constituted of three sub-modules that enable to take routing decision based on power and delay. They are explained as follows:

(1) Network sensing
The primary objective of this sub-module is to sense all the one-hop neighboring nodes' status. Here, every node broadcasts its health status over the network. Here, the innovative part is that we embedded any node's status message with the periodic control information, rather than sending a special purpose message. This is not going to put any additional overhead on network.

(2) Database handler
This sub-module stores the status message received over the air of all neighboring entities. After the reception of the status message, delay from a neighbor is calculated from the packet generation time stamp. The members of the database are defined to store various status-related parameters such as energy, hello message, and delay.

(3) Routing decision

This part of our design aims to select suitable balanced route through which packets can travel with minimal loss. It is time to decide here, which intermediate nodes will actually play the role of forwarding the traffic toward destination. For this purpose, we consulted the database handler sub-module, which records power and delay information of any peer as an integrated load balancing metric. The higher cost links indicate better route, and routing decision is taken accordingly.

4.2 RT-AODV Development

Now, we are going to extend the already built load-balanced framework for real-time AODV procedure. The provision to entertain good QoS was achieved by two phases, namely packet filtering and route determination.

(1) Packet filtering

We envisaged the existence of diversified packet types in our target network. So, it was essential to refine delay-sensitive packets from their generic counterparts for further processing. The filter was materialized by defining a flag variable. This filter was set in the following cases:

i. If a real-time packet was originated by the source node.
ii. If any other node receives information from a RREQ packet, that route request is for real-time traffic.
iii. If any intermediate node receives RREP containing route reply for managing a real-time data.

This flag is declared as a global static variable inside NS2-aodv.cc file. Figure 3 reveals the piece of code which sets the flag, if any real-time voice packet is generated by a source node. Now, the presence of such delay-sensitive data in the network needs to be intimated to all other intermediate nodes toward destination. This information will help in-between relay nodes to take special care of real-time traffic. That is why real-time flag info is also sent with the sendRequest () function.

(2) Route determination

This portion of our algorithm acts as a path selector for all types of packets. The value of the real-time flag is checked to segregate real-time packets from any normal data packets.

5 Comparative Analysis

After design of real-time protocol, we have compared our protocol performance with other three important protocols which are based on different concepts but efficient real-time protocols and their concept are described in this section. A motivating proposal is given in [15], called real-time multihop protocol (RT-WMP) [15] along with its expansion version for QoS management which is executed over many nodal devices with hardware of specific configuration. It allows proficient end-to-end voice communication during the QoS extended module, and the projected system works better for specific situation for definite topology. This application was tested and validated successfully in a real submission at Somport tunnel of Canfranc, Spain. In [16], a suitable proposal has been offered as per suitability of multimedia communication in mobile ad hoc networks. It proposes an innovative strategy of using multicast tree structure protocol for mobile ad hoc networks. The methodology implements the MCT (multicast tree) format for MANETs and assesses the performance with other two protocols serial MDTMR and parallel MNTMR. In [17], a methodology is adopted for better distribution of spectrum in MANETs, which is called MRFM (MANET real-time frequency management). This scheme uses a centralized control mechanism to stay away from confliction with electromagnetic environment. This function runs as an external application to the system that gathers the radio state information in real-time case, relates rules about the spectrum allocation, and controls the frequency levels received by MANETs. Use of the proposed system helps to trigger capacity of DSA (dynamic spectrum access) to radio signals of MANET, which is not directly existing. Verification of MRFM was done in a prototype system where numerous tests were carried out with stimulating results.

6 Simulation and Results

This section elaborates the simulation scenarios deployed through NS2 simulator to proof our concept of providing decent QoS for multimedia packets. Further analysis was done to compare the performance of the proposed routing protocol with best-effort PDO-AODV technique. Our network model speculated the total area of simulation as 1000 m x 1000 m. Numbers of mobile nodes that roam around this area were 60. We used random waypoint mobility model for users' movement. The underlying MAC and wireless PHY protocol was 802_11. Table 1 shows the simulation parameters used during simulation of our proposed protocol.

The following section describes the results of performance evaluated during our protocol design and simulation.

Characteristics of real-time traffic are very critical than normal traffic. Because this traffic is generated through a code that carries out sampling of a continuous real world environment (image, voice etc.) and transmission of constant renew of these data to regenerate a visually diagrammatic or audio type outcome. Therefore, it is the demand of the application to utilize constant bandwidth till the entire transmission period. Sensitivity to delay is another important characteristic of real-time applications due to the fact that sampling and regenerating a continuous incident are

Table 1 Simulation parameters

Parameter name	Parameter value
Channel type	Wireless channel
Radio propagation model	Two-ray ground
Network interface type	Wireless PHY
Type of traffic	V B R
Simulation time	3 min
MAC type	Mac/802_11
Max speed	40-50 ms
Network size	1600 × 1600
Mobile nodes	120
Packet size	512 Kb
Interface queue type	Queue/droptail
Protocol	PDO-AODV, RT-AODV
Simulator	Ns2.35

regularly done by a real-time stream such as in a voice stream every piece of packet segment should reach at the destination end to be executed or played at the same and correct time. In case, a single packet segment arrives late, then there will be an interval and gap in between two consecutive segments to be played.

So the continuous flow of the speech or audio (may be a music) will be disturbed resulting in degradation of audio quality. The level of degradation of quality is important in real-time applications depending on packet delay time and packet loss rate. Due to the importance of packet arrival time, it is difficult for the transport-level protocol to frequently retransmit a lost packet and wait for significant period of time. The reverse trip to source and to wait for re-transmission is much lengthy, and by the time it reaches, it is already too late with missing its play window. Because TCP does not consider the issue, these are carried out with the UDP, which does not employ any recovery mechanism for lost packets. So the normal routing flow starts with packet sent by the sender to the network and they are delivered to the destination in time or with delay due to congestion or sometimes lost during transition.

Figures 1 and 2 show the packet delivery ratio of voice and video packets. It can be observed that in voice, the PDR is approximately 92% in RT-AODV with a better performance than PDO-AODV and AODV, whereas in video stream, the PDR is approximately 70% in RT-AODV with a better performance than

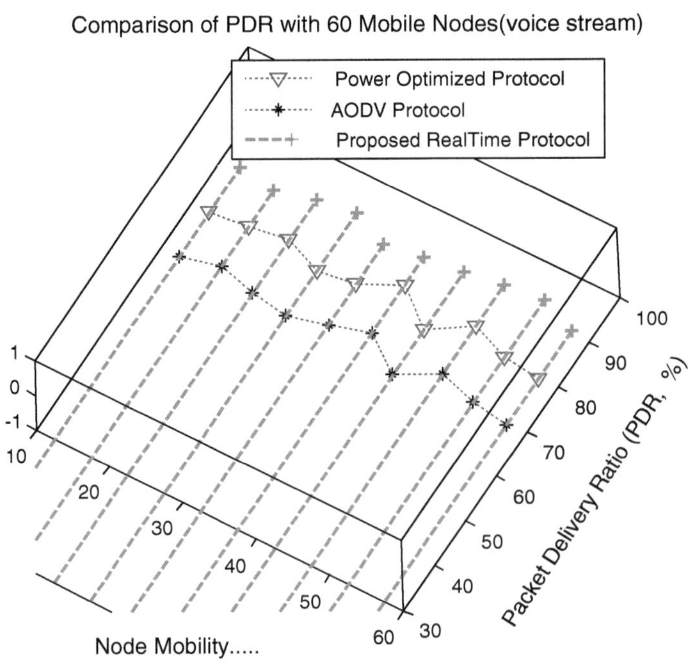

Fig. 1 PDR analysis in voice

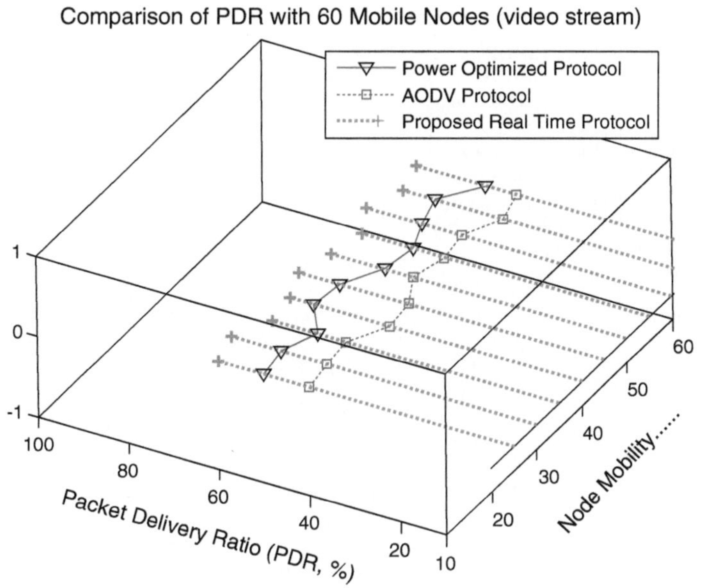

Fig. 2 PDR analysis in video

PDO-AODV and AODV. Due to high density of image and multimedia stuff in video traffic, there is reduction of PDR in video in comparison with audio traffic.

Figure 3 shows the end-to-end delay comparison between different proposed real-time protocols as described above. It can be observed that our proposed RT-AODV exhibits comparatively less delay than the other real-time protocols proposed.

Figure 4 shows the comparison of average bandwidth consumption between various real-time protocols and PDO-AODV. It can be observed that our proposed RT-AODV consumes comparatively lesser bandwidth than the other real-time protocols proposed. PDO-AODV consumes less bandwidth than RT-AODV due to the fact that in RT-AODV, real-time data such as multimedia-rich applications are given priority which consumes higher bandwidth, whereas normal PDO-AODV protocol uses a power-optimized path for sending packets in a load-balanced way.

Figure 5 depicts the data drop analysis between the discussed protocols. It can be seen that there is minimum data drop of packets in our proposed real-time protocol.

Figure 6 shows the throughput analysis between AODV, PDO-AODV, and RT-AODV. It can be seen that there is maximum throughput, approximately 70000 Mbps/s on average, in our proposed real-time protocol which is a better figure than throughput of AODV and power delay-optimized AODV protocol [9].

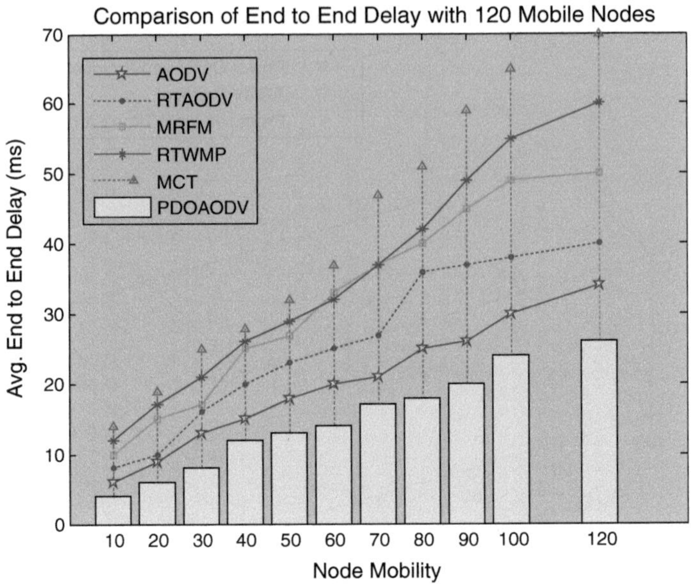

Fig. 3 Delay comparison (video)

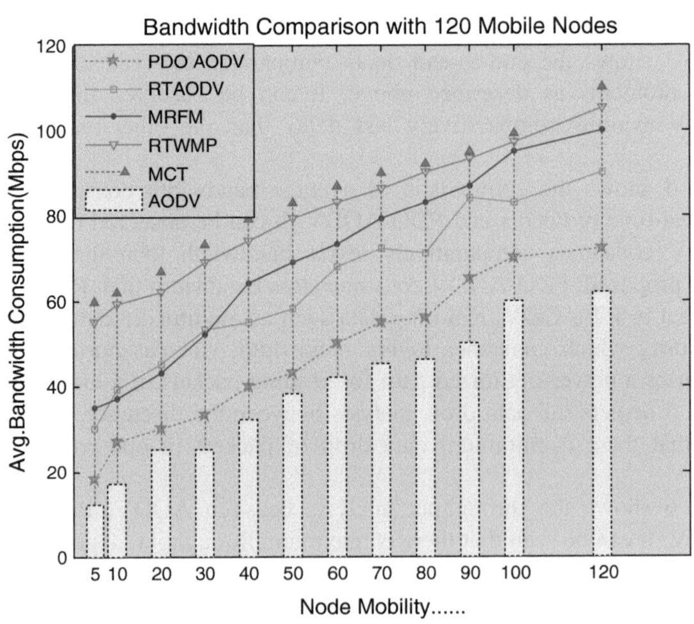

Fig. 4 Bandwidth comparison (video)

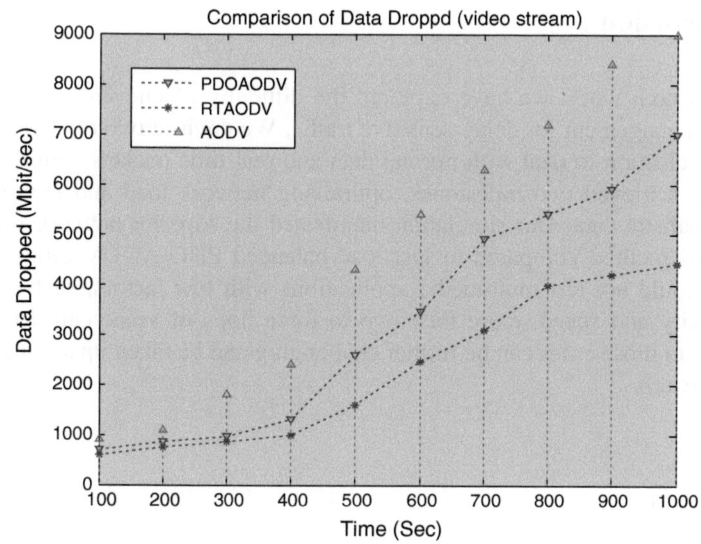

Fig. 5 Data drop analysis (video)

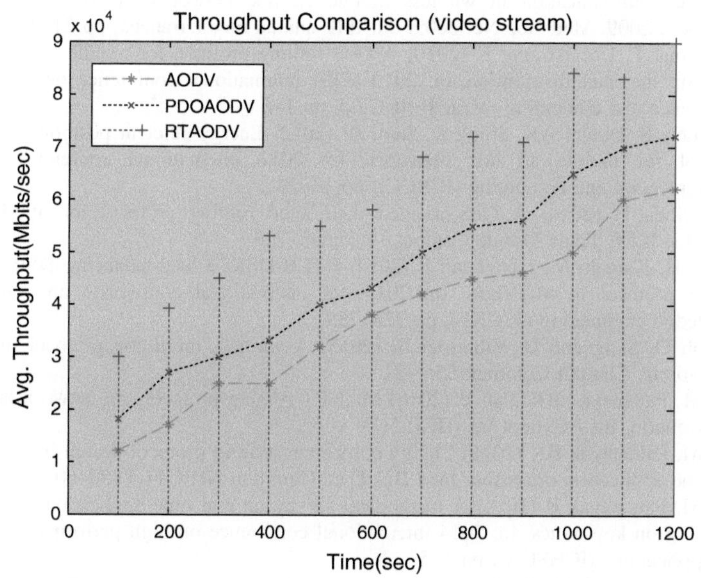

Fig. 6 Throughput analysis (video)

7 Conclusion

In this research work, we have explored the difficulty of uneven load scheduling and QoS management for delay-sensitive traffic. We devised two different formulas for route selection to deal with normal data and real-time packets. Our experiment eventually achieved two milestones, optimizing network load and nourishing the need of realistic data. Our simulation manifested the superior network function of RT-AODV method compared to just load-balanced PDO-AODV protocol. However, we could not test multimedia applications with few factors, such as varying node density and speed, more than two to three hops of voice call. Maintaining good QoS in those cases can be further challenging and be taken up as an activity of future research.

References

1. Vinod Kumar R, Wahida Banu RSD (2012) E2AODV protocol for load balancing in ad-hoc networks. J Comput Sci 8(7):1198–1204
2. Agbaria A, Gershinsky G, Naaman N, Shagin K (2009) Extrapolation-based and QoS-aware real-time communication in wireless mobile ad hoc networks. In: Ad Hoc networking workshop, 2009. Med-Hoc-Net 2009. 8th IFIP annual mediterranean, pp 21–26
3. SivaKumar P, Duraiswamy K (2010) A QoS routing protocol for mobile ad hoc networks based on the load distribution. In: 2010 IEEE international conference on computational intelligence and computing research (ICCIC), pp 1–6
4. Srivastava S, Daniel AK, Singh R, Saini JP (2012) Energy-efficient position based routing protocol for mobile ad hoc networks. In: 2012 international conference on radar, communication and computing (ICRCC), pp 18–23
5. Li Z, Shen H (2014) A QoS-oriented distributed routing protocol for hybrid wireless networks. IEEE Trans Mobile Comput 13(3):693–708
6. Maleki H, Kargahi M, Jabbehdari S (2014) RTLB-DSR: a load-balancing DSR based QoS routing protocol in MANETs. In: 2014 4th international conference on computer and knowledge engineering (ICCKE), pp 728–735
7. Tardioli D, Sicignano D, Villarroel JL (2015) A wireless multi-hop protocol for real-time applications. Comput Commun 55:4–21
8. Rath M, Pattanayak BK, Pati B (2016) MANET routing protocols on network layer in real time scenario. Int J Cybern Inf (IJCI) 5(1)
9. Rath M, Pattanayak BK (2015) Energy competent routing protocol design in MANET with real time application provision. Int J Bus Data Commun Netw 11(1):50–60
10. Rath M, Pattanayak B (2014) A methodical survey on real time applications in MANETS: focussing on key issues. In: 2014 international conference on high performance computing and applications (ICHPCA), pp 1–5
11. Pattanayak B, Rath M (2014) A mobile agent based intrusion detection system architecture for mobile ad hoc networks. J Comput Sci 10:970–975
12. Rath M, Pattanayak B, Pati B (2016) A contemporary survey and analysis of delay and power based routing protocols in MANET. ARPN J Eng Appl Sci 11(1)
13. Rath M, Pattanayak BK, Pati B (2016) Energy efficient MANET protocol using cross layer design for military applications. Def Sci J 66(2)

14. Sharma L, Priti D (2015) An improved AODV with QoS support in mobile ad-hoc network. In: 2015 2nd international conference on computing for sustainable global development (INDIACom), pp 2052–2056
15. Sicignano D, Tardioli D, Cabrero S, Villarroel JL (2013) Real-time wireless multi hop protocol in underground voice communication. Ad Hoc Netw 11(4):1484–1496
16. Vijaya Kumar PDR, Ravichandran T (2013) A real time multimedia streaming in mobile ad hoc networks using multicast tree structure. Res J Inf Technol 5(1):24–34
17. Boksiner J, Posherstnik Y, May B, Saltzman M, Kamal S (2013) Centrally controlled dynamic spectrum access for MANETs. In: Military communications conference, MILCOM 2013. IEEE, pp 641–64
18. Lal C, Laxmi V, Gaur MS (2011) Performance analysis of MANET routing protocols for multimedia traffic. In: 2011 2nd international conference on computer and communication technology (ICCCT), pp 595–600
19. Rath M, Pattanayak, B, Rout U (2015) Study of challenges and survey on protocols based on multiple issues in mobile adhoc network. Int J Appl Eng Res 10:36042–36045
20. Design and implementation of the extended routing information protocol for mobile ad-hoc networks in Linux. http://www.grin.com. Accessed 31 Jan 2016

Securing Network Communication Between Motes Using Hierarchical Group Key Management Scheme Using Threshold Cryptography in Smart Home Using Internet of Things

Gagandeep Kaur and Er. Kamaljit Singh Saini

Abstract IoT is assumed to offer higher connectivity of smart devices, systems, and services that move beyond machine-to-machine communication by using some protocols (RPL, Wi-fi, ZigBee) and techniques (RFID and GSM). Physical objects communicate with each other. Objects can be monitored and controlled through the Internet. IoT has the capability to make home and our life smarter. It can be implemented in different areas such as smart cities, agriculture, energy, healthcare, home automation, and business models. When networks are organized at the large scale, then the security of motes is the main concern. In this paper, we focused specifically on security issues, challenges, and mechanisms of IoT. To secure the network, it is important to establish secure links for the end-to-end communication of motes by using cryptographic mechanism over a home network. We give a solution to secure network communication between motes by purposing Hierarchical Group Key Management Scheme using threshold cryptography in smart homes using IoT.

Keywords IoT · Protocols · Techniques · Security issues · Attacks · Mechanisms

1 Introduction

Internet of Things is a network of objects such as home devices and buildings embedded in hardware, software, sensors, and actuators to exchange data. This is an environment in which things and people communicate with each other to transfer

G. Kaur (✉) · Er.K.S. Saini
Department of Computer Science & Engineering, Chandigarh University, Gharuan, Punjab, India
e-mail: Kaur.deepgagan.gagandeep@gmail.com

Er.K.S. Saini
e-mail: Saini.kamaljitsingh@gmail.com

H.R. Vishwakarma and S. Akashe (eds.), *Computing and Network Sustainability*, Lecture Notes in Networks and Systems 12, DOI 10.1007/978-981-10-3935-5_21

data. It may also be called as the Internet of Everything, the cloud of Things and Internet of relating to Things. Things can be a person, animal, physical devices, everyday objects, and things like a gas valve, home lightening, an air conditioner that are linked intelligently between things and people through the Internet. These objects are called as smart objects. Anyone from anywhere at any time can have connectivity, for anything with IP connectivity can accessing and controlling the devices.

The Internet of Things refers to the ever-growing network of physical objects that feature an IP address of Internet connectivity and the communication that occurs between these objects and other Internet-enabled devices and system [Webopedia].

By embedding short-range mobile transceivers into a wide array of additional gadgets and everyday items, enabling new forms of communication between people and things and between themselves [WSIS 2005] (Fig. 1).

To transfer the data, we are using motes in IoT. Mote is an end device with sensor and actuator while data/information transmission occurs; protection of data is the main concern in the IoT. We are not sure that our connecting devices are secure or not. Data flow must be free from any risk or danger. Identifiable IoT devices have IP addresses they can be discovered and hacked easily. RFID tags and sensors can be read easily. To prevent data from third-party authorization and to secure the network, we are using key management. It plays an important role in protecting group communication of motes in the network. It improves the rebounding against the attacks like node capture. Secure data transmission can be gained by encryption and decryption techniques.

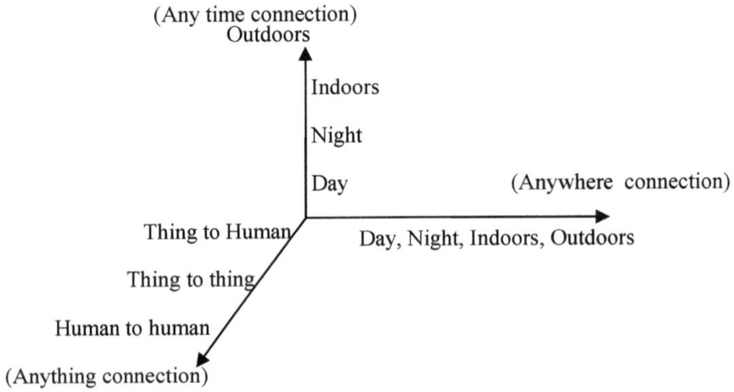

Fig. 1 Internet of Things

2 History

The term IoT is coined by Kevin Ashton, David Brock, Dr. Daniel Engles, and Sanjay Sarma by center of AutoID (automatic identification) (Fig. 2).

3 Architecture of Internet of Things

3.1 Layers of Internet of Things

1. Physical layer: It includes Wi-fi, IEEE 802.15.4 protocols. Processes in this layer are modulation, data rate, transmission mode, and channel encoding. IoT applications need very low data rate demands.
   For example, sensors with limited data size, we use these protocols to make a good performance and link quality on either uplink or downlink channel.
2. MAC layer: It consists of 6LowPAN protocol. Following are the characteristics of MAC are:
   a. Multiple Access: TDMA is used in IoT for low-power operation.
   b. Network topology: Star topology is proper for IoT applications.
3. Network layer: IoT applications required the support of Internet Protocol. UDP/TCP could also provide flexibility on devices. Network services transmit the information between motes using IP routers, to overcome the end-to-end error and sequence control to give reliable service.
4. Transport layer: To handle the TCP connections, this layer is used. It reduces the complexity of implementation. Retransmission also occurs in this layer.
5. Application layer: HTTP and COAP are used in this layer. COAP is network-oriented protocol. HTTP is long-term standard. HTTP based on TCP using point-to-point communication. COAP allows IP multicast.

3.2 Protocols

See Table 1.

Fig. 2 History of IoT

Table 1 Protocols used to communicate to the objects through Internet

Protocol name	Description
UDP/TCP	1. UDP/TCP both work at the transport layer 2. UDP/TCP is transportation protocol that is one of the core protocols of Internet Protocols 3. User Datagram Protocol is connectionless protocol
ROLL	1. Routing over Low power and Lossy network (ROLL) has recently produced the routing protocols
IPv6	1. Also known as IPng (next generation) 2. It uses 128-bit addresses 3. Features are: 1. End-to-end connectivity 2. Faster forwarding/routing Basic support for data integrity, authentication, payload encryption
HTTP	1. Hypertext Transfer Protocol (HTTP) is an application protocol for distributing information used by WWW 2. HTTP defines how messages are formatted and transmitted 3. It is designed to permit intermediate network elements to improve communication between clients and servers
6LowPAN	1. IPv6 Low Personal Area Network 2. It is the simple low-cost communication network that allows wireless connectivity in an application with limited power 3. It is characterized by short range, low bit rate, low power, and low memory usage
COAP	1. Constrained Application Protocol is Web transfer protocol 2. This protocol is designed for machine-to-machine applications such as smart energy
RPL	1. It is routing protocol 2. It is an end-to-end IP-based solution which does not require translation gateways in order to address nodes within the network
ZigBee	1. It is a non-IP stack protocol

3.3 Techniques

See Table 2.

3.4 Communication System in IoT

The user or human directly can connect with physical entities such as objects and machines through protocols using techniques to make our life comfort. Internet of Things decreases energy consumption (Fig. 3).

Table 2 Techniques used to connect to smart objects

Techniques	Features
MOSFET technology	1. Metal-oxide-semiconductor field-effect transistor (MOSFET) is used to amplify or switch the electronic signals 2. This is a four-terminal device: source (S), gate (G), drain (D), and body (B). 3. It requires very little current to turn on
RFID	Radio frequency identification used to transfer data 1. The main purpose of it is to identify and track the tags attached to objects 2. Tags contain electronically stored information 3. RFID tags can be attached to cash, clothing, implanted in animals and people
IPv6	1. It is the recent version of Internet Protocol 2. It uses 128-bit addresses 3. It provides an identification and location system for computers on networks and routes
Cloud computing	1. It is on-demand computing 2. We can share resources, data, and information from the cloud directly by using the Internet 3. It is a pay-per-use model 4. The main advantage is low cost with high availability
LTE	1. Long-term evolution is a mobile communication standard 3GPP Release 9 2. It supports 1.4, 3.0, 5, 10 MHz downlink and uplink 3. It is used in mobile broadband and VOIP
LTE-A	1. Long-term evolution advanced (LTE-A) is a mobile communication standard 3GPP Release 10 2. It supports 70 MHz downlink (DL) and 40 MHz uplink (UL)

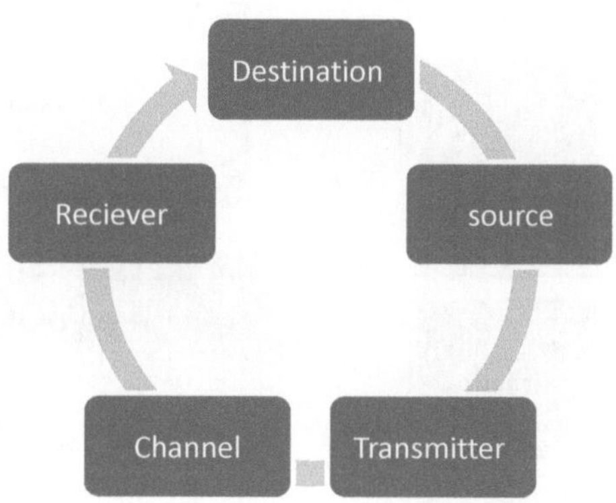

Fig. 3 General communication system

Components in Internet of Things (IoT) communication model are as follows:

(1) Application node (AP),
(2) Control point (CP),
(3) Gateway (GW),
(4) Data sink/data end point (DS), and
(5) Data processor (DP).

(1) Application node: An application does not connect directly with sensors and actuators. It requires communication with control points and data sinks. Application node can communicate with them.
(2) Control point: It is a software executor that containments actuators and sensors and sends commands to sensors and actuators. It will communicate with sensors and actuators and data proceeding, sending them configuration and control messages. It can control bidirectional communication with application node.
(3) Gateway: It is a forwarding element. It is used to connect with local networks. It can communicate with other gateways and forward traffic from control points, data end points, data processors, and application nodes.
(4) Data end point: Data end point is a software executor that receives data from sensors and actuators directly.
(5) Data processor: It is a software executor receiving data from data processors. It performs activities such as filtering, aggregation before sending data to end point.

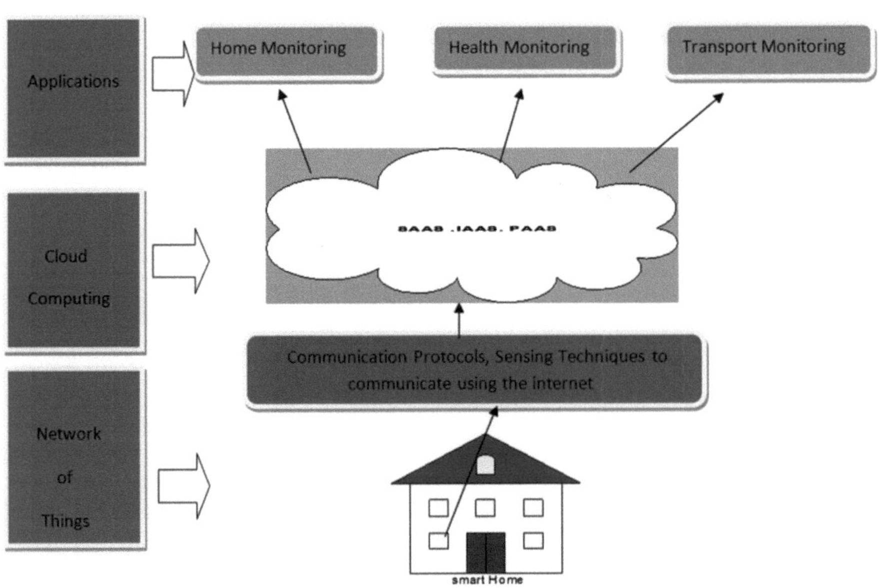

Fig. 4 IoT framework

3.5 *IoT Framework with Cloud Computing*

Cloud computing is tightly coupled in Internet of Things. Cloud computing is a prototype for big data storage. Cloud computing and Internet of Things can enable sensing services. In it, data to be stored allowed by cloud computing and it used for smart controlling and monitoring the smart devices. Cloud computing offers applications to users on demand anytime, anywhere, and anyplace (Fig. 4).

4 Applications of IoT

See Fig. 5.

5 Security Issues

- **Weak passwords in IoT devices**: Passwords are essential to secure the devices. We have to choose strong passwords in our devices (front door, CCTV camera. etc.). If our password is weak, then the attacker can gain access easily to our personal information. Weak passwords are dictionary terms, common phrases, your name or birthday.
- **Poor physical security**: Physical security is the protection of hardware, software, networks, and data from physical events such as a natural disaster, accidents, theft, and terrorism. Poor physical security occurs when hacker disassembles a device to the storage medium.
  For example, USB port can be used to compromise the device; from it, data could be stolen easily.

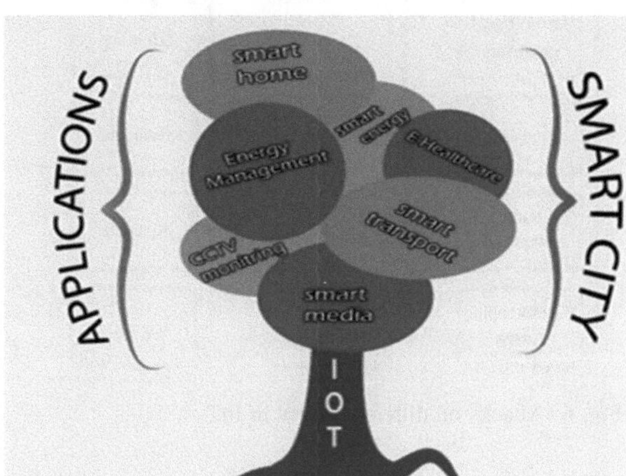

Fig. 5 Applications of IoT

- **Insecure network services**: While motes transmitting in the radio environment, MITM attack occurs due to manipulation of information. Data can be sniffed easily by using insecure network protocols.
- **Insufficient authentication**: Insufficient authentication occurs when any Web site gives directly permission to the hacker to access data without having authenticated to review route traffic and guessing passwords easily to capture the secret information by the hacker. Lack of access control is the main cause of insufficient authentication.
- **Insecure cloud interface**: Insecure cloud interface is present when connection is simply reviewed by an attacker. An unsecured cloud interface could lead to compromise data and control the device.

6 Security Attacks on Different Layer

See Fig. 6.

7 Mechanisms

Mechanisms may include the following reducing vulnerabilities:

Hash functions: These functions can be used to map data of exacting size to data of stable. Values extracted from these functions are called hash value and hash

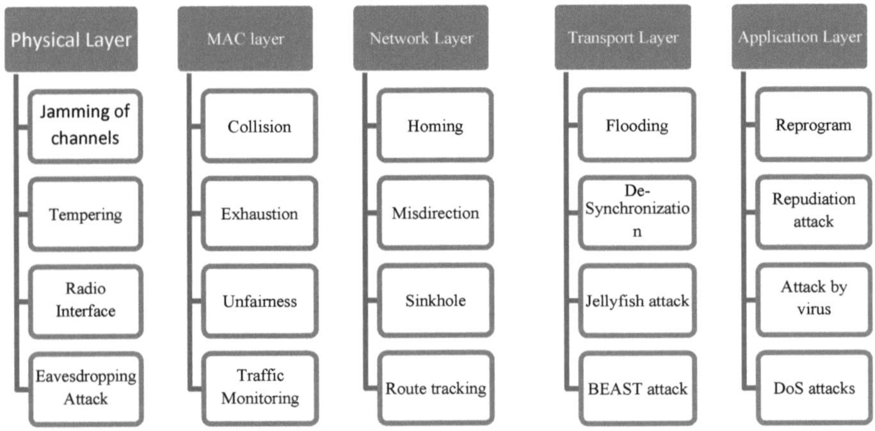

Fig. 6 Attacks on different layers in IoT

codes. For indexing and searching items in the database, we use hashing. It is also used to encrypt and decrypt digital signature to authenticate sender and receiver the message. It offers low risk of collision. MD2, MD4, and MD5 are message digest hash functions.

Symmetric algorithms: When keys are same for encryption and decryption, then this key is called symmetric key. This is also called secret key algorithms. Keys must be known to both the sender and receiver.

Random number generators: The random number has application in cryptography, sampling, etc. It is designed to generate a sequence of numbers or symbols. In short time number produces, when defining single numbers, a random number is one that is drawn from set values.

End-to-End security: It is a secure communication, prevents third parties from accessing the information while in motion state from one end of the device to another. Data is encrypted by sender and receiver is able to decrypt it. Cryptographic keys are used to encrypt and decrypt the data that is stored on the end points.

Security protocols: Security is concerned with integrity. Security protocols are abstract protocols that perform functions of security, and it applies cryptographic methods. It tells how algorithms should be used. TLS (transport layer security) is a main security protocol that used to secure the Web connection.

Key management: It includes lightweight cryptography. It is a simple encryption technique used to implement in RFID, sensors, and RFID tags. It contributes to the security of smart devices. Elliptic curve cryptography to secure the network in it. We use blocks, steams, and ciphers to overcome the security problems. Threshold cryptography is a group-based public key cryptography to enhance the security of network.

8 Final Results

Node Capture Attack: It is a most dreadful security attack. An attacker theft cryptographic key information from captured mote to compromise whole network. It is a combination of passive, active, and physical attacks. The term mote is sensor node in Internet of Things. To prevent the motes in network from being captured and modify by attacker in smart home using Hierarchical Group Key Management Scheme using threshold cryptography using Internet of Things.

In this scheme, there is radio environment in simulator where motes transmit the data. Network considers that forwarding motes are connected to base motes and skymotes in which information sends to each other with authenticity.

8.1 Broadcasting Message Between Motes

Broadcasting is a method of transferring data to all recipients. It is a high-level operation. Each sender transmits the message to all receivers within a group (Fig. 7).

8.2 Data of Source and Destination is Captured

we have captured the data from the interface. COOJA simulates the wireless links. It manages to save them in pcap file so we use Wireshark to read packets. We can select and view packets. We can view individual packets in separate window (Fig. 8).

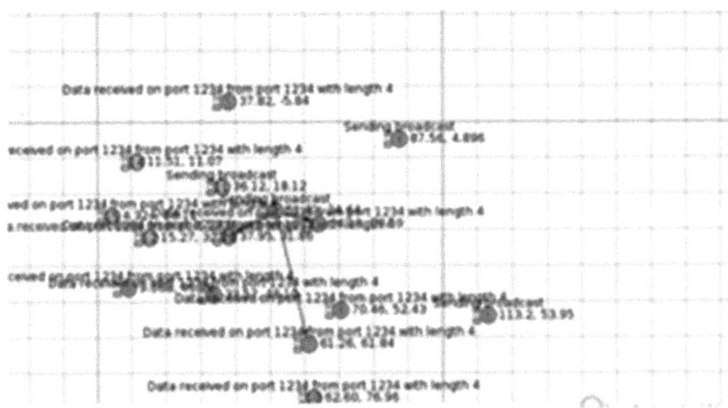

Fig. 7 Message broadcast to base motes and sensor motes

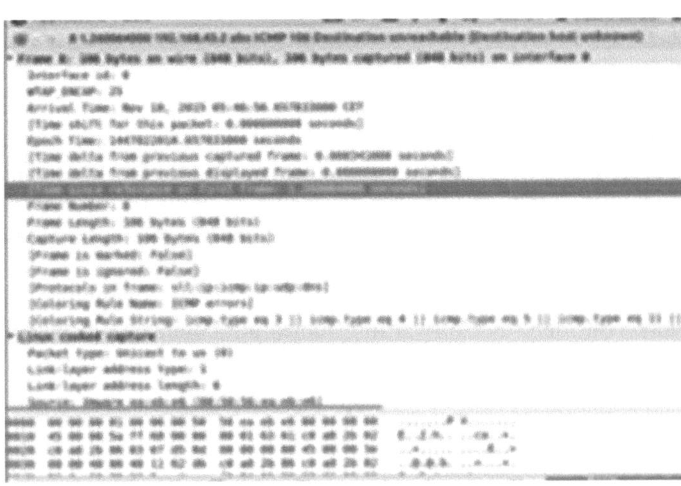

Fig. 8 Mote captured

When mote is captured, we can use the following steps to prevent the attack:

$$\text{Forward mote} \xrightarrow{\text{Request}} \text{Base Mote}$$

$$\text{Forward mote} \xleftarrow{\text{Authenticates}} \text{Base Mote}$$

$$\text{Forward mote} \xleftarrow{\text{Reply}} \text{Base Mote}$$

Forward mote gets group key using threshold cryptography that includes secret sharing key. Forward mote transmits group key to each skymote. This key is divided into two types of keys: data encryption and decryption.

8.3 Algorithm

Step 1: Two sensor motes A and B are located in same cluster.
Step 2: Each mote, base mote, and forward mote are preinstalled with secret key and public and private certificates ($Cert_{pu,}$ $Cert_{pr}$).

Step 3: Motes are established in radio network.
Step 4: Forward motes form clusters by joining motes.
Step 5: Forward mote forwards secret request to the base mote.
Step 6: If key is successfully generated, then

(1) decrypt the nonce value generated by forward mote and
(2) Transmit back secret reply message to forward mote.

Step 7: If validation is failure, repeat Steps 1 and 2.

9 Conclusion

Internet of Things is networked interconnection of objects such as sensors, actuators, and physical objects. Anything can be connected at anytime. Wi-fi, cellular networks, etc. technologies are used for connectivity to access the system through sending commands. It is a wireless network between human and machines. IoT requires security solutions. In IoT, security mechanisms also defined to ignore complex problems in smart environment. Security is the main concern of IoT. Security is critical to any network. We surveyed in this paper security attacks on different layers. Scheme could be used for research era. We removed the mote capture problem by using new scheme that is called Hierarchical Group Key Management Scheme using threshold cryptography in smart home.

References

1. Alsaadi E, Tubaishat A (2015) Internet of things features, challenges, vulnerabilities. Int J Adv Comput Sci Inf Technol (IJACSIT) 3(1):1–13
2. Bhanot R, Hans R (2015) A review and comparative analysis of various encryption algorithms. Int J Secur Appl 9(4):289–306
3. Chen S, Xu H, Liu D, Hu B, Wang H (2014) A vision of internet of things: applications, challenges, and opportunities with china perspective: IEEE Internet Things J 1(4)
4. Cirani S, Picone M, Gonizzi P, Veltri L, Ferrari G (2015) IoT—OAS: OAuth-based authorization service architecture for secure services in IoT scenarios. IEEE Sens J 15(2)
5. Dieter S, Kelly T, Suryadevara NK, Mukhopdhayay SC (2013) Towards the implementation of IoT for environmental condition monitoring in homes. IEEE Sens J 13(10)
6. Gaur A, Scotney B, Paar G, Mcclean S (2015) Smart city architecture and its applications based on internet of things. In: The fifth international symposium on internet of ubiquitous and pervasive things (IUPT) Procedia Computer Science, vol 52, pp 1089–1094
7. Hernández-Ramos JL, Pawlowski MP, Jara AJ, Skarmeta AF, Ladid L (2015) Toward a lightweight authentication and authorization framework for smart objects. IEEE J Sel Areas Commun 33(4)
8. Jin J, Gubbi J, Marusic S, Palaniswami M (2014) An Information framework for creating a smart city through IoT. IEEE IoT J 1(2)
9. Kim JT (2015) Requirement of security for internet of things (IoT) application based on gateway system. Int J Secur Appl 9(10):201–208
10. Kumar S (2014) Ubiquitous: smart home system using android application. Int J Comput Netw Commun (IJCNC) 6(1)
11. Masram R, Shahare V, Abraham J, Mooha R (2014) Analysis and comparison of symmetric key cryptographic algorithms based on various file features. Int J Netw Secur Appl (IJNSA) 6 (4)
12. Sarkar C, Nambi SNAU, Prasad RV, Rahim A, Neisse R, Baldini G (2015) DIAT: a scalable distributed architecture for IoT. IEEE Internet Things J 2(3)
13. Singh K, Sharma L (2013) Hierarchical group key management using threshold cryptography in wireless networks. Int J Comput Appl (IJCA) 63(4):0975–8887
14. Vetrive RS, Pandi Kumar S (2014) Internet of things based architecture of web and smart home interface using GSM. Int J Innov Res Sci Eng Technol 3(3)
15. Zanella A, Bui N, Vangelista L, Zarzi M (2014) Internet of things for smart cities. IEEE Internet Things J 1(1)

A Framework for Recyclable Household Waste Management System in Smart Home Using IoT

Manpreet Kaur and Er. Kamaljit Singh Saini

Abstract In today's world, increasing population density changes the need of the human beings. Today that number has swelled up to 450 and will continue to grow in the future. The need for controlling the devices has made people in developing new technologies like IoT. In today's era, IoT used in many applications such as smart city, retail, smart agriculture, waste management, household waste, and smart home. Household waste is difficult to manage in all over the world. In the waste management system, many types of waste include solid waste, construction waste, industrial waste, agriculture waste and household waste. To overcome the problem of disposing and managing recycled household waste in a smart city, a weight-based billing system is proposed in this framework that will help to clean the environment. This paper also presents information related to IoT, smart city, waste management system problems, and solutions.

Keywords IoT · Applications · Smart city · Waste management · Waste issues · Waste solutions

1 Introduction

Interconnected network objects can collect, analyze, and manage the data. That world is Internet of Things. In this scenario, around 4.9 billion people in all over the world use the internet for sending, receiving, and accessing multimedia and their services, using the social networking sites and for other tasks. The IoT is a new prototype that spread rapidly in the era of wired and wireless telecommunications where people and things connect and communicate with each other. The termi-

M. Kaur (✉) · Er.K.S. Saini
Department of Computer Science & Engineering, Chandigarh University, Gharuan, Punjab, India
e-mail: hanjrahharpreet13@gmail.com

Er.K.S. Saini
e-mail: sainikamaljitsingh@gmail.com

H.R. Vishwakarma and S. Akashe (eds.), *Computing and Network Sustainability*, Lecture Notes in Networks and Systems 12, DOI 10.1007/978-981-10-3935-5_22

nology IoT was introduced by Kevin Ashton, who was the director of Auto-ID centre of Matthew present IoT platform performs actuating, sensing, information gathering, data storage, and processing by connecting it to the internet through physical devices. In Internet of Things, various types of technologies are used such as RFID, WSN, Wi-fi, and LTE (Long Term Evolution). The RFID technology is used as identification and tracking of items or gadgets and person by using radio waves by transmitted the data and the information of a person or an object. RFID is a small chip just like a small piece of rice. The RFID is attached to the person or object. There are many fields in which IoT was used such in industry, hospitals, intelligent home, network, transportation, smart cities, shopping, and much more. The Internet of Things may sound like a futuristic term, but it is already here and improving our lives [Anne Bouverot]. IoT is used rapidly in the present and in the coming future because it becomes the more relevant and inevitable for people. IoT refers to the idea where a thing that is readable, recognizable via the internet, and controlled by using the internet for communication means. The communication can be done with RFID, WSN, WAN, etc. The object connected through people and locations of objects. In 2010, the number of everyday physical objects that are connected to the internet was around 12.5 billion, and it will increase up to 50 billion by 2020. IoT helps the people in different ways i.e., choosing favorite holiday spot and buying some food items.

In smart city, SMART stands for Standardizing Monitoring Accounting Rethinking Transforming. In Smart city, people live a luxurious and comfortable life which is helpful in the development of the society and the country. Various types of facilities are there such as surroundings, living, human beings, and administration. In Smart city, people live a good quality of life. In smart city, sensors and wireless communication technologies are used to achieve the data and the information is 3G, Wi-fi, and 4G. The aim of developing a smart city is for better infrastructure and providing facilities to old-age people. It provides an intelligent way to manage components such as Smart urban lighting, Smart taxi, Smart hospitals, Healthcare, Smart administration, and Waste management. Top ten smart cities in the world are Vienna, Toronto, Paris, New York, London, Tokyo, Berlin, Copenhagen, Hong Kong, and Barcelona. Technology changes too fast and too many stakeholders are there.

2 History

The concept of networks of smart devices was talk about in early 1982. The computer of the twenty-first century, as well as academic venues i.e., UbiComp and PerCom, produced the contemporary vision of Internet of Things. Between 1993 and 1996, most of the company's proposed solutions such as Microsoft's at Work or Novell's NEST. The Internet of Things is a term first coined by Kevin Ashton David Brock Dr. Daniel Engels and Sanjay Sarma in 1999 during a PowerPoint presentation he made while working for P&G. The MIT Auto-ID laboratory creates the IoT using the RFID and WSN. RFID was seen as a prerequisite for the Internet of Things in early days. Besides using RFID, the tagging of things may be achieved through these technologies such as barcodes and message hiding. Sensors, actuators, and other smart technologies were used for enabling communication between person-to-things and things-to-things (Tables 1, 2, 3, 4, 5, 6 and 7).

In the last decade, it is calculated that 4.9 billion things connected will use IoT in the last year and it will reach up to 50 billion by 2020.

Table 1 In year 1830–1835

Year	Invention	Application
1832	Electronic telegraph	Human-to-human transmission of code messages
1833	Coding	For communication over a distance of 1200 m

Table 2 In Late 1940s–1970s, Norman Joseph Woodland obtains the idea of barcode when he drew four lines in the sand at a beach in Miami

Year	Invention	Application
1942	Frequency hopping	Multiple access method in FHCDMA
1946	2-way radio (transceiver)	For transmit and receive content
1950	Sensorama	Head-mounted display
1952	Barcode	Industrial
1955	Wearable computer	Predicting roulette wheels
1960	Stereoscopic 3D television	For enhancing or creating the illusion of depth in an image
1967	Wearable computer having eyeglass	It is used to display the lip reading
1969	ARPANET	DEPT. of Defence

Table 3 In 1971–1980

Year	Invention	Application
1971	Email (electronic mail)	Sending and receiving message
1971	TCP (Transmission Control Protocol)	Computer can connect across the globe
1973	Read–Write RFID tag	RFID tag revolutionizes retailing
1974	UPC (Universal Product Code)	Process purchases at a supermarket
1980	Micro-switches	Coke vending machine to monitor number of bottles

Table 4 In 1981–1990. In Late 1990s Sanjay Sarma, David Brock and Kevin Ashton started connecting objects together using RFID technology at Auto-ID center at MIT

Year	Invention	Application
1981	Wearable personal computer	For military and entertainment
1982	Internet	Communication
1989	WWW (World Wide Web)	Opensource information space where documents and other web resources are identified by URLs
1990	Using a customized computer	It head-up display as a wearable
1990	Internet device Toaster	Information base (SNMP) for turn ON/OFF

Table 5 In 1991–2000

Year	Invention	Application
1991	WWW (World Wide Web)	Accessible for internet connection
1993	Mosaic web browser	Used by commercial organizations
1994	"Forget-Me-Not" a wearable device	wireless transmitters to communicate
1994	Wireless wearable camera	For capturing images
1995	M1	machine-to-machine (M2M) communication
1995	MIT published an article on wearable computing	For wearable computing things
1998	Bluetooth	For transferring files and media
1999	Internet of things	Smart cities
2000	Smart fridge	Manages grocery and food items

Table 6 In 2001–2010

Year	Invention	Application
2001	EPC	Propose a unified directory of identification number
2003	"Project JXTA-C: Enabling a Web of Things"	An open source protocols for peer-to-peer
2004	Publications referring to the potential of the Internet of Things	For configuring home lights, home healthcare and shipping monitoring
2005	"Arduino"	An inexpensive and user-friendly microcontroller to help in interaction of two objects
2008–2009	CISCO internet business Solutions group claimed "The Internet of Things"	There were more objects connected to the internet than the people
2010	Self-driving vehicle	Milestone in development of connected cars

Table 7 In 2011–2015

Year	Invention	Application
2011	IPv6	Objects that connect to the web
2013	Google glass	Use by public
2014	Arduino Wi-Fi shield	Rapid prototyping of IoT applications for makers
2014	AllSeen Alliance	Open framework for IoT
2015	Smart cities	Enhance the level of life of people

3 Applications of IoT

Field	Typical application
Smart city	Comfortable homes, infrastructure, smart road, waste management
Industry	Temperature monitoring, ozone presence
Transportation and logistics	Item location, quality of shipment conditions
Smart environment	Smart museums, gym, forest fire detection
Social sensing	Social networking, thefts
Smarter devices	Watches
Healthcare	Data collection, tracking, sensing
Retail	Intelligent buying at apps, smartness production direction
Smart water	Portable water monitoring, river floods, water leakages
Smart metering	Smart grid, water flow
Smart agriculture	Greenhouses, wine quality enhancing
Smart animal farming	Animal tracking, toxic gas levels
Domotic automation	Intrusion detecting systems, water system use
eHealth	Fall detection, patients surveillance

4 Waste Management

Waste is the main problem in all over the world. People have to face many problems for managing the waste. Waste management is the main aspect in managing and disposal of waste. There are many types of waste such as solid, industrial, household, and much more. The waste is managed in many countries by using 3R formula i.e., Reduce, Reuse, and Recycle.

Waste management is all those activities required to manage waste until its final disposal. Waste includes many types of waste such as a household, Industrial and hospital. In the waste management, it relates to all kinds of waste. Waste management is processing of the raw materials into intermediate then the final products. Types of waste management are as follows:

- Construction waste
- Agriculture waste
- Domestic waste
- Factory waste
- Food processing waste
- Bio-medical waste
- Electronic waste
- Nuclear waste
- Household waste

Solid waste is divided into two categories (Fig. 1):

4.1 *Life cycle of Waste Management*

In the life cycle of waste, it consists of 6 steps as follows:

- Design: In the design, select the type of waste i.e., it is plastic, paper, and aluminum (Fig. 2).
- Manufacture: When selecting the type of waste, it sends to the industry for making new products from them.
- Distribution: In the 3rd step of lifecycle, waste material which sends to the industry is distributed category wise.
- Use: Then, the waste material is molded for further use.
- Reuse, Recovery, and Recycling: In the 5th step, material is reuse, recover, and send for the recycling process.
- Disposal: Left waste material which is not in use is disposed of. For example, landfill

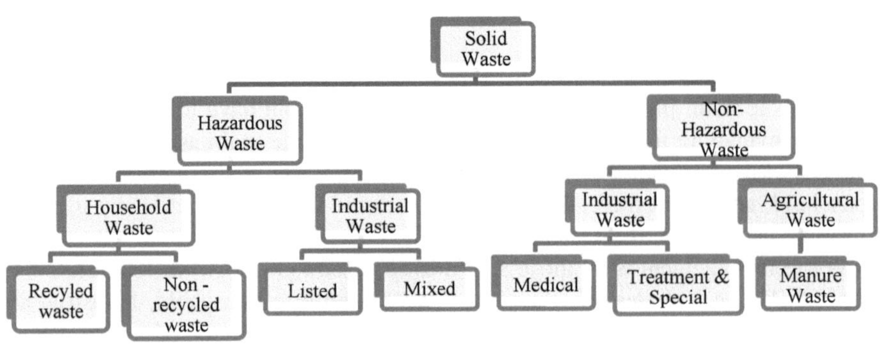

Fig. 1 Types of solid waste

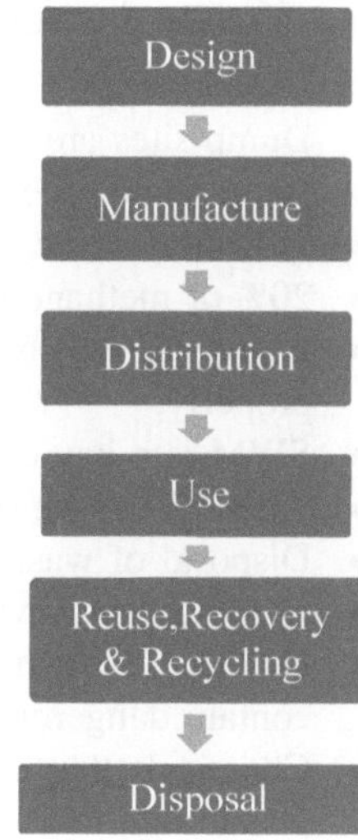

Fig. 2 Life cycle of waste management

5 Disposal Solutions of Waste

- Landfill: In this, put or hide the waste in underground.
- Incineration: Incineration is a method of disposal of solid organic waste which is subjected to combustion to convert into gaseous products.
- Recycling: Recycling is the process of collection and reuse of waste material. Materials are collected from homes, industrial and also from kerbside. By recycling, the metal can save lot of energy that can be used to:
  - Watch TV for 2 h.
  - For 3 h powering the computer
  - 100 W light bulb for 20 h for lightening
- Reuse: Reuse is a method by which reuse the used things by different methods such as biological reprocessing, energy recovery, pyrolysis, resource recovery, and sustainability.
- One Container multiple benefits:
  - Reduce Traffic: A single bin reduces traffic on highway or road rather than use more trucks or bin.
  - Saves Times: By using a bin it saves enough time.
  - Sized to Fit: due to different use of bin, it comes in different sizes and shapes, people can choose the size of bin according to their requirement.
  - Less Waste: By using recycling, it gives new life to waste material.
  - Conserves Energy: It saves more energy i.e., in the manufacturing phase, less energy is used for making new product.
  - Easy to Use: A single container is easy to use for separating the waste.

6 Waste Issues

- Dump sites are not good for local residents
- No space to accommodate fresh garbage waste because it is already overflowing.
- Industrial waste is also dumped into landfills.
- 20% of methane gas emissions in India caused by landfills.
- Segregation of biodegradable waste from non-biodegradable waste is not done properly.
- SWM is a key challenge for local municipal bodies & state governments.
- About 271.7 kg of garbage per person generated per year.
- Disposal of waste is difficult in many cities.
- In India, just 8 waste-to-energy plants are there as compared to other countries.
- Due to the grown in chemical industry products that are used and thrown away contain dangerous chemicals.
- Chemicals that cause health such as BPA (Biphenyl–A), commonly found in plastic like toys.
- Packaging is the largest and most rapidly growing category of solid waste. 30% packaging and 40% plastic waste.
- Current waste disposal system is flawed.

7 Results

The following algorithm is used for designing a framework for recyclable household waste.

Step 1: Collect the household waste
Step 2: Check whether the waste is recyclable or not that is collected from household
Step 3: If waste is Recyclable, then
Step 4: Calculate the weight of waste material which is recyclable by using weight-based billing system
Step 5: Select the waste collector vendor and then send waste for recycling
Step 6: If waste is not recyclable,
Step 7: Then the waste material is throw in the dustbin
Step 8: Select another waste material
Step 9: After that Repeat step 3 and 4
Step 10: Stop (Fig. 3)

Formula for Calculating Recyclable Household Waste:

Recyclable Household Waste = Recycled Household Waste ÷ Total Waste Generated *100 (Calculate in %).

In the formula, the Total waste generated = Disposal waste + Recycled waste (Fig. 4; Table 8).

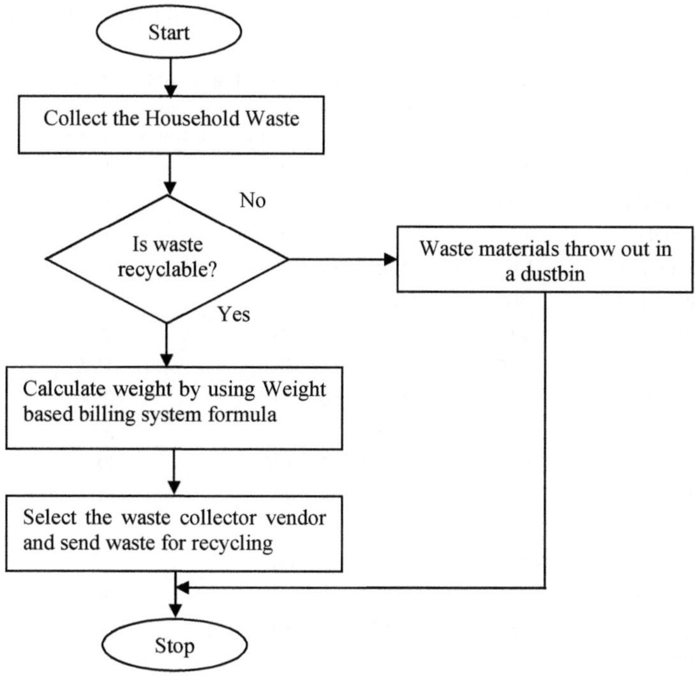

Fig. 3 Flowchart for recyclable household waste management system

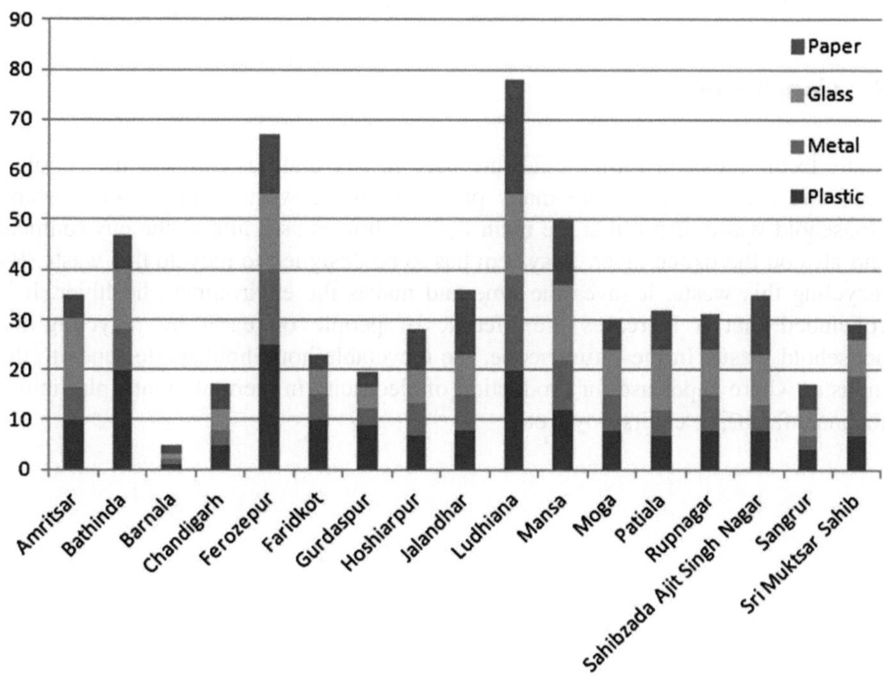

Fig. 4 Represent calculation of recyclable household waste of 17 main districts of Punjab

Table 8 Total household waste generated and recyclable household waste of 17 main districts of Punjab

S. no.	City	Total area in Km^2	Number of households	Total household waste generated in a month tones	Recyclable household waste in a month in tons
1	Amritsar	2,683	2,490,891	60	35
2	Bathinda	3,344	285,788	90.5	47
3	Barnala	26	116,449	10	5
4	Chandigarh	114	1,054,686	30	17
5	Ferozepur	5,305	2,2029,074	110	67
6	Faridkot	18.14	87,695	40	23
7	Gurdaspur	2,610	2,104,011	35	19.5
8	Hoshiarpur	3,365	1,579,160	45	27
9	Jalandhar	3401	862,196	78	37
10	Ludhiana	310	713,493	120	78
11	Mansa	2,174	768,808	89	50
12	Moga	2,230	125,573	54	30
13	Patiala	339.9	406,192	60	32
14	Rupnagar	1,440	684,627	45	31
15	SahibzadaAjit Singh Nagar	1093	176,152	50	35
16	Sangrur	3,685	1,654,408	40	17
17	Sri Muktsar Sahib	32.79	117,085	63	29

8 Conclusion

Many techniques and technologies are used in IoT that has changed the people's need and lifestyle. There are many problems in the waste management system. Household waste disposal is the main issue in homes as it affects the environment and also on the ozone layer. A system has to be designed to recycle this waste. By recycling this waste, it saves the time and makes the environment healthier. It is concluded that it increases the lifetime of people on earth by recycling the household waste. In the future scope, the recyclable household waste sends to the industry where paper use for production of electricity in thermal plants, plastic use for manufacturing chairs, toys, etc.

References

1. Alsaadi E, Tubshit A (2015) Internet of things features, challenges, vulnerabilities. Int J Adv Comput Sci Inf Technol 3(1):1–13
2. Atzori L, Lera A, Morabito G (2010) The internet of things: a survey. Comput Netw 54:2787–2805
3. Catania V, Ventura D (2014) An approach for monitoring and smart planning of urban solid waste management using smart-M3 platform. In: IEEE proceeding of the 15th conference of Fruct association, pp 24–31
4. Caytiles RD, Park B (2012) Mobile IP-based architecture for smart homes. Int J Smart Home 6:29–36
5. Chen S, Xu H, Liu D, Hu B, Wang H (2014) A vision of internet of things: applications, challenges, and opportunities with China perspective. IEEE Internet Things J 1:349–359
6. Chowdhury B, Chowdhury MU (2007) RFID-based real-time smart waste management system. Telecommunication networks and applications conference (ATNAC), Australasian, pp 175–180
7. Dahlén L, Lagerkvist A (2010) Pay as you throw strengths and weaknesses of weight-based billing in household waste collection systems in Sweden. Elsevier Waste Manag 30:23–31
8. Gaur A, Scotney B, Parr G, McClean S (2015) Smart city architecture and its applications based on IoT. Proc Comput Sci 52:1089–1094
9. Gaura A, Scotney B, Parr G, McClean S (2015) Smart city architecture and its applications based on IoT. Proc Comput Sci 52:1089–1094
10. Gubbia J, Buyaa R, Marusic S, Palaniswami M (2013) Internet of things (IoT): a vision, architectural elements and future directions. Future Gener Comput Syst 29:1645–1660
11. Hong I, Park S, Lee B, Lee J, Jeong D, Park S (2014) IoT-based smart garbage system for efficient food waste management. Sci World J 1–13
12. Jin J, Gubbi J, Marusic S, Palaniswami M (2014) An information framework for creating a smart city through internet of things. IEEE Internet Things J 2:112–121
13. Korteum G, Kansar F, Sundramoorthy V (2010) Smart objects as building blocks for the internet of things. IEEE Internet Comput 14:44–51
14. Mantas G, Lymberopolous D, Komninos N (2011) Security in smart home environment: wireless technologies for ambient assisted living and healthcare. Syst Appl 170–191
15. Prajakta G, Kalyani J, Snehal M (2015) Smart garbage collection system in residential area. IJRET Int J Res Eng Technol 4:122–124
16. Robles RJ, Kim T (2010) A review of security in smart home development. Int J Adv Sci Technol 15:13–21
17. Vetrive RS, Pandi Kumar S (2014) Internet of things based architecture of web and smart home interface using GSM. Int J Innov Res Sci Eng Technol
18. Wahab MHA, Kadir AA, Tomari MR, Jabbar MH (2014) Smart recycle bin: a conceptual approach of smart waste management with integrated web based system. IEEE 1–4
19. Weber RH (2010) Internet of things-new security and privacy challenges: Comput Law Secur 26:23–30
20. Zanella A, Vangelista L (2014) Internet of things for smart cities. IEEE Internet Things J 1:21–32

Feature Extraction in Permanent Human Dentition Radiographs

Kanika Lakhani, Bhawna Minocha and Neeraj Gugnani

Abstract Feature extraction in dental images in the form of radiographs involves the identification of major defect areas. While analyzing complex radiograph images, one of the major problems stems from the types of defects present. Analysis with a large number of defects present generally requires a large amount of memory and computational power. Feature extraction applied over the radiographs, once the edge detection process is accomplished, derives combinations of the defects to get around the problems while still describing the problem areas with sufficient accuracy. The process has been implemented over a set of 20 extracted human dentitions for the identification of similar features to actualize the presence of defects in the dentition.

Keywords Feature extraction · Edge detection · Dental radiographs · SOM

1 Introduction

Dental radiograph is an image that is recorded using X-ray radiation. Dental radiographs are actually X-ray radiation consisting of teeth, tooth bones, and soft tissues surrounding the oral cavity. Feature extraction, collection, and analysis for dental clinical diagnostics are the chief requirements nowadays for dental science. In this realm of dental image analysis, most of the research done is crucial for the purpose of human biometric identification. Moving further with the realm is the diagnosis of dental diseases from radiographs that eases the job of a dentist.

K. Lakhani (✉) · B. Minocha
Amity University, Noida, India
e-mail: kanikalakhani@yahoo.co.in

B. Minocha
e-mail: bminocha@amity.edu

N. Gugnani
DAV Dental College, Yamunanagar, India
e-mail: drgugnani@gmail.com

H.R. Vishwakarma and S. Akashe (eds.), *Computing and Network Sustainability*, Lecture Notes in Networks and Systems 12, DOI 10.1007/978-981-10-3935-5_23

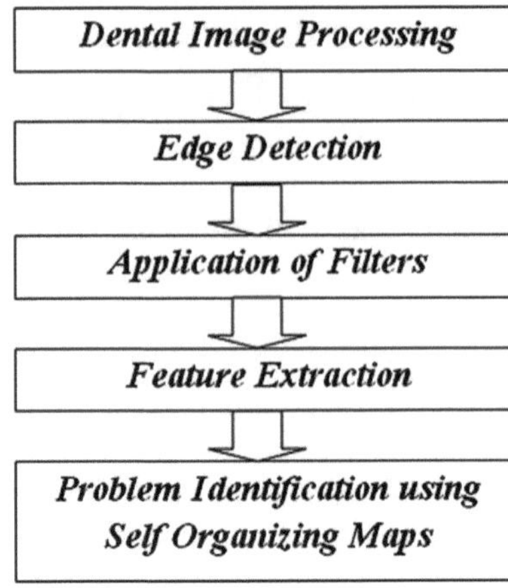

Fig. 1 Steps for the identification of problems related to permanent dentition in human beings

Different branches in dentistry deal with different types of dental problems such as tooth decay, mouth sores, and tooth erosion. Some of these problems have common symptoms that can be diagnosed easily. The idea is to detect these problems and be aware of the true picture before undergoing appropriate treatment. Furthermore, this can be used as a preliminary diagnostic aid by the dentists. The process to be followed has been depicted in Fig. 1.

The images retained from radiographic devices produce a relatively blurred image. A physician's or dentist's involvement in the diagnosis is of attern importance. It is not possible to detect fine details with ease in such radiographs. Latest technology with high-contrast radiograph readers is available but that sounds heavily on a physician's pocket. This paper discusses an alternate technique for the same. It includes the processing of radiographs for identifying the exact location and depth of damage in affected tooth.

Edge detection process identifies and locates the lack of continuity, inequalities, and varied orientations in an image [1]. This discontinuity describes the sudden changes in the pixel intensity. The discontinuities that occur in image intensity can either be step edge or line edge [2]. These discontinuities are rare in real images because instant changes rarely occur.

This paper has been organized as follows: Section 2 discusses edge extraction using various edge extraction techniques. Section 3 shows the comparison of various techniques. Section 4 focuses on the observations and findings. Finally, this paper concludes with describing the scope for future work.

2 Conventional Edge Extraction Operators in Dental X-Ray Images

Sobel Operator: Most of the edge detection methods are based on the assumption that edges are found in the image where there is discontinuity. Based on this assumption, the derivative is taken for image intensity value and the points are located where intensity derivatives have maximum value so as to locate the edges [3].
Prewitt Operator: [4] It is computationally less expensive and faster method for edge detection. It is only appropriate for noiseless and well-contrasted images [5]. Prewitt approximation is applied on the derivatives of intensity function. It results in edges where gradient of intensity function has maximum value. Prewitt operator detects two types of images: horizontal edges and vertical edges. The difference between the corresponding pixel intensities of the image results in edges. Derivative masks are used for edge detection technique. Prewitt operator generates two masks, one for detecting edges in horizontal direction and other for vertical direction.
Canny Edge Extraction Algorithm for Dental X-Ray Images: Canny edge extraction process is a multistage algorithm that is popularly known as the optimal edge detector. The regions are represented by the local maxima which are marked as the edges in the gradient image. A non-maximal suppression is used to find the local maximum points in the gradient edge map. The weak edge areas are suppressed by double thresholding. The algorithm produces over-segmented images from which none of the root features can be identified. Generally, the Canny edge extraction algorithm is assumed to provide optimum results, but is proving over segmented edge maps in the case of these dental X-ray images which have illumination variations, noise and different gradient angles.

3 Comparison of Edge Detection Techniques

The above techniques were implemented on a sample space of 20 teeth. The radiographs were taken for a set of 20 extracted teeth. Further, the radiographs were converted to JPEG format, and the number of black and white pixels was calculated using Sobel operator, Prewitt operator, and Canny edge detection technique. Table 1 represents the difference between the numbers of black and white pixels of each tooth in the sample space (Figs. 2, 3, 4 and 5).

It can be concluded from the implemented data set that the features extracted from the proposed technique can describe the exact shape and the accuracy in classifying the dental image. The effect of pretreatment of the image so produced is not idyllic, and the proposed technique is still relatively lacking perfection that still leaves a scope for further improvement (Fig. 6).

Table 1 Comparison of number of pixels

Tooth no.	Sobel		Prewitt		Canny	
(JPEG)	No. of black pixels	No. of white pixels	No. of black pixels	No. of white pixels	No. of black pixels	No. of white pixels
1	1,654,642	4358	1,654,643	4357	1,602,577	56,423
2	1,652,873	6127	1,652,931	6069	1,592,284	66,716
3	1,655,807	3193	1,655,805	3195	1,599,029	59,971
4	1,654,293	4707	1,654,329	4671	1,590,427	68,573
5	1,653,300	5700	1,653,350	5650	1,607,227	51,773
6	1,652,825	6175	1,652,847	6153	1,600,715	58,285
7	1,654,162	4838	1,654,147	4853	1,618,285	40,715
8	1,655,245	3755	1,655,272	3728	1,600,768	582,232
9	1,655,550	3450	1,655,557	3443	1,597,341	61,659
10	1,654,731	4269	1,654,734	4266	1,602,566	56,434
11	1,654,290	4710	1,654,271	4729	1,593,289	65,711
12	1,653,874	5126	1,653,907	5093	1,598,133	60,867
13	1,653,652	5348	1,653,668	5332	1,621,309	37,691
14	1,654,329	4671	1,654,359	4641	1,614,303	44,697
15	1,653,948	5052	1,653,989	5011	1,614,303	44,697
16	1,654,502	4498	1,654,515	4485	1,609,869	49,131
17	1,653,824	5176	1,653,793	5207	1,623,816	35,184
18	1,654,397	4603	1,654,433	4567	1,625,303	33,697
19	1,652,901	6099	1,652,917	6083	1,621,286	37,714
20	1,653,490	5510	1,653,605	5395	1,615,613	43,387

Fig. 2 Tooth no. 6 radiograph image converted to JPEG format

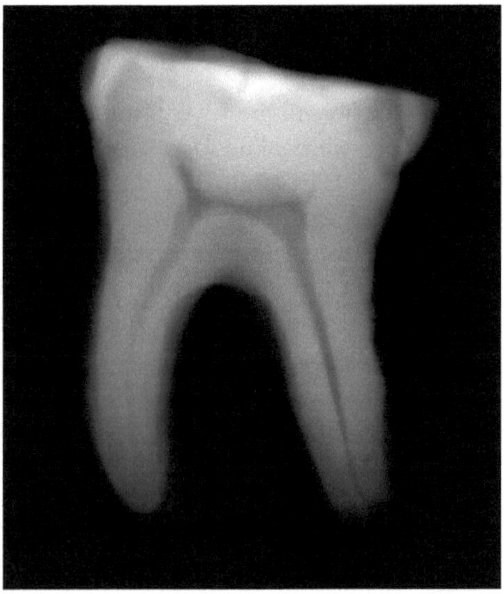

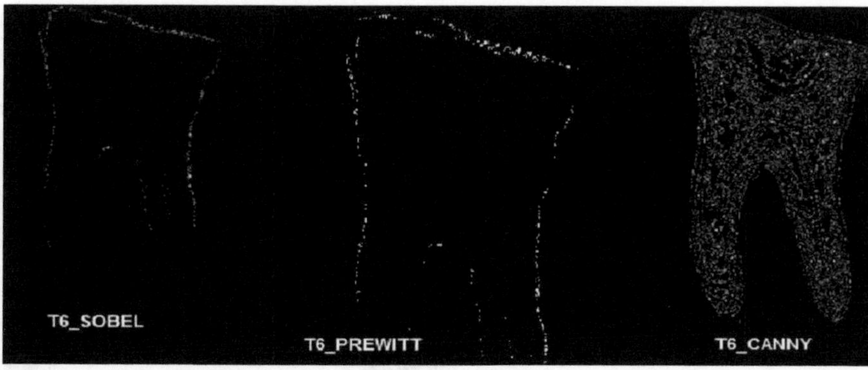

Fig. 3 Comparison of various techniques on tooth no. 6

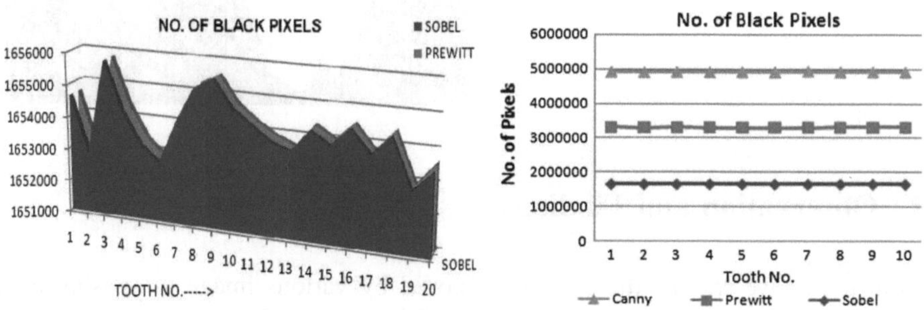

Fig. 4 Comparative graph of number of black pixels

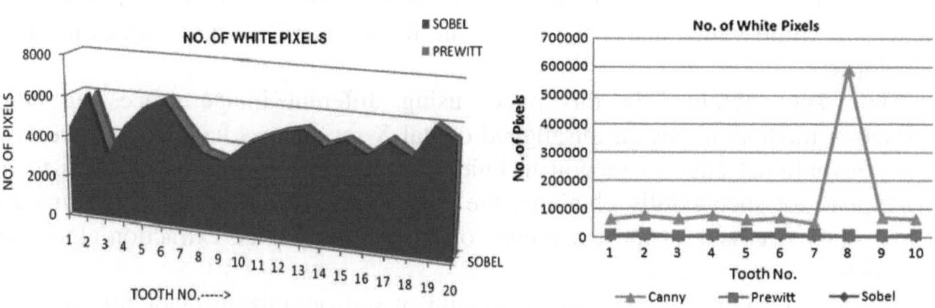

Fig. 5 Comparative graph of number of white pixels

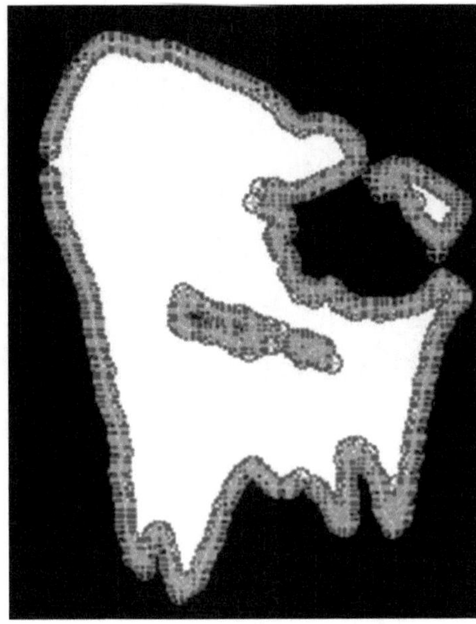

Fig. 6 Feature extraction on tooth no. 6

4 Observation and Findings

From the above results, the challenges posed by various image processing edge extraction techniques on non-uniform dental images can be evaluated. Sobel and Prewitt operators provide edges which are not complete and sufficient. The Canny algorithm, known as the optimal edge extraction algorithm, is producing over-segmented edge map containing edges of unwanted noise areas also. Thus, a better algorithm which can make use of the linear character of the edges must be developed.

The issues obtained in this paper using different image processing edge extraction methodologies on misaligned dental X-ray images have been discussed. The conventional edge extraction techniques mentioned in this paper seem to be inadequate for successfully obtaining the edge features from dental X-ray images. Thus, study necessitates the emergence of an improved edge extraction algorithm over dental X-ray images.

Future work includes the implementation of self-organizing maps for the identification of problems related to permanent dentition. The extracted features in the radiographs shall be grouped together to form clusters, and thus, the classification of diseases shall be done.

References

1. Thanki R, Trivedi D (2012) Introduction of novel tooth for human identification based on dental image matching. Int J Emerg Technol Adv Eng 2(10). ISSN 2250-2459
2. Hema (2013) Edge extraction algorithm using linear prediction model on dental X-ray images. IJCA 100
3. Harandi AA, Pourghassem H (2011) A semi automatic algorithm based on morphology features for measuring of root canal length. In: Proceedings of IEEE international conference on communication software and networks (ICCSN)
4. Li H, Chutatape O (2004) Automated feature extraction in color retinal images by a model based approach. IEEE Trans Biomed Eng 51(2):246–254
5. Solanki AJ, Jain KR, Desai NP (2013) ISEF based identification of dental caries in decayed tooth. In: Proceedings of international conference on advances in information technology and mobile communication
6. Sharma DK, Gaur L, Okunbor D (2007) Image compression and feature extraction with neural network. Proc Acad Inf Manag Sci 11(1):33–37
7. Sushma Sri M, Narayana M (2013) Edge detection by using look-up table. IJRET: Int J Res Eng Technol 02(10). eISSN: 2319-1163|pISSN: 2321-7308
8. Oprea S, Marinescu C, LiŃă I, Jurianu M, Visan DA, Cioc IB (2008) Image processing techniques used for dental X-ray image analysis. Electron Technol ISSE 125–129
9. Said EH, Nassar DEM, Fahmy G, Ammar HH (2006) Teeth segmentation in digitized dental X-ray films using mathematical morphology. IEEE Trans Inf Forensic Secur 1(2):178–189
10. Lain AK, Chen H Matching of dental X-ray images for human identification. Pattern Recognit 37(7):1519–1532
11. Abood ZM (2013) Edges enhancement of medical color images using add images. IOSR J Res Method Educ (IOSR-JRME) 2(4):52–60. e-ISSN: 2320–7388, p-ISSN: 2320–737X
12. Patel A, Patel P Analysis of dental image processing. IJERT 1(10)

Index-Based Image Retrieval-Analyzed Methodologies in CBIR

B. Prasanthi, P. Suresh and D. Vasumathi

Abstract Quality, efficiency, and scalabilities are main focusing concepts in image retrieval from various image databases. These are major topics in image retrieval from image databases with preferable analysis. Image retrieval from large image databases is a complex task in present days because of image color, shape, and visual features of images in CBIR (content-based image retrieval). So in this paper, we analyze already presented different index-based image retrieval techniques with procedure of each technique for processing quality-based image retrieval from various image databases. Each technique follows basically features such as color, shape, and image length and width in image retrieval from different image data warehouses. We analyze each method implementation procedure for retrieving efficient image retrieval based on query, relevancy with index-based structures.

Keywords Index-based structures • Color • Shape • Query relevancy • Content-based image retrieval

1 Introduction

Dynamic improvement of Web image data retrieval in last presented years may concern complex task for retrieve images from large amount of image data. Past years have been achieved to define advanced techniques in content-based image retrieval. Image retrieval applications such as medicine, education, entertainment,

B. Prasanthi (✉)
Department of CSE, MGIT, Hyderabad, Telangana, India
e-mail: biraliprashanthi19@gmail.com

P. Suresh
Department of IT, CBIT, Hyderabad, Telangana, India
e-mail: plpsuresh@gmail.com

D. Vasumathi
Department of CSE, JNTUH, Hyderabad, Telangana, India
e-mail: vasukumar_devara@yahoo.co.in

H.R. Vishwakarma and S. Akashe (eds.), *Computing and Network Sustainability*, Lecture Notes in Networks and Systems 12, DOI 10.1007/978-981-10-3935-5_24

and manufacturing make use large amount of visual data in the form of images to process efficient problem solutions with their respective operations processed in visual image data retrieval. Growing popularity of the Internet, new consumers may achieve digital image, video, and the emergence of digital standards for efficient storage and retrieval of multimedia data-like image. In this paper, we analyze different index-based structure methodologies such as efficient content-based image retrieval (CBIR), robust color object analysis approach, fast and semantics-tailored image retrieval procedure, clustering-based approach, and feature-based adaptive tolerance tree) for efficient image retrieval in real-time configurations in multimedia data retrieval for processing effective data assurance in image retrieval. Remaining of this paper may explain above-mentioned methodologies.

2 Efficient CBIR for Image Retrieval

Image color performs a very part in picture fetching from large image data present in real-time image retrieval. For example, we consider two pictures; their structure is likeness, but their combinations are different in different features. In the event that the recuperation depends just on shading, these photographs are distinctive, else they are similarity [1]. CBIR procedure may shown in following Fig. 1 with architectural implementation with image retrieval perspective operations.

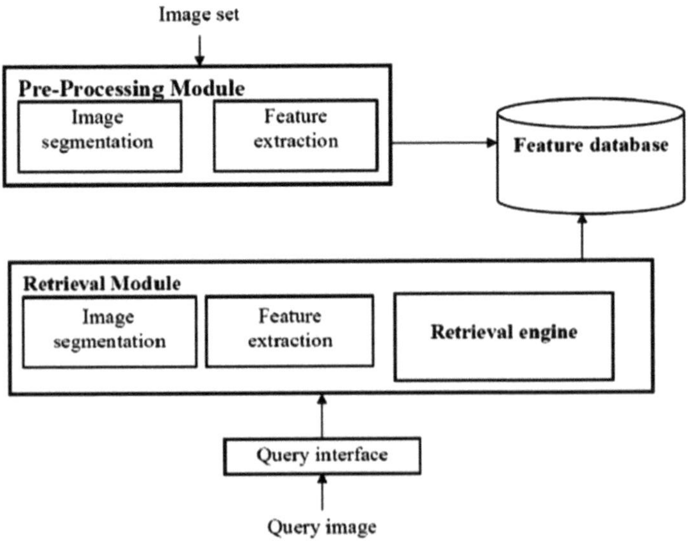

Fig. 1 Procedure of CBIR in relevant image data retrieval

The framework incorporates two primary modules: pre-preparing and recovery module. The pre-handling subsystem is responsible for getting proper components from pictures and sparing them into the photo-information source [2].

In this technique, performance procedure is as follows: The recovery efficiency was analyzed using a test data source of 8000 pictures. This picture data source will be used to indicate the efficiency of technique [2, 3]. Keeping in mind the end goal to guarantee the proficiency of looking system, three concerns were taken out and every question has utilized spatial color features (SCF). Give X a chance to imply an arrangement of pictures identified with picture information source, A means an arrangement of significance pictures bought picture information source, and X_A implies an arrangement of significance pictures from A.

$$recall = \frac{area(X_A)}{area(X)}$$

$$precision = \frac{area(X_A)}{area(A)}$$

The perspective results of efficient content-based methodology are as follows (Table 1).

Fig. 2 shows the procedure in real-time image retrieval for proceedings efficient image retrieval with comparison SCF and co-occurrence-based color (CC) in warehouse databases.

Table 1 Recall with precision in image retrieval from various image databases

Recall values	SCF	CC
2	4.8	3.5
4	4.3	3.2
6	3.5	2.8
8	3.4	1.8

Fig. 2 Recall with precision in image retrieval from image data warehouses

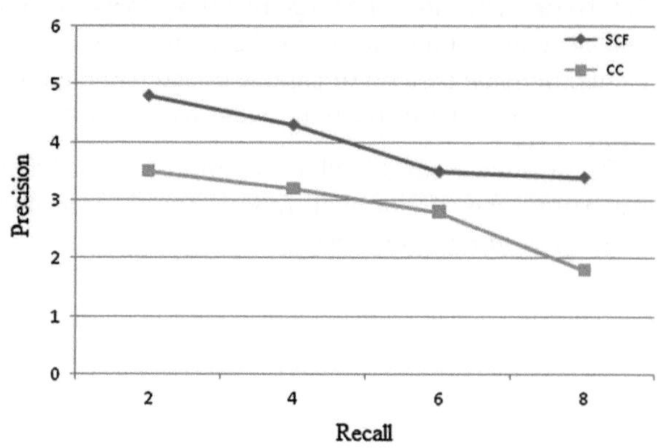

Figure 2 shows that caused by the SCF method better than the CC in image data retrieval for efficient communication for classification with recall and precision in image data retrieval. Efficient CBIR techniques give overall conclusion with recall and precision response with spatial color features and color histogram features in relevant image retrieval.

3 Robust Color Object-Based Image Retrieval

For efficient image retrieval which consists fuzzy color features with image segmentation in real search data from different image Web databases. In this model, we propose productive, bunching-based and fuzzified highlighted representation to the broadly useful of substance-based picture recovery [4]. In this approach, we process to integrate intensive clustering-based segmentation with fuzzy color histogram with index databases. So we call this method as CLEAR for fuzzy index-based image retrieval, and it contains following modules in real-time image retrieval proceedings.

3.1. Image Segmentation: The inquiry picture and all photographs in the database are initially portioned into ranges. The indistinct capacity of shade, composition, and structure are created to be the trademark of every district in one picture. Segmentation may appear superior implementation for image retrieval proceedings.

3.2. Fuzzy Histogram for Each Region in Image Retrieval: Along with reflection would be harsh and dark on the off chance that we essentially draw out shade capacity of one keep (the delegate square) to be the shade trademark of every zone as Wang et al. did. Shape is excellent viewer in image display in real-time process. The key reason why we cure shade property along these lines is twofold: (i) We need to characterize the neighborhood property of hues precisely and heartily and (ii) shade part in the region capacities is delivered superior to anything composition and structure and it is more proficient to clarify the semantics of pictures.

3.3. Representation of Shape in Fuzzy Way: To bolster the dark picture division and instability of human comprehension, we offer adulterate every area delivered by picture division by a set of parameterized participation works [5]. Fuzzy Orientation may appear shape in fuzzy-oriented map of images.

These are the basic modules in image retrieval from various fuzzy-related image databases, and if we will follow above procedure, then it provides efficient robustness in real-time image retrieval.

4 Fast-Based Image Retrieval

In this section, we implement FAST in a prototype system for fuzzy-based indexed image retrieval with relevancy and other features such as shape color and fuzzy histogram analysis in image retrieval. In this approach, compare previously discussed approach CLEAR in terms of accuracy of image retrieval with the increase number of images [6]. This approach also follows CLEAR approach procedure with the extension of homogeneous image region for relevance.

4.1. FAST Accuracy with CLEAT: FAST is investigated against one of the best in class CBIR frameworks, CLEAR [5, 7], with respect to productivity. Recovery viability is figured by recollecting and flawlessness examination. For a given inquiry and a given assortment of pictures recuperated, accuracy gives the rate between the measure of fitting pictures recovered and the measure of recouped pictures in complete. Review gives the rate between the measure of proper pictures recovered and the inclusion of suitable pictures the accumulation.

Accuracy in FAST with respect to image retrieval for processing efficient data storage and other configurations processed in image segmentation and homogeneous fuzzy region selection shows in Table 2 [2]. The perspective procedure

Table 2 Accuracy of FAST with respect to CLEAR in image retrieval

Images retrieval	FAST	CLEAR
10	4.8	3.5
20	4.3	3.2
30	3.5	2.8
40	3.4	2.7
50	2.5	1.9
60	1.9	1.3
70	1.4	0.9
80	0.9	0.4

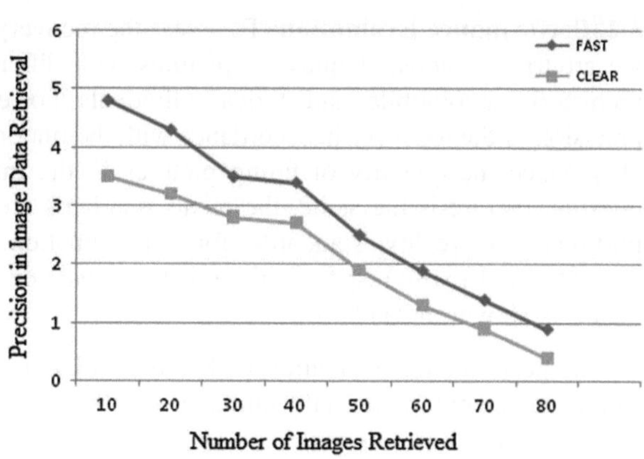

Fig. 3 Accuracy of FAST with CLEAR in image retrieval

achieves recent contributions of image retrieval precision accuracy as shown in Fig. 3 with respect to image retrieval in relevancy procedure.

To ponder the versatility of FAST, we incrementally test the first 10,000-picture database to create two littler information sources, one with 3,000 pictures and the other with 6,000 pictures [8]. Accuracy FAST may increase in precision of image retrieval.

5 Cluster-Based Image Retrieval

In this section, we analyze unsupervised segmentation technique, i.e., FUZZY CLUB for image retrieval which is followed by index features with fuzzy color histogram. FUZZYCLUB follows following procedure to retrieve relevant image retrievals [9, 10]. The main principles of FUZZY CLUB region-based tracking are as follows.

5.1. Feature-Based Regions: Within each area, we determine three kinds of features: shade, structure, and form, along with the traditional geometrical details as the function vector for picture listing. Color is the most famously used kind of functions in picture listing [3, 11]. We determine a Gaussian operator to be the fuzzy similarity function with respect to Euclidean distance as follows:

$$\mu_c(c') = \frac{1}{\sqrt{2\pi\sigma^2}} \exp\left\{ -\frac{d^2(c, c')}{2\sigma^2} \right\}$$

where d is the Euclidean extent between shade c and c′ in LAB region, and σ is the normal reach between shading c and c′. This fluffy shading outline permits to grow the impact of an offered shade to its adjacent shades, as indicated by the uncertainty idea and the perceptual likeness.

To estimate the gap between two areas, we apply L2 range measurement to unclear shade histogram, texture vector, and form vector, respectively.

5.2. Performance Evaluation: To assess the recovery efficiency of FUZZYCLUB, we arbitrarily choose 4 question pictures with different semantics, namely plant, prehistoric, automobile, and Africa individuals. For each question, we analyze the precision of the recovery in accordance with the importance of the picture semantics [12]. Since the quantity of fitting pictures in the information hot spot for every question picture is the same, the recall standards are not figured as they are proportionate to reality standards for this situation. Evaluated performance of FUZZY CLUB with respect to time with comparison of IRM (information retrieval mean) is shown in Table 3.

The performance evaluation of FUZZYCLUB may effective with the comparison of registered traditional technique, i.e., IRM with respect to time evaluation as shown in Fig. 4.

Table 3 Experimental evaluation of FUZZY CLUB with IRM with respect to time

Image retrieval	IRM	FUZZY CLUB
1	4.8	3.5
2	4.3	3.2
3	3.5	2.8
4	3.4	2.7
5	2.5	1.9
6	1.9	1.3
7	1.4	0.9
8	0.9	0.4

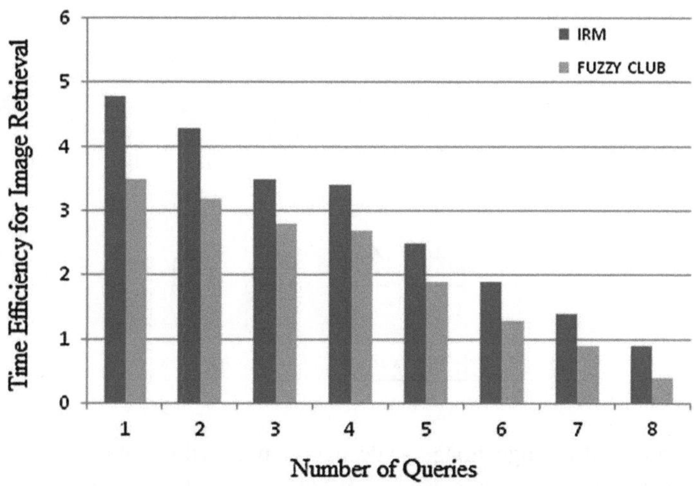

Fig. 4 Comparison results of FUZZYCLUB with IRM with respect to time evaluation

The common reaction time decrease of FUZZYCLUB to IRM [7] is 29.56%. With the dimensions of the database improve, the recovery performance of FUZZYCLUB improves proportionally.

6 FATT-Based Image Retrieval

We present FATT comprises of main node (grandparent) is having the highest possible equivalent mother or father nodes of 256 and detail of the shrub relies upon the amount of functions regarded and size of the shrub relies upon the person catalog value [13, 14]. The FATT is designed based on the catalog rule developed. The framework of the shrub can be modified based upon the pictures found. The general framework (4 levels, m = 4, N = 256) of the FATT is shown in Fig. 5.

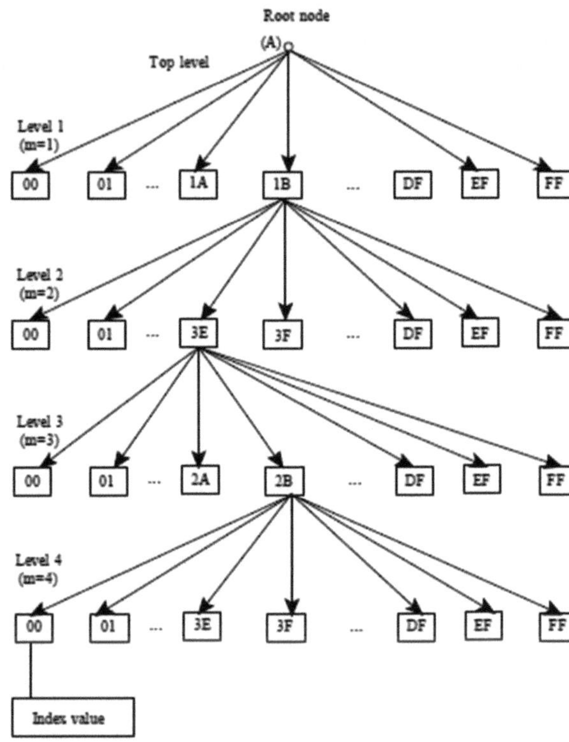

Fig. 5 General procedure for FATT for image retrieval

FATT can catalog huge image-scale data source using catalog code generated and euclidean likeness evaluate used as distance metric Index (nextImage, levelNo)

- If the parentnode > root node then do; Go to the lchild of the main node Until no lchild is available Check whether the listing value is obtained.
- Elseif parentnode < main node Return the catalog value in [15].
- Else not discovered navigate phase # 1 and 2.
- Do it again the actions 1, 2 and 3 until catalog value is discovered.

The listing criteria index (nextImage, levelNo) first performs the listing with main node: Case (1) if parentnode > root node listing starting with lchild node until no leftmost kid is available. Check whether catalog value is available. Case (2) if parentnode < main node, then come back the catalog value otherwise catalog for leftmost kid node is available.

We have examined the FATT to catalog and recover the images with three different data sources through a question procedure such as query by example (QBE). Concluding the overall procedure discussed in FATT, it may give better image retrieval with query-based search as shown in Fig. 6.

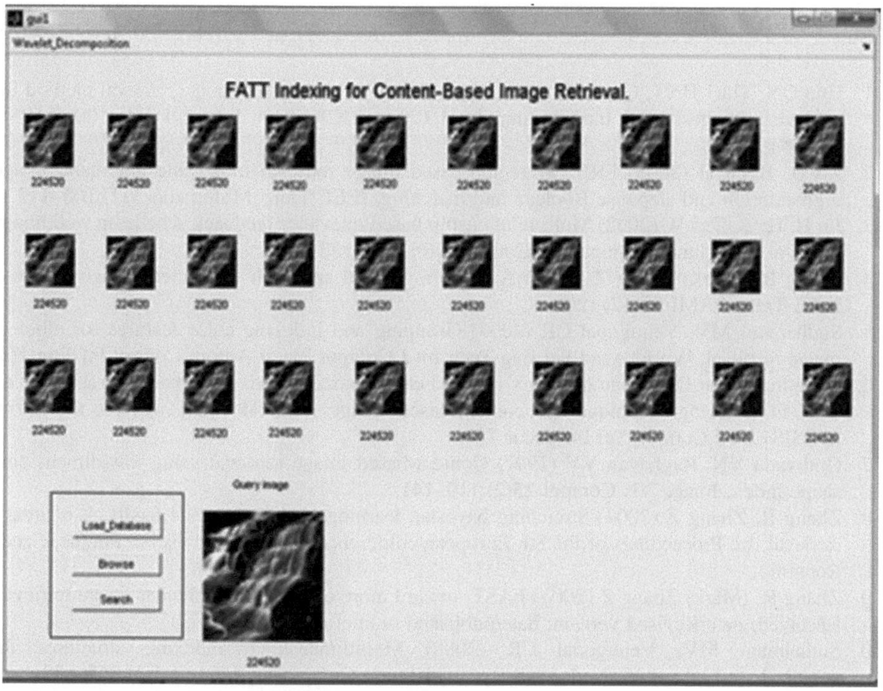

Fig. 6 Image retrieval from already stored database

7 Conclusion

In this paper, we analyze five different and effective and efficient image retrieval techniques. By concluding these techniques, following points were analyzed. (A) query-based image retrieval with proceedings of relevant images stored in image databases; (B) attributes is the main focusing index features in the image, so object-based analysis; (C) incorporate new indexing, a hierarchal indexing structure with a hierarchical search-based image using FAST image retrieval prototype methodology; (D) in order to improve the query response time and color feature inaccuracy problem, introduce secondary clustering schema such as FUZZYCLUB; it achieves global and regional matching and pre-organizes for image retrieval efficiency; (E) a novel listing strategy for material-based image recovery is developed particularly for quick placement, searching, and listing, moreover to deal with the issue of overlap between the nodes.

References

1. Huu QN, Thul HNT, Quoc TN (2012) An efficient content based image retrieval method for retrieving images. Int J Innov Comput Inf Control ICIC 8(4). International c 2012 ISSN 1349-4198
2. Ko B, Byun H (2005) FRIP: A region-based image retrieval tool using automatic image segmentation and stepwise Boolean and matching. IEEE Trans Multimedia 7(1):105–113
3. Jin H, He R, Tao W (2008) Multi-relationship based relevance feedback scheme in web image retrieval. Int J Innov Comput Inf Control 4(6):1315–1324
4. Zhang R, (Mark) Zhang Z (2000) A clustering based approach to efficient image retrieval. IEEE Trans PAMI 22(12) (2000)
5. Sudhamani MV, Venugopal CR (2007) Grouping and indexing color features for efficient image retrieval. World Acad Sci Eng Tech Int J Comput Electr Autom Control Inf Eng 1(3)
6. AnandhaKumar DP, Balamurugan V (2010) Feature-based adaptive tolerance tree (FATT): an efficient indexing technique for content-based image retrieval using wavelet transform. (IJCSIS) Int J Comput Sci Inf Secur 7(3)
7. Gudivada VN, Raghavan VV (1997) Content-based image retrieval using low-dimensional shape index. Image Vis Comput 15(2):119–141
8. Zhang R, Zhang Z (2004) Stretching bayesian learning in the relevance feedback of image retrieval. In: Proceedings of the 8th European conference on computer vision, Prague, Czech Republic
9. Zhang R, (Mark) Zhang Z (2005) FAST: toward more effective and efficient image retrieval. Received: date/Revised version: date/published online: date_c. Springer
10. Sudhamani MV, Venugopal CR (2008) Multidimensional indexing structures for content-based image retrieval: a survey. Int J Innov Comput Inf Control 4(4):867–882
11. Quynh NH, Tao NQ (2010) A novel content based image retrieval method based on splitting the image into homogeneous regions. Int J Innov Comput Inf Control 6(10):4029–4041
12. Zhang R, Zhang Z (2003) Addressing CBIR efficiency, effectiveness, and retrieval subjectivity simultaneously. In: 5th ACM international workshop on multimedia information retrieval. In conjunction with ACM Multimedia, Berkeley, CA
13. Zhang R (2004) A robust color object analysis approach to efficient image retrieval. EURASIP J Appl Signal Process 2004(6):871–885
14. Jing F, Li M, Zhang H, Zhang B (2002) An effective regionbased image retrieval framework. In: Proceedings 10th ACM multimedia. Juan-les-Pins, France, pp 456–465
15. Traina C, Traina AJM, Seeger B, Faloutsos C (2000) Slim-trees: high performance metric trees minimizing overlap between nodes. In: Proceedings of Oedbt 2000, Konstanz, Germany, Mar 2000, pp 51–65
16. Vertan C, Boujemaa N (2000) Embedding fuzzy logic in content based image retrieval. In: 19th international meeting of the North American fuzzy information processing society (NAFIPS '00), Atlanta, Ga, USA, July 2000
17. Chen Y, Wang JZ (2002) A region-based fuzzy feature matching approach to content-based image retrieval. IEEE Trans Pattern Anal Mach Intell 24(9):1252–1267
18. Traina C, Traina AJM, Faloutsos C, Seeger B (2002) Fast indexing and visualization of metric data sets using slim-trees. IEEE Trans Knowl Data Eng 14(2):244–260
19. Zhang R, Zhang Z (2004) Hidden semantic concept discovery in region based image retrieval. In: IEEE international conference on computer vision and pattern recognition (CVPR), Washington, DC

Advanced Cyber Hiding of Key Logging During Cyber Communication

D. Veeraiah and D. Vasumathi

Abstract Software program key loggers were used to spy on laptop users to accumulate sensitive statistics for decades. Their primary attention has been to seize keystroke statistics from keyboards. Cyber assaults targeting critical infrastructure using this information could bring about tremendous catastrophic structure failures. It is important to shield society in this age of modern technology. Online security risks manipulate the extended complexness and connection of critical facilities components, linked to a country's security, economic climate, and public security. Prior techniques have focused on using dedication techniques that are cryptographic primitives only protecting data stored in RREP and RREQ message formats as part of a sequential execution of tasks. These techniques proven to be effective, might be used along with attack recognition techniques for determining affected routers to increase overall system security by marginalizing the working limitations of an attacker, thus jeopardizing their visibility. A realistic execution validates our claim.

Keywords Malware · Software security · Cyber security · Cyber infrastructure framework · Key logger

1 Introduction

Human computer interfacing is an ever evolving field wherein numerous new and modern technologies have recently emerges as mainstream in society. Touch screens, speech popularity, gesture management, and swipe passwords have made their way into the hands of consumers over the last decade. However, these features frequently affect our capacity to improve our ability to satisfy risks whilst using

D. Veeraiah (✉)
Lakireddy Bali Reddy College of Engineering, Mylavaram, Andhra Pradesh, India
e-mail: veeraiahdvc@gmail.com

D. Vasumathi
JNTUH College of Engineering, Hyderabad, Telangana, India
e-mail: rochan44@gmail.com

H.R. Vishwakarma and S. Akashe (eds.), *Computing and Network Sustainability*, Lecture Notes in Networks and Systems 12, DOI 10.1007/978-981-10-3935-5_25

prime logger systems to satisfy real-time network storage environments. Key loggers have long been a severe threat to PC systems across the world. In many cases key loggers have been the medium by which passwords have been obtained [1]. In Verizon's (2014) information breach investigations record, key logger malware became one of the top 10 threats leading to intrusions and crimeware, at 2% and 13%, respectively. Key loggers also are closely utilized in cyber-espionage and have been linked to 38% of associated statistical breaches (Fig. 1).

The above figure shows the steps in an efficient key logger activity in a real time cyber security situation for processing efficient data hiding. We must turn our attention to the unique threat that key loggers will pose to the terminals providing direct access to the foundations that smart cities will be built upon. This danger gained widespread momentum with the inclusion of an on-display screen keyboard in Microsoft's Windows 8 machines. Some researchers implement a theoretical analysis of smart city development in recent applications protecting users from other services. The majority of anti-key logger methods take advantage of variety.

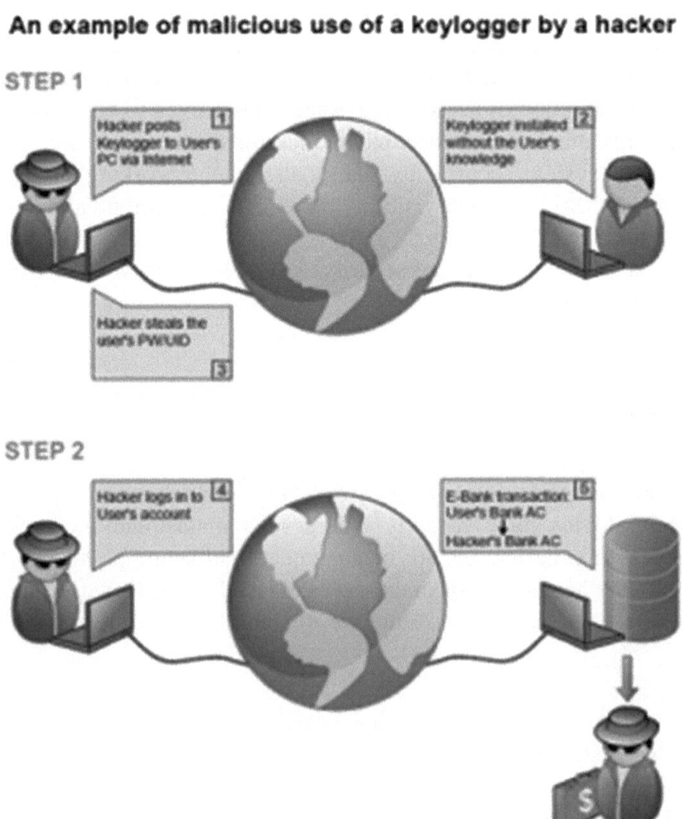

Fig. 1 Key logger steps for malicious use by hackers

For example, anti-Key logger methods may employ several regularity groups, different MAC programs, or several redirecting routes. Such variety helps to control the effects of performing strikes by demanding the key logger act on several sources at the same time. In this chapter, we recommend using them along with strike recognition methods for determining affected routers. This protects the overall network by marginalizing the working limitations of an attacker, thus mishandling visibility. To create an effective use of this redirecting variety, however, each source node must be able to create a brilliant throughput of traffic across all available routes while considering the potential effect of performing this on the required data throughput.

1.1 Organization of This Chapter

This chapter is organized as follows: Sect. 2 formalizes the problem statement for malware detection proceedings in real-time cyber communication systems. Section 3 introduces a proposed methodology for real-time cyber security. Section 4 defines an implementation procedure with experimental evaluation for processing malware risk management analysis. Section 5 concludes the chapter, considering cyber security.

2 Problem Statement

In order to discover the capacity key loggers already have to seize contact display keystroke statistics, we decided to look at and test a sample of five software key loggers. In order to identify the ramifications of the programs, we selected one business key logger, two freeware key loggers, one JavaScript browser primarily based on key logging, and a malware key logger. We additionally tested a hardware key logger to confirm our assumption that hardware key loggers are unable to seize digital keyboard facts.

Free Key Logger Version 3.95—represented software to be had from multiple utility web hosting net websites.

Real Key Logger model 3.2—business software to be had from *actualkeylogger.com*. We utilized the trial model, which limits key logging consultation time to 40 min.

Meta Split Meter Preterm Key Logger—a key logger constructed into the meter preterm malware payload available through Meta Split. We used Meta Split Version 4.11.1-2015032401 on a far flung Kali Linux 1.1.0 system to create the meter preterm payload. Meta split was then installed to concentrate on incoming meter preterm connections and the malicious payload became hooked up on the test gadget via a USB flash drive. Once the meter preterm changed into mounted it relayed to the far away Meta Split server. We then commenced use of the built-in key logger and were able to view the keystrokes made on the test system on the far away Kali Linux server.

Metasploit JavaScript Key Logger—a browser-based key logger designed to seize keystrokes of users browsing the internet.

- This was a Metasploit module written by Marcus Carey that creates a website with a malicious script that captures the keystrokes of all people visiting the internet site. Once more, the Kali Linux far flung host was used as the server to host this malicious internet site. The web browser utilized in our test was Internet Explorer 11.

3 Proposed Approach

We process aggregator utilization in generation of key logger with prescribed events like data usage between users with real-time data communication. This can be performed, for instance, with the utilization of multi-radio handsets. Additionally, the assailant has online reception apparatus that let the wedding festivity of a sign from one hub and performing of the same sign at another. For exploration reasons, we surmise that an aggressor can effectively key log various pieces just beneath the ECC capacity towards the beginning of transmission. They can then decide to irretrievably harm a visited site by performing the last symbol. It has been affirmed that specific damage can be performed with far less sources. A key lumberjack arranged with a stand out, half-duplex handset will suffice to sort and key log visited bundles. In any case, our outline gets a more successful assailant that can be compelling even at high transmitting rates of rate (Fig. 2).

In the first set of tests, we installed a bandwidth of 3 MB for information files between two customers A and B linked via a multi-hop path. The TCP method was used to effectively transport their requested information files. For the MAC part, RTS/CTS procedure was allowed. The transmitting rate was set to 11 MB per second at each web link. The key logger was placed within the vicinity of one of the advanced trips in the TCP relationship. Four performing techniques were considered: (a) particular performing of collective TCP-ACKs, (b) particular performing

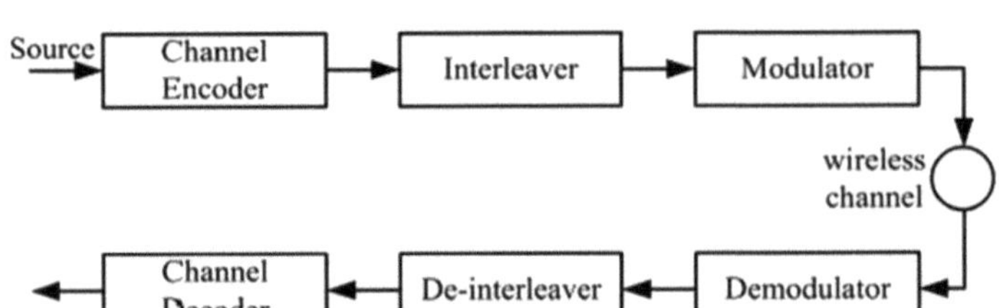

Fig. 2 Generic secure communication systems in cyber networks

of RTS/CTS information, (c) particular performing of information packages, and (d) unique performing of any bundle. In each of the techniques, a portion *p* of the focused packages was packed.

We examined the recent applications for key logger systems with data security via data transmission between users in a network application. Key logger procedure difficulties in transfer control protocol with RTS/CTS in real time protection with proceedings of data users. User's data protection in real time configuration with security acknowledgement in TCP for data communication. Structure of the communication in key-logger system procedure which enables aggregative full duplex enhanced cyber security with communicative information packages in packet data transmission.

4 Experimental Evaluation

In this section we analyze and observe traditional approaches with comparison to our proposed approach with respect to time. We use java framework for data protection in real-time data transmission with load balancing and effective throughput issues in data for wireless sensor networks.

The proposed techniques may achieve real-time data transmission for advanced encryption systems with reliable packet delivery with respect to time. Other configurations may perform efficient data communications. Such functions can be applied in components very efficiently. Symmetric encryption such as AES can be applied at rates of speeds of 10s of Gbps/s when noticed with Program Specific Incorporated Tours (PSITs) or Field Automated Checkpoint Arrays (FACA). These handling rates of speed are purchases of scales higher than the transmitting rates of speed of most current Wi-Fi technological innovations, and hence, do not show any delay. In the test the consumer asked for a 1 MB computer file from a server. We analyzed the effects of bundle concealing by calculating the effective throughput of the TCP relationship in the following scenarios: (a) no bundle concealing (N.H.); (b) MAC-layer security with a fixed key (M.E.); (c) SHCS (C.S.); (d) time-lock CPHS (T.P.); (e) Hash-based CPHS (H.P.); (f) straight line AONT-HS (L.T.); and (g) AONT-HS depending on the program converter (P.T.) (Table 1).

A graphical representation of relative dimensionality with secure accuracy is shown in Fig. 3.

Table 1 Comparison analysis of security accuracy with increasing nodes in cyber networks

Number of nodes	Key logger	Hiding cyber security
1	1	2.4
2	2.2	3.5
3	3.5	4.5
4	4.5	5.4
5	4.8	6.4

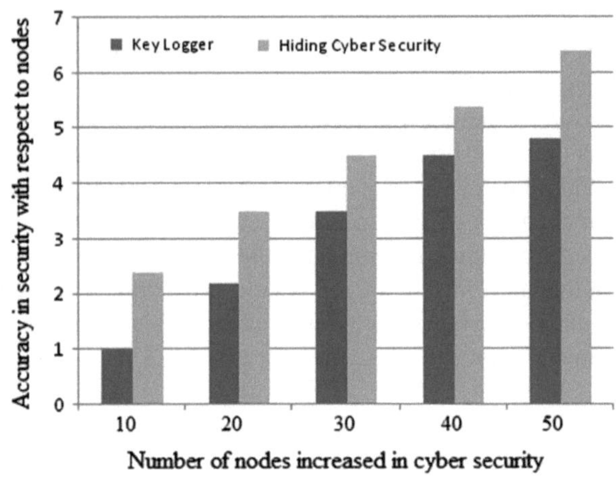

Fig. 3 Graphical representation of security analysis in cyber networks

This is accepted by the moderately small association cost of each hiding technique and the small hold up at cutting-edge remote switches due to the absence of any cross activity streams regarding security that appeared in Fig. 3. The AONT-HS, relying upon the project change over, acquired a slightly lower throughput, since it occurs for every parcel at a cost of 128 pieces instead of 56 pieces required for SHCS. We additionally see that disguising systems, relying upon cryptographic inquiries, diminish the compelling throughput of the TCP relationship by one half, in contrast with the "no covering" circumstance. This effectiveness is normal since much of the time required to settle a test, after a group has gained the MAC part, is equivalent to the transmitting length of every pack. While this contains a diminishing imperative effectiveness, we highlight that cryptographic enquiries are recommended as a candidate arrangement only when image measurement is small, to the point that more compelling covering methods don't give sufficient levels of security.

An aggregate of 20 customer/server sets traded messages of size 1 KB using the TCP strategy, at self-picked start times. The measurements of the messages traded between sets of hubs were kept small to abstain from skewing the street discovering proficiency because of framework blockage.

4.1 Comparison Results with respect to Time

If the number of clients increased in real-time cyber communication systems then the key logger system did not follow an efficient time complexity in node communication. Comparison results shown in Table 2.

This wait is described as plenty of your time difference between the transmitting of the first RREQ from a source and the wedding celebration of the corresponding

Table 2 Time efficiency results in real-time data cyber communication systems

Number of nodes	Hiding cyber security	Key logger system
1	1	2.4
2	2.2	3.5
3	3.5	4.5
4	4.5	5.4
5	4.8	6.4

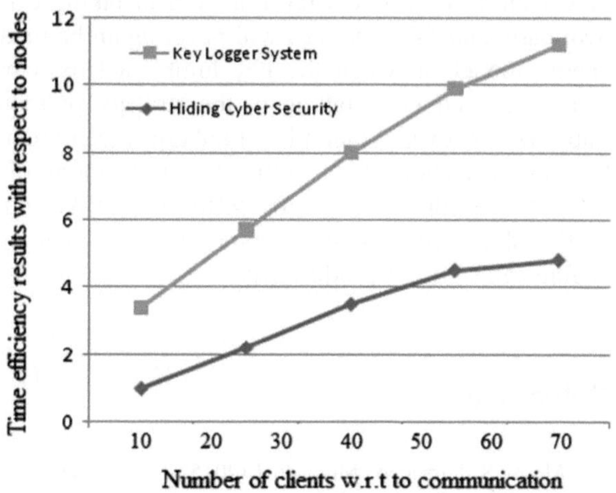

Fig. 4 Time efficiency results with comparison to data communication

RREP from the destination. We notice that the effect of bundle concealing on the road-finding wait was small in comparison to the situation where no bundle concealing was employed in Fig. 4.

We see that powerful packet concealing strategies, for example, SHCS and AONT-HS have a moderately small impact on the general throughput. This is on the grounds that in a swarmed framework, the productivity is generally dependent on the queuing mishaps at the hubs. The communication cost, made by transmitting the pack covering elements, is small, consequently, it does not fundamentally impact the throughput. Then again, for CPHS, we see a productivity loss of 25–30% contrasted with the circumstances requiring no parcel cover up. This decline is identified with the sitting tight made by CPHS for the wedding festivity of every pack. Note that in the swarmed framework circumstance, the throughput loss of CPHS is smaller in contrast to the non-congested one since hubs can exploit the lining difficulties to alter questions.

5 Conclusion

The budgetary administrations business sector throws a spotlight on the critical need for money related organizations to assemble to confront malware strikes thus sparing monetary misfortune, harm to prominence, diminished client assets, research breaks, directing administrations, and/or absence of administration control over creation property.

Financial institutions need to be aware that malware suppliers depend upon a solid monetary market in order that they can exploit criminal movement. They are unrealistic to objective vital arrangement taking care of parts for stress that their own particular fake dealings will never again be readied. We respected an encased enemy model in which the key lumberjack is a piece of the framework under assault, consequently taking in the strategy prerequisites and shared framework traps. We uncovered that a key lumberjack can order visited bundles progressively by comprehension of the initial few indications of a constant transmission. We assessed the impact of specific performing strikes on framework conventions, e.g., TCP and diverting. Our outcomes uncover that a specific key lumberjack can significantly impact productivity.

References

1. Moses S, Larson A, Mercado J (2015) Touch interface and key logging malware. In: 2015 11th international conference on innovations in information technology (IIT)
2. Business Verizon (2014) 2014 data breach investigations report. Verizon Bus J 2014(1):1–60
3. V. E. Solutions (2015) 2015 DBIR contributors. Verizon Bus J (2015)
4. Sagiroglu S, Canbek G (2009) Key loggers: increasing threats to computer security and privacy. IEEE Technol Soc Mag 28(3):10–17
5. Wood CA, Raj RK (2015) Key loggers in cyber security education
6. Aron J (2015) Smartphone jiggles reveal your private data. New Sci 211(2825)
7. Demopoulos D, Kambourakis G, Grizzlies S (2013) From key loggers to touch loggers: take the rough with the smooth. Comput Secur 32:102–114
8. Holz T, Engel Berth M, Freiling F (2009) Learning more about the underground economy: a case-study of key loggers and drop zones. In: Lecture notes in computer science (including subseries Lecture notes in artificial intelligence and Lecture notes in bioinformatics), LNCS, vol 5789, pp 1–18
9. Framework for Improving Critical Infrastructure Cyber security. Version 1.0 National Institute of Standards and Technology, 12 Feb 2014
10. PWC (2015) Managing cyber risks in an interconnected world, Sept 2014
11. ICS-CERT (2014) Alert (ICS-ALERT-11-343-01A)
12. Jacobson M, Ramzan Z (eds) (2008) Crime ware: understanding new attacks and defenses. Safari Technical Books
13. Coffman S (2011) USSS malware update for FS/ISAC, 14 Mar 2011
14. Panda Labs (2010) Panda labs annual report 2010. http://www.pandasecurity.com
15. FS-ISAC (2011) Threat viewpoint, advanced persistent threat
16. Krebs B (2010) 'Stuxnet' worm far more sophisticated than previously thought. KregsOnSecurity.com, 22 Sept 2010

17. Winfield N, Worthier B (2010) Microsoft battles cyber criminals. Wall Street J
18. Profane A, Lazes L (2011) Packet-hiding methods for preventing selective jamming attacks. Proc IEEE Trans Dependable Secure Comput 9(1)
19. Perkins C, Belding-Royer E, Das S (2003) RFC 3561: ad hoc on demand distance vector (AODV) routing. Internet RFCs
20. Popper C, Strasser M, ˇCapkun S (2009) Jamming-resistant broadcast communication without shared keys. In: Proceedings of the USENIX security symposium
21. Rivest R (1997) All-or-nothing encryption and the package transform. Lecture notes in computer science, pp 210–218
22. Rivest R, Shamir A, Wagner D (1996) Time-lock puzzles and timed release crypto. Massachusetts Institute of Technology
23. Schneider B (2007) Applied cryptography: protocols, algorithms, and source code in C. Wiley
24. Stinginess (2010) Break DES in less than a single day. http://www.sciengines.com

Development of Path Loss Models for Localization and Creation of Wi-Fi Map in a Wireless Mesh Test Bed

Moirangthem Sailash Singh, Pramod Jayaram, R.K. Nikshitha, P. Prerna, Meghana Deepak and Viswanath Talasila

Abstract In order to localize mobile nodes, know the correct positions for the placement of static nodes and reduce the time taken to scan the Wi-Fi SSIDs. Wi-Fi path loss models are developed from the setup test bed. The three models can then be used to localize mobile nodes using triangulation method. With the three models, Wi-Fi heat maps have been created. The maps give us complete information for the signal strengths in the given test bed.

Keywords RSSI · OpenWrt · Wi-Fi heat maps · Path loss models · Localization

1 Introduction

A wireless mesh network based on 802.11b/g may consist of both indoor and outdoor environments. The indoor environment is more complex than the outdoor environment because of presence of numerous objects that can scatter, diffract, reflect, and absorb radiation. Moreover, a mesh environment consists of static nodes

M.S. Singh (✉) · P. Jayaram (✉) · R.K. Nikshitha · P. Prerna · M. Deepak · V. Talasila
MS. Ramaiah Institute of Technology, Bangalore, India
e-mail: sailashm@gmail.com

P. Jayaram
e-mail: pramod.j94@gmail.com

R.K. Nikshitha
e-mail: nikshithark@gmail.com

P. Prerna
e-mail: prerna210195@gmail.com

M. Deepak
e-mail: meghanad10@gmail.com

V. Talasila
e-mail: viswanath.talasila@msrit.edu

H.R. Vishwakarma and S. Akashe (eds.), *Computing and Network Sustainability*, Lecture Notes in Networks and Systems 12, DOI 10.1007/978-981-10-3935-5_26

as well as mobile nodes. For the mobile sensor nodes, it is always advantageous to know the position of a mobile mesh node with respect to the static nodes. The existing handoff process in a mesh network when a mobile node moves out of the coverage area of one access point and re-associates with a different access point may consist of four phases viz. Scanning Phase, Choosing the access point, Authentication, and Re-association phase [1]. The delay due to handoff process is mainly contributed by the scanning phase [2]. Moreover, the continuous scanning of the access points when a node tries to join a mesh network introduces overhead. If the delay during the scanning phase can be reduced, the handoff process efficiency can be increased and the overhead can be reduced.

Unlike the wired network, the wireless network employs radio channels at various frequencies, which make it much complicated than the wired network. The geographical layout, the frequency of operation, velocities of the mobile clients, interference sources, and other factors need to be considered while dealing with wireless communication as these factors can bring high differences in the performance of the network. The primary factors that need to be considered for the design and deployment of any wireless communication systems, and how the wireless medium operates are scattering, diffraction, reflection, and transmission [3].

The main focus of this work is to localize mobile nodes in a mesh network, improve the handoff efficiency, and to have a knowledge of which access point has the best signal strength at which location by the help mathematical models.

The experiment consists of three static mesh nodes and one mobile node. The received signal strengths from the three static nodes are noted down for each location of the test bed. Mathematical path loss models are then developed using the readings with respect to the three static nodes.

These models can then be used for localization of mobile nodes in the wireless mesh test bed. The models will determine the best access point at any location in the test bed, thereby eliminating the process of scanning during the handoff process. Moreover, the correct positions for the placements of access points that can maximize the efficiency can also be known.

1.1 Path Loss Models for Pico-Cellular Indoor Areas

Pico-cells refer to the base stations that cover small areas such as offices, restaurants, and shopping malls. Pico-cells often cover area in the span between 30 m and 100 m. It is often employed for WLANs and WMNs. The most common models for pico-cellular are as follows:

Log Distance Path Loss Model [4]: This model predicts the path loss in an indoor environment. The path loss equation is given by

$$P_L = U_L + 10 \cdot n \cdot \log(d) \quad (1)$$

P_L = path loss
U_L = power loss (dB) at 1 m distance (30 dB)
N = path loss coefficient factor
d = distance between transmitter and receiver

The JTC model [5]: This model is an improvement to the above model. In the above model, relation between the number of floors and path loss is linear, which does not comply with most of the experiments. Therefore, the JTC model describes the path loss as a nonlinear function of the number of floors. It is given by the equation:

$$Lp = A + Lf(n) + B \log(d) + X \quad (2)$$

Lf(n) = function defining the nonlinear relation between the path loss and the number of floors, X = log normally distributed random variable representing the shadow fading.

Empirical Channel Model for 2.4 GHz IEEE 802.11 WLAN [6]:

In this paper, a received signal strength equation based on distance was developed in an office environment that somehow resembles the test bed used in this experiment. The model is given by:

$$RSSI = 2.1 * 10 * \log(d) + 7 \quad (3)$$

1.2 Hardware and Software Requirements

Hardware requirements consist of three tp-link routers, one raspberry pi board along with the Wi-Fi adapter. The details are shown in Table 1.

Table 1 Hardware and software requirements

Hardware platform	TP-Link WR1043ND (3 numbers)
Wi-Fi	802.11b/g/n; 2.4 GHz, Tx. Power: 20 dBm
Mobile node	Raspberry Pi B+
Software requirements	OpenWrt Firmware for Raspberry Pi B+ and TP-Link
WR1043ND chipset	QUALCOMM Atheros QCA 9558
WR1043ND RAM, flash	64 MB, 8 MB
WR1043ND WLAN driver	Ath9k

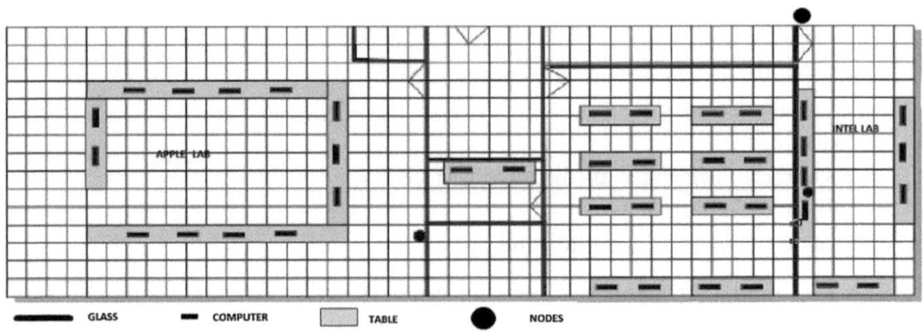

Fig. 1 Topology of the test bed

2 Experimental Setup

The topology for the experimental test bed is shown below:

The entire test bed covering an area of 213.3 m^2 is divided into 15 $\times$ 45 tiles, with each tile having a dimension of 55.6 $\times$ 55.6 cm. The test bed is a large computer room partitioned by three glass walls as shown in Fig. 1. The presence of tables, computers, chairs, etc. adds to the scattering and diffraction effects in the test bed resulting in path loss.

The three tp-link routers are set as static access points with SSIDs Node1, Node2, and Node3 to form a mesh network. They are placed at the locations shown by the black dots in the topology shown in Fig. 1. The mobile nodes are placed at each locations of the tiles.

3 Procedure

All the static nodes as well as the mobile nodes were flashed with openWrt firmware. WiPi, wireless card for the raspberry pi board was used as the transceiver for the pi board.

The test bed consists of 690 tiles. 100 readings per tile with respect to Node1 were taken by placing each mobile mode at each locations of the tiles.

$$\mathrm{R_N} = \frac{\sum_{i=1}^{100} \mathcal{R}i}{100} \tag{4}$$

Here $\mathrm{R_N}$ represents the signal strength received by the mobile node with respect to the N (Node1, Node2, or Node3) static node.

Three heat maps were generated based on the received signal strengths from Node1, Node2, and Node3 (Table 2).

Table 2 RSS color distribution

RSS	Color
If RSS > 65 dB	Yellow
55 dB < RSS < 65 dB	Greenish yellow
45 dB < RSS < 55 dB	Bluish green
35 dB < RSS < 45 dB	Cyan
25 dB < RSS < 35 dB	Light blue
RSS < 25 dB	Blue

The received signal strengths at different locations but equidistant from Node1 were considered and a graph of distance vs RSS was plotted. Same procedure was followed for Node2 and Node3. Mathematical models that fit the curve of the graphs were then developed and compared with the reference model given by Eq. (3).

4 Results

The heat maps that show the variations of received signal strengths from one tile to another with respect to Node1, Node2, and Node3 are shown in Figs. 2, 3 and 4.

With Node1 as center, concentric circles with radii increased by 1 unit are drawn on the entire topology. The tiles covered by the first circle, second circle, till the last circle are marked and their readings are taken down.

Figure 5 shows the variation of received signal strength at every radius with Node1 as center.

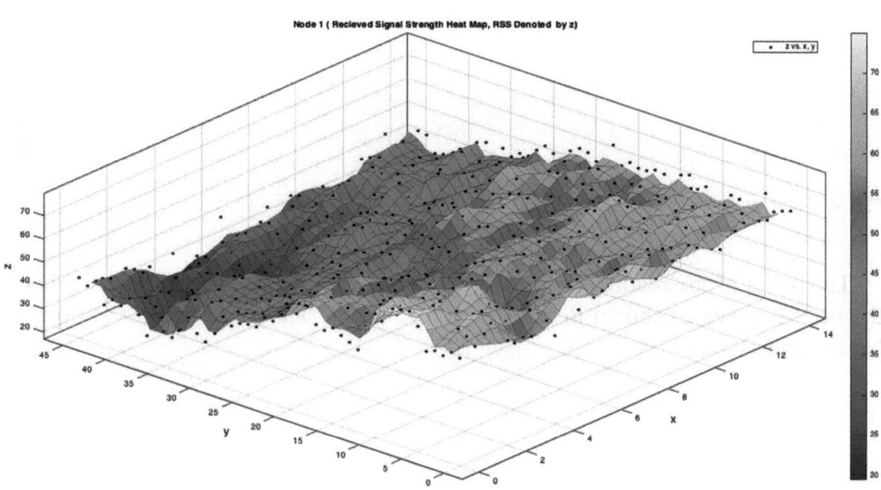

Fig. 2 Heat map with respect to Node1

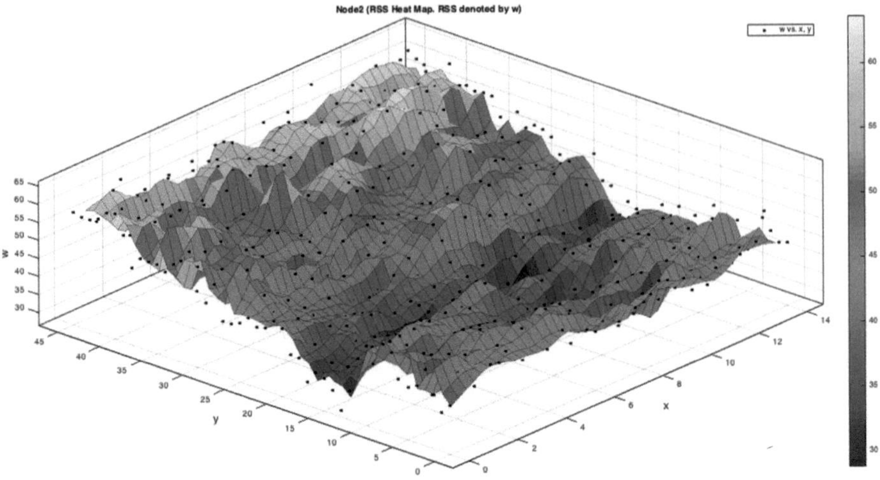

Fig. 3 Heat map with respect to Node2

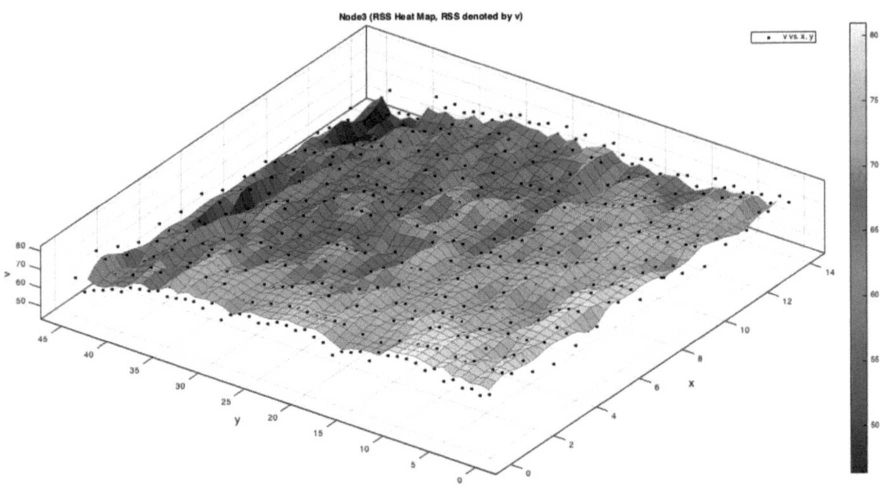

Fig. 4 Heat map with respect to Node3

The mean of the readings for every circle is taken down and is used to generate the best fit curve as shown in the Fig. 6 for Node1. The same process is done for Node2 and Node3.

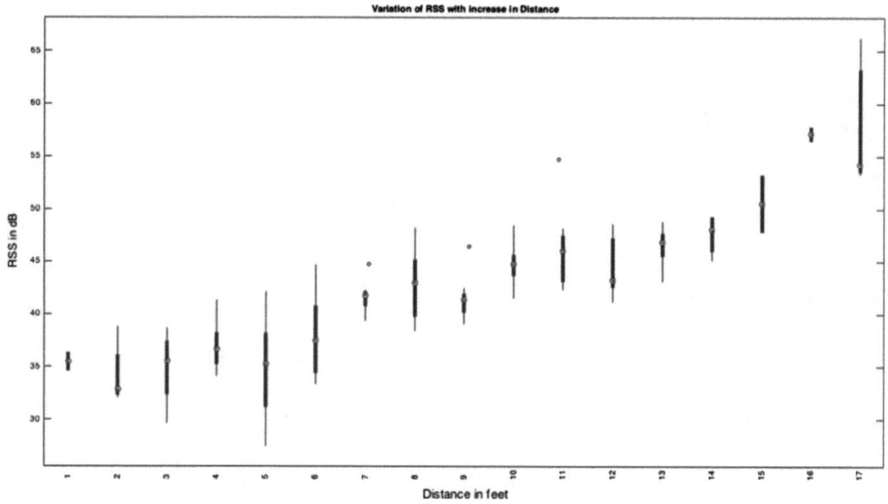

Fig. 5 RSS versus distance variation

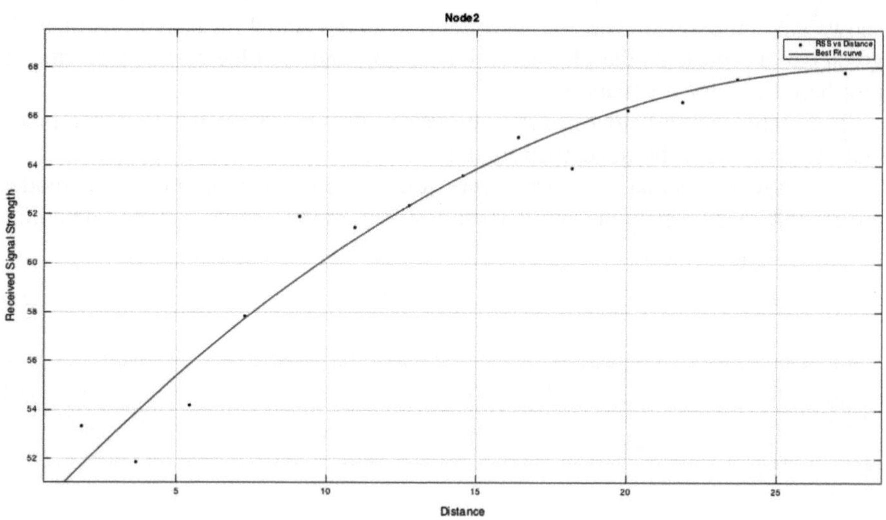

Fig. 6 Curve fit with respect to Node2

5 Mathematical Models

Based on the plots shown above, the following mathematical models were developed using MATLAB with 'd' representing the distance.

- Model with respect to Node1

$$RSS = -0.006511 * x^2 + 0.9588 * x + 27.97 \tag{5}$$

- Model with respect to Node2

$$RSS = -0.0229 * x^2 + 1.305 * x + 49.44 \tag{6}$$

- Model with respect to Node3

$$RSS = -0.0007883 * x^2 + 0.4932 * x + 39.88 \tag{7}$$

In this section, a brief comparison between the developed Eqs. (5), (6), and (7) and the model from reference [1] given by Eq. (3) is discussed. The graph in Fig. 7 shows the comparison. The red curve represents the actual measured readings, the green curve represents the plot based on Eq. (3), and the blue curve represents the plot based on above equations.

The graph shown below in Fig. 8 compares the error from the reference model and the error from the developed model.

The above mathematical models were verified with real experiment. The models are 80–90% accurate. In the above mesh test bed, mobile nodes can then determine

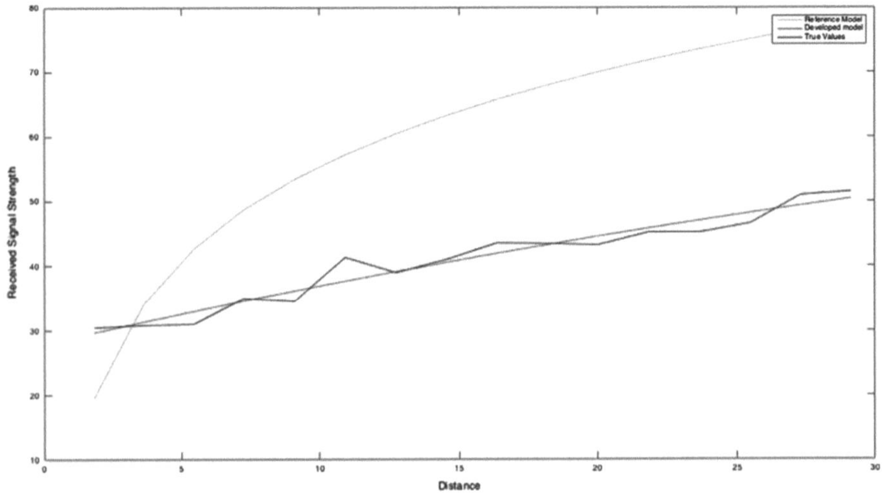

Fig. 7 Comparison between (3) and (4)

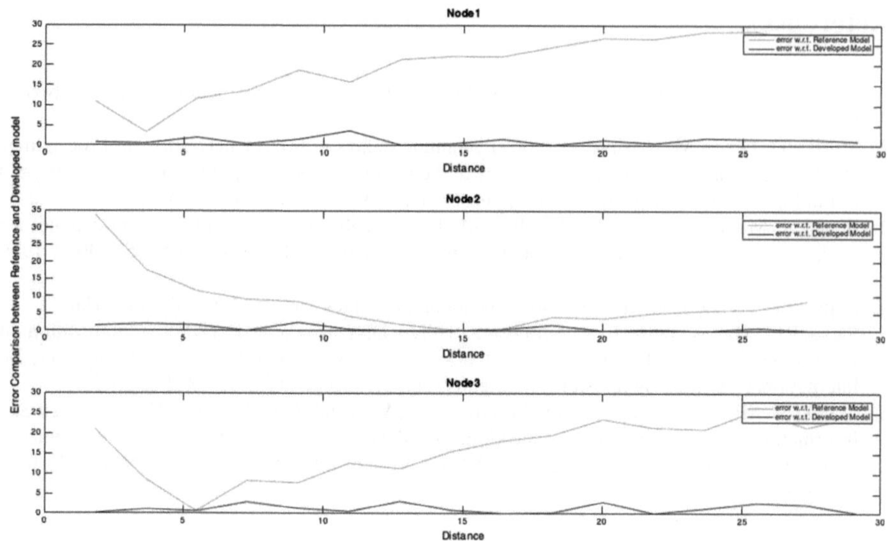

Fig. 8 Error comparison between the reference and developed model

its distance from Node1, Node2, and Node3. The intersection of the arcs drawn with these three distances will give an approximate location of the nodes within the mesh test bed.

In the future, we are developing proactive or table driven mesh protocols that are based on these received signal strength tables. The idea of this protocol is to allow the mobile nodes connect automatically to the static nodes based on the highest signal strength.

6 Conclusion and Future Works

The above mathematical models were verified with real experiment. The models are 80–90% accurate. In the above mesh test bed, mobile nodes can then determine its distance from Node1, Node2, and Node3. The intersection of the arcs drawn with these three distances will give an approximate location of the nodes within the mesh test bed.

In the future, we are developing proactive or table driven mesh protocols that are based on these received signal strength tables. The idea of this protocol is to allow the mobile nodes connect automatically to the static nodes based on the highest signal strength.

Acknowledgements This work in this paper has been partly funded by ABB, under a MoU between MSRIT and ABB.

References

1. Jaydip Sen Tata Consultancy Services. Mobility and handoff management in wireless networks India
2. Murray D, Dixon M, Koziniec T (2007) Scanning delays in 802.11 networks. In: Proceedings of the international conference on next generation mobile applications, services and technologies, 12–14 Sept. IEEE Computer Society, Washington, DC, USA
3. Beneat J, Marku M, Pahlavan K (1999) Modeling of the wideband indoor radio channel for geolocation applications in residential areas P. Krishnamurthy, (0-7803-5565-2/99 0 1999 IEEE)
4. Rappaport TS (2002) Wireless communications principles and practices. Prentice-Hall
5. Safna RF, Manoshantha EJN, Suraweera SATS, Dissanayake MB (2015) Optimization of wireless path loss model JTC for access point placement in wireless local area network. In: International research symposium on engineering advancements 2015 (RSEA 2015)
6. Cebula SL III, Ahmad A, Graham JM, Hinds CV, Wahsheh LA, Williams AT, DeLoatch SJ Information assurance research, education, and development institute (IA-REDI.: Empirical Channel Model for 2.4 GHz IEEE 802.11 WLAN). Norfolk State University (NSU), Norfolk, VA, USA

Clustering Algorithms: Experiment and Improvements

Anand Khandare and A.S. Alvi

Abstract Clustering is data mining method to divide the data objects into n number of clusters. Clustering algorithms can be used in domains such as e-commerce, bio-informatics, image segmentation, speech recognition, financial analysis, and fraud detection. There is abandon knowledge in the clustering research and applications and also various improvements are done on various clustering algorithms. This paper includes the study and survey of various concepts and clustering algorithms by experimenting on it on some data sets and then analyzed gaps and scope for enhancement and scalability of algorithms. Then improved k-means is proposed to minimize these gaps. This improved algorithm automatically finds value of number of clusters and calculates initial centroids in better way rather random selection. From the experimentation, it is found that numbers of iterations are reduced; clusters quality increased and also minimized empty clusters in proposed algorithm.

Keywords Data object · Data mining algorithm · Clustering algorithm · K-means

1 Introduction

Extraction of hidden data and knowledge form large data is called data mining [1]. The clustering is unsupervised method of grouping data objects into the cluster which are more similar to each other. It is a main task of data mining and used for data analysis in different domains listed in the survey. This paper provides the study and analysis of the various modern clustering.

A. Khandare (✉)
Department of CSE, SGB Amravati University, Amravati, India
e-mail: anand.khandare1983@gmail.com

A.S. Alvi
Department of CSE, PRMIT & R, Badnera, Amravati, India
e-mail: abrar_alvi@rediffmail.com

H.R. Vishwakarma and S. Akashe (eds.), *Computing and Network Sustainability*, Lecture Notes in Networks and Systems 12, DOI 10.1007/978-981-10-3935-5_27

1.1 Types of Clustering Algorithms

There are various types of clustering algorithms [1–6]. First is, partitioned method, in which, input data objects are given and partitioned into a number of clusters. In this, relocation objects are done by moving data from one cluster to other clusters from an initial partitioning. So due to the checking of all possible solutions, this method may suffer from computational complexity problems. Also this method requires that number of clusters should be given by the user at the beginning only. The k-means and k-medoids/PAM belong to this type of clustering. Second type is grid based, in which a cluster is converted into a number of cells or grids. Then it combines grids together to form grid format. Grid is used to perform all clustering operations. Number of grids will be the time required to perform clustering. This method is not sensitive to number of data objects. The STING and CLIQUE belong to this category. Third type is hierarchical clustering, in which, hierarchy of clusters is formed to group the data objects using some standard criteria. Dendrogram data structure is used to represent, how clusters are related to each other. There are two main types of it, first is agglomerative clustering, in which, object represents a cluster and then clusters are merged until the required cluster structure is formed. Second type is divisive clustering in which, all objects represent cluster, and then the cluster is divided into subclusters. Process continues until the required clusters are formed. The CURES and BIRCH are the examples this type. In the fourth type, Density-based clustering, high dense regions of input data are merged to form required clusters. This method scans complete data in one scan and also handles the noisy data. Examples are DBSCAN and OPTICS. Next type is Model based which finds descriptions of each cluster, based on some mathematical model. Decision trees, neural networks, EM, and SOM are the examples. Last type is soft computing, and it is an advanced clustering method in which fuzzy logic is used to cluster data objects.

1.2 Applications, Requirements and Problems of Clustering Algorithms

Applications of clustering algorithms

- For market-basket research, recognition of patterns, large data analysis, processing images, and for multimedia [7–9].
- Clustering methods help to discover distinct clusters of customers or people [10].
- It is also used to create animal/plant taxonomies and genes categorization [11].
- Clustering also help in earth observation database for land identification.
- In software, web and wireless sensors network, clustering can be used [7, 8].
- Clustering is used in bank applications such as credit card fraud detection and risk analysis [1].
- Clustering is also used in decision making of new product development and recognition system [12, 13].

Requirements of clustering algorithms

- Algorithms must be highly scalable to deal with large- and high-dimensional data [14, 15].
- Algorithms must handle all types of data [16].
- Algorithms must handle data with noise or missing value [1].
- Algorithms must accurately find required number of clusters and initial centroids [4, 5, 8, 11].
- Algorithms must able to produce clusters with quality and efficiency [2, 16, 17].

Problems of clustering algorithms

- No algorithm addresses all the requirements [17].
- There are chances to increase complexity when dealing with large and dimensions data items [14, 15, 17].
- Process of clustering highly depends upon initial centroids, value of k, and the distance measures [10, 11, 17, 18].
- Algorithms are also sensitive to noisy data [1, 17].
- Single algorithm cannot handle all types of data.

2 Related Work

This paper [14] presents method to characterize attributes of the big data and proposes the novel techniques for processing the data using data mining. Paper suggests that clustering algorithms need to be enhanced for big data applications. Paper [19] presents a self-optimal clustering, and it is the advanced version of enhanced mountain clustering method. The proposed clustering is done with major changes and modifications in old algorithm. Quality of results may be improved further. In paper [4], implementation of Lloyd's k-means clustering was presented. This algorithm is also known as the filtering method. It is easily implemented by using a KD tree data structure. There is a scope to improve overall performance of proposed algorithm. In paper [5], method is suggested to use multiple values of k if quality clustering results are required. But this method is computationally expensive. The Quad tree-based k-means algorithm [8] is applied for finding the initial cluster centroids in software fault detection application. The clustering gain is used to determine the quality of clusters formed. Paper [15] introduces efficient method for clustering of large data sets. This is known as the sketch and validate techniques, and it consists of two algorithms that depend on k-means in per iteration on the reduced number of dimensions. The first operates in a batch fashion and second in a sequential way with maintaining computational efficiency. But the performance of algorithm can be improved further. In [20] paper, new clustering method is proposed to handle the spatial similarity between node readings in underwater sensor network. Also, two-tier data aggregation technique is proposed. Paper [21] proposed the coding aided k-means algorithm for the blind transceiver for the space shift, keying multiple input, multiple output systems to minimize detection complexity of receiver.

Paper [22] proposes a new clustering technique using in the quaternion domain representation for qualitative classification of electronic nose data. The proposed technique is similar to the basic k-means algorithm. In this technique, pools of cluster centers are created by subjecting the center to a fixed rotation in quaternion domain space. There is a further scope for reducing the number of computations.

Paper [23] presents cluster analysis to help the researchers to develop profiles of learners for accessing tasks and information in education. In this, author has used hierarchical clustering and k-means clustering. Paper [24] proposes the fault detection clustering algorithm which is based on the incremental clustering technique. This algorithm finds wafer faults in class distribution and also efficiently processes large sensor data with minimum storage. Paper [18] presents new clustering algorithm, like k-means, called enhanced k-means algorithm. Paper [10] proposes modified k-means clustering algorithm to effectively mine emotional intelligence data. This algorithm can be enhanced again. In paper [25], adaptive fuzzy k-means clustering algorithm for the operation on images like segmentation is purposed. It is based on fuzzy belonging, adaptive and iterative concepts to obtain the optimum value of centroids to segment the image in better way. But there is no enhancement in the processing time of algorithm. Paper [26] presents the extensions of k-means by including intracluster compactness and intercluster separation. The new algorithms are extended k-means (Ek-means), extended weighting k-means (EW k-means) and extended attribute weighting k-means (E AWA k-means). But the complexity of the proposed algorithms is more and improvements are required when used for other applications.

3 Experimentation

3.1 Standard K-means

The k-means is simple unsupervised type of clustering algorithm used for the various applications studied in above survey. The k-means is divided into three steps [10] and its working is as follows:

1. Select initial mean or centroids from the input data objects.
2. Then assign data objects into the cluster based on closest distance.
3. Refine centroids by repeating step 2 and 3 up to no change in clusters.

Random selection creates bad quality clusters, also empty clusters. The effectiveness of an algorithm is also depends on distance measures used. Various methods are proposed in above literatures to enhance the accuracy and efficiency of the k-means clustering algorithm. This paper is presenting algorithm to find value of k and method for finding the proper initial centroids for assigning data to required clusters with less time complexity and more clusters quality.

3.2 Improved K-means

From above survey, it is observed that proposed methods of selecting the number of clusters and initial means have further scope for improvements. A new method to decide value of k in k-means algorithm is proposed in this paper. This method also calculates the initial centroids using arithmetic mean method instead of random selection. Due to this, the quality and number of iterations required for creating final clusters has improved to a great extent. The proposed algorithm is tested on simple numerical numbers of various ranges and it is found that algorithm is scalable and also efficient. The algorithm consists of three phases given as follows:

Phase 1: *Deciding value of K (Number of clusters):*
Input: n data objects
Output: value of k

1. Find distinct number digits in each data object(parameter-1)
2. Find range of data objects in input(parameter-2) and apply k=Roundup(Max/2)
3. If number of digits in data deviates then calculate difference (parameter-3) of first and second parameters.
4. Find number of digits after decimal point if input is real number

Phase-2: *Initial Clusters*:
Input: n data objects and k
Output: initial clusters

5. Read Input array
6. Split up array into sub array using s= n/k where s is clusters, n is elements and k is number of clusters.
7. Use sub arrays as a initial clusters

Phase-3: *Final Clusters*:
Input: Initial Clusters
Output: Final Clusters

8. Compute centroids based on distance between remaining data objects.
9. Check if distance of data object less or equal distance then it will remains in the same cluster otherwise they will be moved to required clusters.
10. Repeate steps 7 and 8 until there is no change in clusters.

Working of improved k-means is shown in Fig. 1.

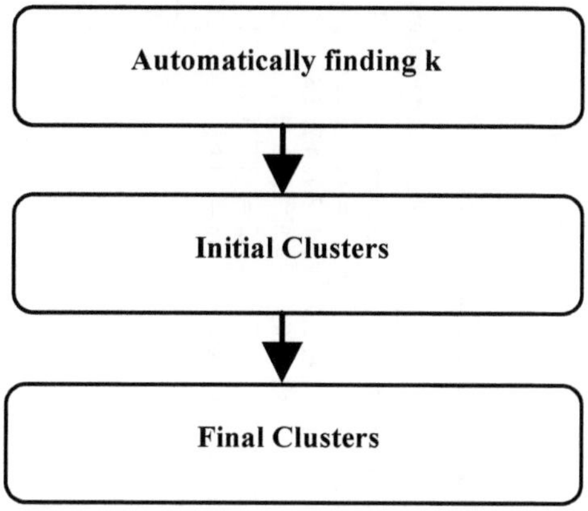

Fig. 1 Improved K-means

Table 1 Working of algorithms

Algorithm	Number of iteration required							Final clusters
Standard K-means **K = 3** D = {1.7, 1.3, 1.6, 2.9, 3.1, 1.4, 3.7, 4.2, 2.8, 4.8, 3.5, 2.3} Let, initial centroids, M1 = 1.7 M2 = 1.3 M3 = 1.6	Iteration 1: A1 = {1.7, 2.9, 3.1, 3.7, 4.2, 2.8, 4.8, 3.5, 2.3} A2 = {1.3, 1.4} A3 = {1.6}	Iteration 2: A1 = {2.9, 3.1, 3.7, 4.2, 2.8, 4.8, 3.5} A2 = {1.3, 1.4} A3 = {1.7, 1.6, 2.3} So, M1 = 3.6 M2 = 1.4 M3 = 1.9	Iteration 3: A1 = {2.9, 3.1, 3.7, 4.2, 2.8, 4.8, 3.5} A2 = {1.3, 1.4, 1.6} A3 = {1.7, 2.3} So, M1 = 3.6 M2 = 1.5 M3 = 2.0	Iteration 4: A1 = {2.9, 3.1, 3.7, 4.2, 2.8, 4.8, 3.5} A2 = {1.3, 1.4, 1.6, 1.7} A3 = {2.3} So, M1 = 3.6 M2 = 1.5 M3 = 2.3	Iteration 5: A1 = {3.1, 3.7, 4.2, 4.8, 3.5} A2 = {1.3, 1.4, 1.6, 1.7} A3 = {2.3, 2.8, 2.9} So, M1 = 3.9 M2 = 1.5 M3 = 2.7	Iteration 6: A1 = {3.7, 4.2, 4.8, 3.5} A2 = {1.3, 1.4, 1.6, 1.7} A3 = {2.3, 2.8, 2.9, 3.1} So, M1 = 4.1 M2 = 1.5 M3 = 2.8	Iteration 7: A1 = {3.7, 4.2, 4.8, 3.5} A2 = {1.3, 1.4, 1.6, 1.7} A3 = {2.3, 2.8, 2.9, 3.1}	A1 = {3.7, 4.2, 4.8, 3.5} A2 = {1.3, 1.4, 1.6, 1.7} A3 = {2.3, 2.8, 2.9, 3.1}
Improved K-means Here we have considered parameter3 K = 4.8/2 = 3 **K = 3** D = {1.7, 1.3, 1.6, 2.9, 3.1, 1.4, 3.7, 4.2, 2.8, 4.8, 3.5, 2.3} **s = 12/3 = 4** number of elements in each initial cluster	Iteration 1: A1 = {1.3, 1.6, 1.7, 2.9} A2 = {1.4, 3.1, 3.7, 4.2} A3 = {2.3, 2.8, 3.5, 4.8} So, M1 = 1.7 M2 = 3.1 M3 = 3.4	A1 = {1.3, 1.4, 1.6, 1.7, 2.3} A2 = {2.8, 2.9, 3.1} A3 = {3.5, 3.7, 4.2, 4.8} So, M1 = 1.7 M2 = 2.9 M3 = 4.1	Iteration 3: A1 = {1.3, 1.4, 1.6, 1.7, 2.3} A2 = {2.8, 2.9, 3.1} A3 = {3.5, 3.7, 4.2, 4.8}	Iteration 4: A1 = {1.3, 1.4, 1.6, 1.7, 2.3} A2 = {2.8, 2.9, 3.1} A3 = {3.5, 3.7, 4.2, 4.8}				A1 = {1.3, 1.4, 1.6, 1.7, 2.3} A2 = {2.8, 2.9, 3.1} A3 = {3.5, 3.7, 4.2, 4.8}

4 Experimental Results

For the experimentation, this paper implemented basic k-means and improved k-means and performed experimentation on it by using some numerical data sets. Sample results of both algorithms are shown in Table 1. For comparisons, this paper is using various parameters such as sum squared error (SSE), number of iterations, quality of clusters, empty cluster, and the applications. Comparison statistics is shown in Table 2.

It can be seen from Table 2 and Fig. 2, the performance of improved k-means with respect to number of iterations and SSE is better than standard k-means. The number of iterations in standard k-means is 7 whereas in improved k-means is 4 for the same data sets. A good algorithm should have minimum SSE value. It can be seen that SSE for k-means is 1.7 and for proposed algorithm is 1.4, i.e., minimum. Quality of cluster is increased and empty clusters are also minimized in proposed algorithm.

Table 2 Comparison of algorithms

Algorithm	Iterations	(SSE)	Empty clusters	Quality of clusters	Applications
Standard K-means	7	1.9	Found	Not good	Used for small data
Improved K-means	4	1.5	Not found	Increased	Used for large data

Fig. 2 Results analysis of algorithms

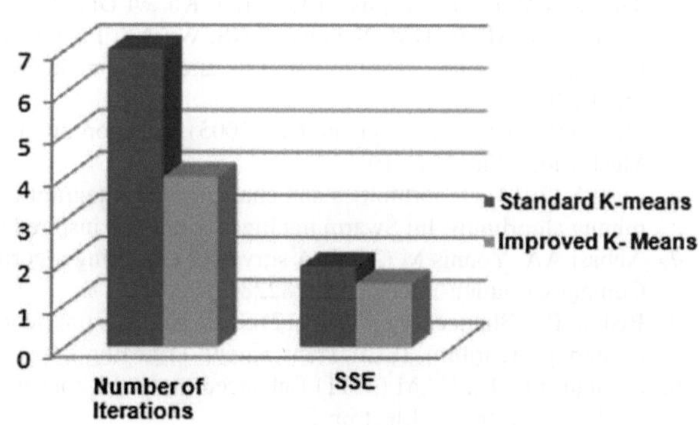

5 Conclusion and Future Work

This paper presents survey and experimentation of clustering algorithms along with its applications in various domains such as medical, software, web, fraud detection, decision making in business. From the survey research gaps and requirements of algorithms are identified. It is found that standard k-means algorithm does not always produce good results as the accuracy of the final clusters depends on proper initial centroids. Selection of random centroids leads to vary time complexity and make inefficient algorithm for large data. Then k-means algorithm is modified accordingly and applied to some data set. This modified k-means algorithm has systematic and efficient method for finding initial centroids and assigning data points to clusters. Using this method, process of clustering is done in lesser iterations without sacrificing the accuracy of clusters. The future works of this paper is like further improvements in k-means and apply it in domains such as search engine optimization, selecting other clustering algorithms, performs experiments, and modify accordingly.

References

1. Dunham MH (2006) Data mining-introduction and advanced concepts. Pearson Education
2. Aggarwal CC, Zhai C (2012) Survey on text clustering algorithms in mining text data. Springer, USA, pp 77–128
3. Mahmood A, Leckie C, Udaya P (2007) An efficient clustering scheme to exploit hierarchical data in NW traffic analysis. IEEE Tran. Knowl Data Eng 20(6):752–767
4. Kanungo T, Mount DM, Netanyahu NS, Wu AY, Piatko CD, Silverman R (2002) An efficient k-means clustering algorithm-analysis and implementation. IEEE Trans Pattern Anal Mach Intell 24(7)
5. Pham DT, Dimov SS, Nguyen CD (2005) Selection of value of k in k-means clustering. Proc Mech Mech Eng Sci 219
6. Fong S (2013) Opportunities and challenges of integrating bio inspired optimization and data mining algorithms. In: Swarm intelligence and bio inspired computation. Elsevier, pp 385–401
7. Abbasi AA, Younis M (2007) A survey of clustering algorithms for wireless sensor networks. Comput Commun 30(14), 15, 2822841
8. Bishnu PS, Bhattacherjee V (2012) Software fault predictions using quad tree based k-means clustering algorithm. IEEE Trans Knowl Data Eng 24(6)
9. Siddiqui FU, Isa NAM (2011) Enhanced moving k-means algorithm for image segmentation. IEEE Tran Consum Electron 57(2)
10. Khandare AD (2015) A modified k-means algorithm for emotional intelligence mining, ICCCI-15, Coimbatore, India, pp 1–3
11. Harrison R, Zhong W, Altun G, Tai PC, Pan Y (2005) Improved k-means clustering algorithm for exploring local protein sequence motifs representing common structural property. IEEE Trans Nanobiosci 4(3)
12. Jaber H, Marle F, Jankovic M (2015) Improving the collaborative decision making in the new products development project using clustering algorithm. IEEE Trans Eng Manag 62(4)
13. Li T-HS, Kao M-C, Kuo P-H (2016) Recognitions system for the home service related sign languages using entropy based kmeans algorithm and the ABC based HMM. IEEE Trans Syst Man Cybern Syst 46(1)

14. Wu X, Zhu X, Wu G-Q, Ding W (2014) Data mining on big data. IEEE Trans Knowl Data Eng 26(1)
15. Traganitis PA, Slavakis K, Giannakis GB (2015) Sketch and validate big data clustering. IEEE J Sel Top Signal Process 9(4)
16. Khandare A, Alvi AS (2016) Survey of improved k-means clustering algorithms-an improvements, shortcoming and scope for further enhancement and scalability, INDIA-2016, vol 434. AISC Springer, pp 495–503
17. Xu R, Wunsch D II (2005) Survey of clustering algorithms. IEEE Trans Neural Netw 16(3)
18. AM Fahim, AM Salem, FATorkey, M.A. Ramadan (2006) An efficient enhance kmeans clustering algorithm. J Zhejiang Univ Sci 7(10):1626–1633
19. Verma NK, Roy A (2014) Self optimal clustering techniques using optimized threshold function. IEEE Syst J 8(4)
20. Harb H, Makhoul A, Couturier R (2015) Enhanced k-means, ANOVA based clustering approach for similarity aggregation in underwater wireless sensor networks. IEEE Sens J 15(10)
21. Liang H-W, Chung W-H, Kuo S-Y (2016) Coding aided k-means clustering blind transceiver for space shift keying mimo system. IEEE Trans Wirel Commun 15(1)
22. Kumar R, Dwivedi R (2016) Quaternion domain *k*means clustering for the improved real time classification of E-Nose data. IEEE Sens J 16(1)
23. Antonenko PD, Toy S, Niederhauser DS (2012) Using cluster analysis for the data mining in educational technology research R&D
24. Kwak J, Lee T, Kim CO (2015) Incremental clustering algorithm based fault detection algorithm for class imbalanced process data. IEEE Trans Semicond Manuf 28(3) (Yonsei University, Seoul, Korea)
25. Sulaiman SN, Isa NAM (2010) Adaptive fuzzy k-means clustering algorithm for image segmentation. IEEE Trans Consum Electron 56(4)
26. Huang X, Ye Y, Zhang H (2014) Extensions of k-means type algorithms: a new clustering framework by integrating intra cluster compactness and inter cluster separation, IEEE Trans Neural Netw Learn Syst 25(8)
27. Xie M, Cui H, Cai Y, Huang X, Liu Y (2014) Cluster validity index for adaptive clustering algorithms. IET Commun 8(13)
28. Bandyopadhyay S, Coyle E (2003) An energy efficient hierarchical clustering algorithm for wireless sensor networks. In: Proceedings of the 22 annual joint conference, IEEE computer and communication societies, San Francisco, California
29. An F, Mattausch HJ (2013) k-means clustering algorithm for multimedia application with flexible hardware and software co-design. J Syst Archit 59(3) (Elsevier)

Urban Traffic State Estimation Techniques Using Probe Vehicles: A Review

Vivek Mehta and Inderveer Chana

Abstract Accurate and economical traffic state estimation is a challenging problem for future smart cities. To curb this problem, fixed roadside sensors are used for traffic data collection traditionally, but their high costs of installation and maintenance has led to the use of probe vehicles or mobile phones containing GPS-based sensors as an alternative cost-effective method for traffic data collection. However, the data collected by the latter method are sparse because the probe vehicles are very randomly distributed over both time and space. This survey paper presents state-of-the-art techniques prevalent in the last few years for traffic state estimation and compares them on the basis of important parameters such as accuracy, running time, and integrity of the data used. The dataset used for the implementation of techniques comes from probe vehicles such as taxis and buses of cities such as San Francisco, Shanghai, and Stockholm with different sampling rates (frequencies) of probes. Finally, it represents the challenges that need to be addressed along with the possible data processing solution.

Keywords Traffic state estimation · Probe vehicles · Data segmentation

1 Introduction

The present demand of traffic state estimation for smart cities requires real-time traffic information which is complete enough and covers a large area of the city. The ultimate aim of getting this information is to build a traffic monitoring system that can be used for the following: 1. Better trip planning. 2. Traffic management. 3. Urban roads and highways engineering, and 4. Urban infrastructure planning.

V. Mehta (✉) · I. Chana
Computer Science and Engineering Department, Thapar University,
Patiala 147004, India
e-mail: vivek_mehta90@yahoo.com

I. Chana
e-mail: inderveer@thapar.edu

H.R. Vishwakarma and S. Akashe (eds.), *Computing and Network Sustainability*,
Lecture Notes in Networks and Systems 12, DOI 10.1007/978-981-10-3935-5_28

Already ongoing approaches for the intelligent transportation systems rely on the real-time data collected by the means of fixed roadside sensors, e.g., loop detectors and video cameras, to detect the traffic state variables such as density, average speed, and travel time [1–3]. However, due to high costs involved in their deployment and maintenance, the authorities cannot cover the entire road network with these devices. Also, as they are fixed they measure only the speed on the spot of their location thus a wider view of the speed is not reflected.

An alternate approach to this static kind of approach is to use GPS receivers which are now coming as embedded in vehicles and mobile phones. This has enabled to get the location and speed dynamically on their path of movement (although it is random). The speed and location updates of the vehicles are known as probes; thus, the corresponding vehicles are known as probe vehicles (taxis, buses, ambulances, etc.). These data probes can be transmitted over a cellular network such as GSM/GPRS to a monitoring center for traffic estimation. As these vehicles cover the entire city, the data are collected over a large area, and due to the low cost of onboard GPS receivers, the overall cost of the system is low. However, this new approach faces some challenges. This paper presents a survey of the techniques which have been used to overcome these challenges. One of the challenges is that the distribution of the vehicles is uneven and incomplete in-between the whole space and time. Figure 1 shows the complete life cycle of data in an intelligent transportation system. As it is shown, firstly the probe measurements are taken with the help of GPS-based mobile phones or probe vehicles. After the collection of raw data, pre-processing techniques such as re-sampling, coordinate

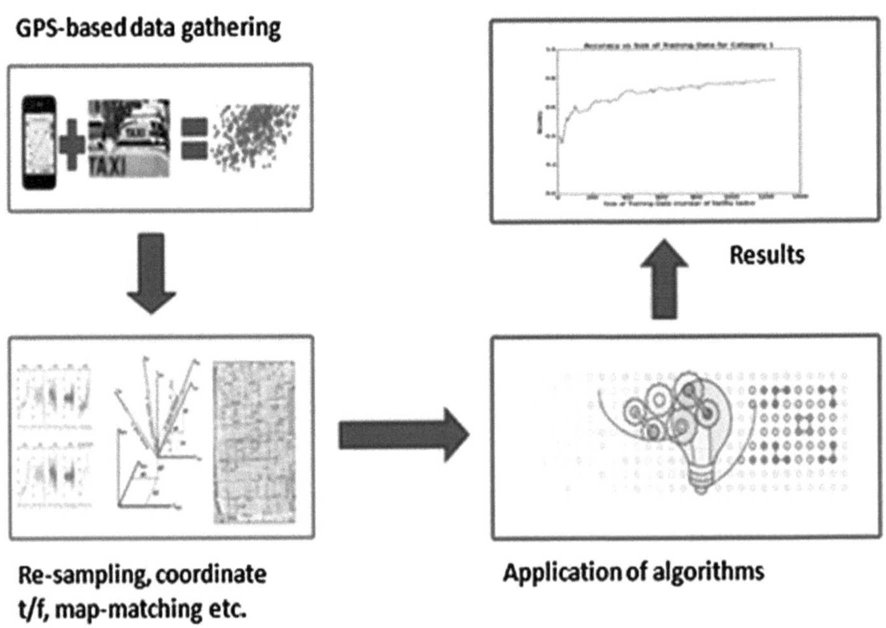

Fig. 1 Data flow in an ITS

transformation, map-matching are applied for suitable application of prediction/ estimation algorithms. Coordinate transformation and map-matching are used to represent the collected GPS measurements on a digital map precisely. Re-sampling of data may also be required in case the data are not real and is thus collected in a simulation framework. After this step, one of the different categories of algorithms such as statistical, Bayesian, and machine learning models are applied to get the estimation of traffic state.

In this survey paper, we present a comprehensive study of all the techniques proposed up till now for the estimation of traffic state. This is done by comparing them on the basis of the attributes such as accuracy, running time, sampling frequency of the dataset, and number of vehicles.

2 Survey Approach

This survey has been done with an objective to describe the current state of the art in intelligent transportation research using GPS-based probe vehicles for traffic data collection. This has been done by covering the following aspects.

1. Introduction to GPS-based traffic monitoring system and its advantages over traditional fixed sensor-based technology.
2. Characteristics of a traffic monitoring system.
3. Different approaches that have been used for traffic estimation in the past few years.
4. Challenges that need to be handled in this technology.
5. Existing projects that have implemented this technology.

2.1 Sources of Information and Search Criteria

The research papers used to perform this survey mainly comes from database IEEE explore along with other databases such as Sciencedirect (www.sciencedirect.com) and Taylor and Francis (www.tandofline.com).

Search criteria involved using relevant combinations from a set of strings such as GPS-based techniques, traffic state estimation, urban traffic monitoring, and probe vehicles.

3 Brief Review of Existing Approaches

In the past few years, several approaches have been proposed to measure the traffic state variables, for example Bayesian, compressive sensing, aggregation, curve-fitting, statistical, neural-network, and K-NN. They are summarized as follows:

3.1 Bayesian Network Approach

In this approach [4], firstly an expectation-maximization approach is used to learn about the state variables of traffic congestion from a historically available large dataset. Traffic state prediction is done in real time with streaming data. Historical training provides more robustness to the model. In order to make an improvement over another approach of *scaling partial link travel time in proportion with length of the partial links,* it [4] uses density modeling to estimate partial link travel times from link travel times.

3.2 Compressive Sensing Approach

This algorithm is inspired by the idea that applying the method of *principal component analysis* discovers some hidden structures in a large dataset of probe data [5]. Now using compressive sensing, these internal structures are exploited for traffic state estimation. This algorithm has been claimed to outperform KNN and MSSA.

3.3 Curve-Fitting-Based Method and Vehicle-Tracking-Based Method

These two algorithms have been simultaneously proposed by a single author [6]. To support the traffic state estimation of the algorithms, a new method to construct an exact GIS-T digital map has been proposed. On comparing the two algorithms with each other, it has been found that the vehicle-tracking-based method provides higher accuracy but takes more than double the time taken by curve-fitting method.

3.4 Statistical Model

In this model, the total travel time on the road network is considered as the sum of travel time on links and delay at traffic signals and intersections [7]. Trip conditions such as time of the day, season, and weather conditions, and the network characteristics are taken into account by expressing the mean and variance of link travel times and turn delays as functions of explanatory variables in combination with fixed effects for groups of segments.

3.5 *Artificial Neural Network*

An artificial neural network of 3 layers has been proposed in [8] which takes vehicle's position, link id's time stamps, and speed as the input. In it, the model has been compared with Hellinga's model and has been found to perform better than that probably due to higher number of parameters used than those of Hellinga's model. Mean absolute percentage error was less than 6% and is inversely proportional to traffic demand. However, in [8], real GPS data were not used.

4 Comparison and Analysis of Results of Different Approaches

Table 1 shows the complete summary of results achieved by different techniques applied for traffic state estimation. In total, 7 different techniques are compared on the basis of 7 different parameters. The important parameters are explained as follows:

1. *Accuracy*: It is defined as:

$$\mathrm{Ac} = 1 - \mathrm{MAPE}$$

where MAPE is mean absolute percentage error and is represented as [6]:

$$\frac{1}{N}\sum_{n=1}^{N}\frac{|\overline{vn} - \overline{va}|}{\overline{va}}$$

Here, N is the total number of estimated values during the experiment, $\overline{ve}$ is the estimated value of the mean speed and $\overline{va}$ is the actual value of the mean speed.

2. *Running Time*: It is the time taken by the algorithm for one cycle of estimation. A cycle is the interval of time for which a new traffic state is estimated each time it gets elapsed (time granularity), for example a new estimation can be made after each interval of 4 min or it may be an interval of 15 min.
3. *Temporal Integrity*: This parameter defines the percentage of the total number of time intervals for which GPS sample points appeared for each link of the network on an average. It is worth mentioning here that the actual temporal integrity can vary from as low as '5%' for some of the links to as high as '90%' for the other links.
4. *Sampling Interval*: It is defined as the time elapsed between two consecutive probe reports. A probe report is the speed and location update sent by the probe vehicle to the monitoring center.
5. *Time Granularity*: The traffic state from the collected data is time when a particular interval of time is elapsed. This time interval is known as the time granularity, for example state can be estimated after each interval of 5 min or after each interval of 15 min.

Table 1 Comparison of techniques used for traffic estimation

Method	Accuracy (%)	Running time (sec)	# links, # nodes	#Probe vehicles	Temporal integrity (%)	Sampling interval	Time granularity (min)
Curve fitting	75.42	34	150000, 17000	16000	68	NA	4
Vehicle tracking	83.08	71	15000, 17000	16000	68	NA	4
Compressive sensing	75	0.827	221, NA	4000	60	30 s–several minutes	15
Naïve KNN	55	<0.1	221, NA	4000	60	30 s–several minutes	15
Correlation-based KNN	40	<0.1	221, NA	4000	60	30 s–several minutes	15
Bayesian	85–90	NA	815, 527	500	NA	1 min	15
ANN	95	NA	NA	Simulated data were used	NA	1 min	NA

In general, the major features of a short-term forecasting system are as follows (originally described by Eleni et al. in [9] and further used by Soufiene et al. in [10]).

1. *Determination of the Scope*: It relates to determining that whether our forecasting model should be implemented as a part of traffic management system (TMS) or a traveler information system (TIS) along with the area of implementation (e.g. freeway, highway, and urban arterials).
2. *Conceptual specification of the output*: Here, we specify the size of the horizon and the step. The forecasting horizon denotes the extent of the time ahead which the traffic state has been predicted for. The forecasting step is the actual time interval upon which traffic state is forecasted and thus gives the frequency of prediction in the forecasting horizon. So, intuitively the larger is the forecasting horizon, lesser will be the accuracy of the model. The shorter is the forecasting step used, the more difficult will be to predict. A 15-min interval of time has been indicated as the best interval by the Highway Capacity Manual (2000).
3. *Methodology used for the modeling of data*: An appropriate selection of the methodology should be made for traffic forecasting. In 1990s, ARIMA models were applied to forecast parameters such as urban traffic volume, bottleneck formulation in a freeway, but ARIMA models have a tendency to concentrate just on means leaving the extremes. Over the last decade, techniques such as artificial neural network, nonparametric regression are being widely used.

The results presented in Table 1 shows that the top-performing techniques are ANN, Bayesian, and vehicle-tracking method having more than 80% of accuracy. In order to get more accurate results, these models should be ensembled into one single model with such credibility factors, so that the resulting ensemble model estimates the traffic state with a greater accuracy. To further improve the accuracy and the execution time of the algorithm, data segmentation approach as mentioned in [11] can be applied.

5 Existing Projects Using Probe Technology

This section discusses the projects that used probe vehicles and mobile phones for getting information about traffic flows.

In California, an experiment named Mobile Century [12] was conducted to demonstrate the feasibility of a traffic monitoring system based on GPS-enabled mobile phones. The experiment used the concept of virtual trip lines [13] as its sampling strategy to collect the measurements and send updates. The experiment was conducted using 100 vehicles running in loop on a 10-mile highway in California and the mobile device used was NOKIA N95. Travel times generated by VTL were compared with those generated by loop detectors, and it was suggested that 'VTL measurements are more likely to be closer to the actual velocity observed on the field' [12].

In Hanoi, Vietnam, Hitachi started a demonstration project [14] in 2011 that collected and processed probe data from 300 vehicles in the first year, 2011, and 800 vehicles in the second year, 2012. The output graphics, data, and other forms of information could be used for the estimation of the traffic state. The accuracy achieved for traffic situation identification was approximately 70%. Another ITS project by Hitachi in the province of Bali, Indonesia, obtained GPS data from 300 vehicles operated by a local taxi company. Travel times, speed for a section of road, and travel time for a choice of route were calculated [15].

In the past decade, the potential of smartphones has been exploited by researchers for carrying out many traffic-related tasks such as road incident detection, traffic crowd-sourcing, and traffic queue length detection [16] gives a comprehensive review of all the endeavors that has been done in this area. After analyzing and comparing the existing systems that exclusively depends on mobile phones, [16] states that it is certainly possible to implement a vehicle monitoring system that provides an adequate performance using smartphone-based sensing especially for a developing country.

6 Conclusion

GPS-equipped probe vehicles have come out to be a very promising medium to collect traffic data as it can cover a larger area of road network as compared to fixed sensors. In this paper, we have exhaustively summarized the latest techniques that exploited this form of data and arrived to the best models among them. Further, we proposed that combining the top-performing models (ANN, Bayesian, vehicle-tracking method) with suitable credibility factors into an ensemble model would result into a more accurate model. In addition, a data segmentation approach can result in an even more accurate model with lesser execution time.

References

1. Bramberger M, Brunner J, Rinner B, Schwabach H (2004) Real-time video analysis on an embedded smart camera for traffic surveillance. In: Real-time and embedded technology and applications symposium, 2004. Proceedings. RTAS 2004. 10th IEEE, IEEE, pp 174–181 (May)
2. Coifman B (2002) Estimating travel times and vehicle trajectories on freeways using dual loop detectors. Trans Res Part A Policy Pract 36(4):351–364
3. Kong QJ, Li Z, Chen Y, Liu Y (2009) An approach to urban traffic state estimation by fusing multisource information. IEEE Trans Intell Trans Syst 10(3):499–511

4. Hofleitner A, Herring R, Abbeel P, Bayen A (2012) Learning the dynamics of arterial traffic from probe data using a dynamic Bayesian network. IEEE Trans Intell Trans Syst 13 (4):1679–1693
5. Zhu Y, Li Z, Zhu H, Li M, Zhang Q (2013) A compressive sensing approach to urban traffic estimation with probe vehicles. IEEE Trans Mob Comput 12(11):2289–2302
6. Kong QJ, Zhao Q, Wei C, Liu Y (2013) Efficient traffic state estimation for large-scale urban road networks. IEEE Trans Intell Trans Syst 14(1):398–407
7. Jenelius E, Koutsopoulos HN (2013) Travel time estimation for urban road networks using low frequency probe vehicle data. Trans Res Part B Methodol 53:64–81
8. Zheng F, Van Zuylen H (2013) Urban link travel time estimation based on sparse probe vehicle data. Trans Res Part C Emerg Technol 31:145–157
9. Vlahogianni EI, Golias JC, Karlaftis MG (2004) Short-term traffic forecasting: overview of objectives and methods. Trans Rev 24(5):533–557
10. Djahel S, Doolan R, Muntean GM, Murphy J (2015) A communications-oriented perspective on traffic management systems for smart cities: challenges and innovative approaches. IEEE Commun Surv Tutor 17(1):125–151
11. Bernas M, Płaczek B, Porwik P, Pamuła T (2015) Segmentation of vehicle detector data for improved k-nearest neighbours-based traffic flow prediction. IET Intell Trans Syst 9(3): 264–274
12. Herrera JC, Work DB, Herring R, Ban XJ, Jacobson Q, Bayen AM (2010) Evaluation of traffic data obtained via GPS-enabled mobile phones: the Mobile Century field experiment. Trans Res Part C Emerg Technol 18(4):568–583
13. Hoh B, Gruteser M, Herring R, Ban J, Work D, Herrera JC, Jacobson Q (2008) Virtual trip lines for distributed privacy-preserving traffic monitoring. In: Proceedings of the 6th international conference on mobile systems, applications, and services. ACM, pp 15–28 (June)
14. Okubo T, Yoshioka K, Nakamura A, Taniguchi N (2014) Realizing smart mobility using probe data. Hitachi Rev 63(6):359
15. Morioka M, Kuramochi K, Mishina Y, Akiyama T, Taniguchi N (2015) City management platform using big data from people and traffic flows. Hitachi Rev 64(1):53
16. Engelbrecht J, Booysen MJ, van Rooyen GJ, Bruwer FJ (2015) Survey of smartphone-based sensing in vehicles for intelligent transportation system applications. IET Intell Trans Syst 9 (10):924–935
17. Engelbrecht J, Booysen MJ, van Rooyen GJ, Bruwer FJ (2015) Survey of smartphone-based sensing in vehicles for intelligent transportation system applications. IET Intell Trans Syst 9 (10):924–935

A Fault Attack for Scalar Multiplication in Elliptic Curve Digital Signature Algorithm

Deepti Jyotiyana and Varun P. Saxena

Abstract The dominant operation in cryptographic scheme of elliptic curve is the multiplication using point on an elliptic curve by an integer. This paper specifically discusses the competent algorithms for scalar multiplication a very tedious process in Elliptic Curve Cryptography that are relevant for systems using constrained resources like smart cards. The taxonomy of the work in the open literature for these devices is not only from security perspectives, but likewise some implementation attack, such as fault attacks, must be considered. We survey different implementation approaches and algorithms with the purpose of providing a valuable reference of implementing scalar multiplication in order to retrieve information with a way to determine secret signing key. In addition, this paper provides a review of injecting different fault attacks in a system constrained environment with Elliptic Curve Cryptography. Finally, some arguments about future scope that should be undertaken are provided.

Keywords Fault attack · Scalar multiplication · Elliptic curve cryptography · Elliptic curve digital signature algorithm

1 Introduction

The necessity to provide a secure and sound approach in the direction to authorize the genuineness of digital sources and contents is now a days an emerging requirement for cutting edge computing frameworks and the fulfilling the demand by cryptographic digital signature protocols. In modern computing environment, the most creative and existing cryptosystem ready to provide a signature scheme

D. Jyotiyana · V.P. Saxena (✉)
Department of Computer Science and Engineering, Government Women Engineering College, Ajmer, India
e-mail: varunsaxena82@gmail.com

D. Jyotiyana
e-mail: deepti12346@gmail.com

H.R. Vishwakarma and S. Akashe (eds.), *Computing and Network Sustainability*, Lecture Notes in Networks and Systems 12, DOI 10.1007/978-981-10-3935-5_29

accomplish the need of authenticity of digital contents are signified using Elliptic Curve Digital Signature Algorithm (ECDSA), standardised by both IEEE and NIST [1] and also be recommended by NSA suite B. Elliptic Curve Cryptographic algorithms, with Discrete Logarithmic Problem (specifically ECDLP) provided the rigid security that are suitable for system with constrained resources like mobile devices or smart cards with storing the secret key confidentially in a tamper-proof device. Without destroying the information, it is considered to be hard to retrieve the key, the decryption or signing process usually done inside the card for security purpose [1]. In a situation where the signature is perform in the device might be seized via fast access to the secret token used for signing holding the secret key for example a smart card and ready to replicate it before sending back to the authentic possessor. Thus, various adversarial attacks have been developed with purpose of regulating the secret signing key. These attacking policies must depend on information with faults that are induced in cryptographic devices. Analyses of fault attack on implementation of cryptography have been studied [2]. Boneh, Lipton, DeMillo in 1997 was proposed the first attack for RSA [3]. A fault based attack For Elliptic Curve Cryptography (ECC), was proposed by Biehl et al. [1]. The fault attack is well known side channel technique try to break the cryptographic system and reveals the secret key. These fault attack are active attack inducing a fault on primitive computation of cryptographic scheme. The opponent formerly observes complete information of entire channels, together with the output, trying to recover entire key information and fundamentally differ from all passive side channel attacks. Regarding security, Cryptographic systems are modelled for public key cryptosystem [4] which will be more efficient. Effective security presented with low processing overhead but it will be reduced, due to computation growth during signature generation and verification. Like in public key cryptosystem, the security using ElGamal Digital Signature Algorithm with the ease of private key environment has been analysed where Hybrid model [5] have advantages of both public key and private key cryptosystem. Different fault effects have been analysed for various algorithms [6, 7]. The speed of calculating scalar multiplication significantly affects the performances of ECDSA and security depends with the hardness of determining the value secretly including the information of the openly accessible parameters and the digital signature. The general execution of ECC depends on point scalar increases, with point on the curve is multiplied by a scalar.

In the present paper we give an well-organized overview of various approaches of scalar multiplication specifically one of the key algorithm of Elliptic Curve Cryptography (ECC) and concentrating on the discussions over the security of standardised cryptographic signature: Elliptic Curve Digital Signature Algorithm (ECDSA) aiming to recover the value scalar k. Various implementations of some fast point multiplication are observed to present the attack. We start by describing mathematical background of Elliptic Curve Cryptography. The subsequent sections examine the algorithm for scalar multiplication. We then summarize some of the viewpoints of ECDSA with its two primitives, and discuss. The paper is structured as follows. Section 2 gives an introduction to well-known theory of Elliptic curve and clarifies different possible finite fields that can be considered while implementing ECC. Accordingly, this section also presents ECC domain parameters and ECC protocol algorithm. Sections 3 and 4 examines, an operation which will have crucial impact in the scalar multiplication and gives details of known active attacks implementing on Elliptic Curve Scalar multiplication (ECSM) which demonstrate how faults can be operated to select the secret key k, respectively. Section 5 gives summary on this topic issue. Paper is concluded in Sect. 6.

2 Background

A brief review of elliptic curve is specified using unique respect to scalar or Point multiplication.

2.1 Elliptic Curve Cryptography

Elliptic Curve Cryptography relies on mathematical structure of elliptic curve defined over finite fields which are binary finite field or Galois Field (GF (2^{m})) and prime fields or Galois Field (GF (p)). In an asymmetric key cryptosystem, an elliptic curve is a finite abelian additive element group with a tremendous prime subgroup [8]. An elliptic curve E over a field K is characterised by a weierstrass equation and can be rearranged by applying change of coordinates.

$$E: y^2 + b_1xy + b_3y = x^3 + b_2x^2 + b_4x + b_6 \quad (1)$$

Where b_1, b_2, b_3, b_4, b_6 $\in$ K and $\Delta \neq 0$. Here Δ is the discriminant of E. When the characteristics of this field $\neq 2$ or 3, a point at infinity O along with the set of solutions describes the algebraic structure of an additive group:

$$E: y^2 = x^3 + ax + b \bmod p \quad (2)$$

The smoothness of curve and distinct roots are guaranteed by $4a^3 - 27b^2 \neq 0$. With the point at infinity O it can show that the set of point on E (Fq) forms a group in doubling rule and a related rule for point addition. Let P = (x_1, y_1) $\in$ E (Fp) and Q = (x_2, y_2) $\in$ E (Fp) be two points such that $x_1 \neq x_2$, Formerly (x_1, y_1) + (x_2, y_2) = (x_3, y_3) i.e. P + Q = R, where

$$X^3 \equiv \lambda^2 - x_1 - x_2; y^3 = \lambda(x_1 - x_3) - y_1 \quad (3)$$

Where $\lambda \equiv (y_2 - y_1)/(x_2 - x_1)$, if P $\neq$ Q and $\lambda \equiv (2(x_1)^2 + a)/2y_1$, if P = Q such elliptic curves with characteristics $\neq 2$ or 3 are prime field elliptic curve. The curve with characteristic 2, is known as binary field elliptic curve as defined by Eq. 4:

$$E: y^2 + xy = x^3 + ax^2 + b \quad (4)$$

The set of points on E (F2 ^{m}) form group with given addition rule and related doubling rule. Then $(x_1, y_1) + (x_2, y_2) = (x_3, y_3)$ i.e. P + Q = R, where

$$x^3 \equiv \lambda^2 + x_1 + x_2 + a, \quad (5)$$

$$y^3 \equiv \lambda^2(x_1 + x_3) + x_3 + y_1, \quad (6)$$

With $\lambda \equiv \frac{y2 + y1}{x2 + x1}$

2.2 ECDSA

ECC implicate certain additional parameters apart from the curve parameters a and b, that must be agreed by all parties involved in trusted and secured communication with domain parameters D to be used in the protocols implementing ECC. D = (Fq, a, b, G, n, h), where Fq: field size; a, b: curve parameters; G: the generator point or base point, n: generator point order such that n.G = O; h: co-factor; where h = #E (Fq)/n.

The ECDSA cryptosystem [8] consists two operations: a sign generation algorithm and signature verification algorithm. Secret values of Digital signature recognized only by the signer and contain authentication token produce by signing algorithm while signature verification algorithm is used to identify that authenticity of signature. Before the digital signature be used in elliptic curve, the parties must accept to all domain parameters particularly, D = (Fq, a, b, G, n, h).

- ECDSA Signature Generation: For entity A to sign a message M (M = m), does following steps:
  Input: m (message), D is Domain Parameters, d (users private key)
  Output: Signature (r, s)
  - Select and pick the random integer k (k = b) follows the interval [1, n − 1].
  - Compute bG = (x_1, y_1).
  - Compute r = x1(mod n). if r = 0 then return step1.
  - Calculate e = H(m), H is cryptographic hash function, for example, SHA-1 or SHA-2.
  - Compute s = b^{-1} (e + rd) mod n. If that s = 0 return to step1.
  - Return the signature for the message m with pair of integer (r, s).

After signature generation if user B want to authenticate the signature, then signature verification algorithm be used. Apart from ECDSA, scalar multiplication is generally utilized as a part of encryption, decryption, key generation, key agreement etc.

- ECDSA Signature Verification: To verify A's signature (r, s) on a message m, entity B do the following:
  - Verify that r and s are integer in the interval [1, n − 1], and compute the hash value H of the message m
  - Compute $u_1 = s^{-1}H \bmod n$ and $u_2 = s^{-1}r \bmod n$
  - Compute $u_1 P + u_2 Q = (x_1, y_1)$
  - Signature is accepted if and only if $r = x_1 \bmod n$

Next section discusses some widely used algorithm of scalar multiplication. Next section discusses some widely used scalar multiplication algorithms.

3 Common Approaches for Scalar Multiplication

The speed of scalar multiplication shows significant role in the effectiveness of entire framework where operation of key generation, key agreement, signature generation and verification take place. The ECC security ECC depends on toughness of the Elliptic Curve Discrete Logarithm Problem (ECDLP) especially, discovering k such that Q = kP for two point P and Q, the kP is called point multiplication or scalar multiplication, means given altered open point P has a place with prime subgroup, figure scalar different k (which is P + P + P = kP). Scalar multiplication in an optimized form is a significant component for complete performance and implementation of ECDSA. To the representation of scalar k or pre-calculation are mostly offered by conventional techniques.

3.1 Binary Method

Binary method [6] is the simplest traditional scalar multiplication method of compute kP based on the binary representation of the scalar k using (0, 1).

The integer k is represented as

That is $\mathbf{k} = \sum_{j=0}^{l-1} kj2^j$, where $k_j \in \{0, 1\}$. This method is called binary method [7] which checks the bits of k either from left-to-right or right- to-left. The binary method for the computing kP is given in the following Algorithm1

Algorithm 1: Binary method

Binary representation of point P and integer k where $P \in E(Fq)$

Input: $k = (k_{n-1} \ldots k_1 k_0)_2$

Output: kP

R = S; P = T

$S \leftarrow T$

For i = 0 to n − 1 do

$S \leftarrow 2S$ (doubling)

If $k_i = 1$ then S = S + T (addition)
$i \leftarrow i - 1$
Return R

The number of non-zero digits represented as hamming weight of scalar multiplication. The binary method requires n − 1/2 additions and n − 1 doublings Above computation is speed up by reducing the number of 1's of scalar multiplication or hamming weight.

3.2 Non Adjacent Form

In 1951, Booth [9] proposed Signed binary technique, another scalar representation and later Rietweisner [10] verified that along with canonical representation each integer could be uniquely signified selected as Non Adjacent Form (NAF) [8] with having a property that from any two integers, at most one is non-zero. The equation show the representation of integer k = Σl−1j = 0 kj2j, where each kj ∈ {−1, 0, 1}. As compare to other algorithm, it requires additional L–bits memory.

3.3 Window Method

Window method [7] depends on size of the windows or blocks involve precomputed points. In this method typical window with size w was selected first. At that point the values of kP are figured for k = 0, 1, 2, 3, 4... 2^{w-1}. The major advantages of this algorithm are that number of point addition is reduced compared to the previous signed binary method. This approach gives faster execution to scalar multiplication.

Traditional method such as binary method with average hamming weight L/2 is used for computing scalar multiplication. Booth [9] develops a signed binary representation as a new scalar representation with attempt to decrease the number of bit to 1. Binary representation of k has an average hamming weight with L/3. The benefit of NAF over binary representation is that it has low average hamming weight. Prior report contrasting the whole three calculations and new proposed optimized algorithm [6] which states that a new scheme for scalar multiplication enhancing the efficiency implementing on ECDSA. A technique proposed in [11] constructing a scalar k by way of sequence arranged periodically and also showing that this role can speed up the process of elliptic curve scalar multiplication apparently.

4 Fault Attacking Details

This section introducing the details of the advancement of different fault attacks on elliptic curve cryptography. Commencing from the investigation with the computation carried out in weak environments, using a technique to recover data not to be known at the exterior to the device. Starting by analysing the computation executed in unreliable condition, various side channel techniques have been implemented to recover information by developing some method that is not made up to be identified outside the device. Specifically, after analyse first false outcome, it will be possible to come to know the key value i.e. d or the transitional value e + rd in signing process, after that, by examining these values subsequently be used to rebuilt the entire secret key d producing the faulty signature result through the propagation of error [12].

Effectively, implementing the fault attack by using the scalar multiplication algorithm with Euclidian Addition Chain (EAC) based on binary sequence computing the private key and scalar k that cracks the whole cryptosystem [13]. The author proposed the capability of Side channel attack to completely recover the secret key by injecting the single bit transient faults, with minimal cost tools and without damaging the device [14]. Furthermore, side channel information can be combining with cryptanalysis techniques to observe the information about secret keys and implementation structure and also inserting the bit faults into the computation of elliptic curve in taper resistant device [1]. Other approach to present the fault attack in the Elliptic curve cryptosystems were studied by Biehl et al. [1]. Adversary select point P from the calculation of dP while different models just require the information of P where they expect that only a limited faulty bits are embedded either at the time of calculation or into P only preceding at the point multiplication. Then, whole key recovery from multiple point multiplications can be explained and countermeasures are also being discussed to keep the leakage of secrets [15].

Another survey explains a fault attack on the elliptic curve digital signature algorithm (ECDSA) with the program flow modification where parts of ephemeral key are needed to be recovered. This key is randomly selected for each signature. Thus, retrieved information are then agree in order to determine secret signing key by performing lattice attack [16].

5 Summary

Maximum attacks and their cryptanalysis are depends on the mathematical security of the cryptosystem. The common approaches of scalar multiplication are used to provide better efficiency and fast computation on elliptic curve cryptography. Introducing the idea of inducing commonly known side channel techniques i.e. fault attack based on information leakage by injecting the faults in the implementation of ciphers. It is also possible by modifying the scalar multiplication in order to exposes the secret signing key and perform with random exponent rather than fixed exponent. These attacks are induced in the way of permanent and transient faults. In Elliptic Curve Cryptography, moving the scalar multiplication from curve towards weaker curve often require fault induction. This survey proposes conclusion from the above discussion, that the side channel attacks against ECDSA exploits the provably secure algorithm by manipulating in an unintended way. Finally, it should be analyse that fault attacks create a real and severe threats on the security of cryptographic schemes.

Currently, we are examining the software implementation of ECDSA over NIST recommended binary fields and prime fields. A vigilant and considerable study of ECC implementation in software and hardware such as smart cards would be valuable. In the best case, attack breaks a basic and differential side channel analysis safe usage with information/yield curve parameters and point validity.

6 Conclusion

Concerning the scheme of attacking techniques for Elliptic curve Cryptography, many perspectives can occur. These attack based on the arithmetic properties and scalar multiplication of ECC. In this paper, we have presented a review on different optimizing techniques used in scalar multiplication on Elliptic Curve Cryptography. While trying to recover the secret signing key, We examine different available side channel attacking techniques by injecting faulty schemes. Implementation of Elliptic Curve Cryptography in a constrained environment, the situation is consistent to deal with utilization of implementing speedups, for instance key expression in NAF form. By representing the cryptographic keys using different scalar bit, it can reduces the amount of hard computations that are required to perform the cryptographic operation, reducing power consumption and by way of inducing the fault allows adversary to decrease the ECDLP which is computationally solvable in short time. This discussion emphasis on showing the feasibility of using the signed bit representation to produce faulty output on optimal scalar multiplication algorithm to retrieve the secret scalar key k. Further methods will be relevant to determine the elliptic curve digital signature algorithm oven binary field.

References

1. Biehl I, Meyer B, Müller V (2000) Differential fault attacks on elliptic curve cryptosytems. In: Advances in cryptology—CRYPTO 2000. Springer, pp 131–146
2. Blömer J, Otto M, Seifert J-P (2006) Sign change fault attacks on elliptic curve crytosystems. In: Fault diagnosis and tolerance in cryptography. Springer, pp 36–52
3. Boneh D, DeMillo RA, Lipton RJ (2001) On the importance of eliminating errors in cryptographic computations. J Cryptology 14(2):101–119

4. Saxena VP, Nalwaya P (2014) A novel cryptographic approach based on feedback mode of elgamal system. Int J Adv Res Sci Eng (IJARSE) 3(2):128–138. ISSN - 23198354
5. Saxena VP, Priya Nalwaya PN (2014) A cryptographic approach based on integrating running key in feedback mode of elgamal system. In: 2014 international conference on computational intelligence and communication networks (CICN). IEEE Computer Society, pp 719–724. http://doi.ieeecomputersociety.org/10.11
6. Biham E, Shamir A (1997) Differential fault analysis of secret key cryptosystems. In: Advances in cryptology—CRYPTO'97. Springer, pp 513–525
7. Barenghi A, Bertoni G, Palomba A, Susella R (2011) A novel fault attack against ECDSA. In: IEEE international symposium on hardware-oriented security and trust (HOST). IEEE, pp 161–166
8. Hankerson D, Menezes AJ, Vanstone S (2006) Guide to elliptic curve cryptography. Springer Science & Business Media
9. Booth AD (1980) A signed binary multiplication technique. Computer arithmetic-benchmark papers in electrical engineering and computer science, vol 21
10. Reitwiesner GW (1960) Binary arithmetic. Adv Comput 1:231–308
11. Li H, Zhang R, Yi J, Lv H (2013) A novel algorithm for scalar multiplication in ECDSA. In: 2013 fifth international conference on computational and information sciences (ICCIS). IEEE, pp 943–946
12. Fan J, Verbauwhede I (2012) An updated survey on secure ecc implementations: attacks, countermeasures and cost. In: Cryptography and security: from theory to applications. Springer, pp 265–282
13. Saxena VP, Anubhav Saxena SM (2015) Implementation of fault attacks on elliptic curve cryptosystems. Reading, vol 3, no 4, p 0
14. Amiel F, Clavier C, Tunstall M (2006) Fault analysis of dpa-resistant algorithms. In: Fault diagnosis and tolerance in cryptography. Springer, pp 223–236
15. Ciet M, Joye M (2005) Elliptic curve cryptosystems in the presence of permanent and transient faults. Des Codes Crypt 36(1):33–43
16. Schmidt J-M, Medwed M (2009) A fault attack on ECDSA. In: 2009 Workshop on fault diagnosis and tolerance in cryptography (FDTC). IEEE, pp 93–99
17. Ling J, King B (2013) Smart card fault attacks on elliptic curve cryptography. In: 2013 IEEE 56th international midwest symposium on circuits and systems (MWSCAS). IEEE, pp 1255–1258

Proposing an Architecture for Scientific Workflow Management System in Cloud

Vahab Samandi and Debajyoti Mukhopadhyay

Abstract With the growth in IT infrastructure and advances in technologies, workflow scheduling poses many challenging issues for complex applications which require many computing resources. Hence, there is a requirement of a workflow management system adaptable with many cloud environments due to the heterogeneity of resources and applications. In this paper, we have proposed a general workflow management system architecture and a scientific workflow model, followed by a model for monitoring tool in the cloud environment, based on a comprehensive study of literature in cloud computing.

Keywords Cloud computing · Workflow · Virtualization · Scheduler · Resource monitoring

1 Introduction

Cloud computing provides three major services required for an individual or any types of enterprise. These services are Infrastructure as a Service (IaaS), Platform as a Service (PaaS), and Software as a Service (SaaS). IaaS cloud provides the user an unlimited pool of virtual resources with the help of virtualization technology. The cloud resources are heterogeneous in nature and distributed over many locations, and they can be accessed by user demand. PaaS cloud provides middleware services, and in addition to supporting application hosting, it offers developing applications on the PaaS cloud to developers. The SaaS cloud provides applications and tools for consumers so that they can access and utilize unlimited cloud services.

Scientific workflows are the set of applications composed of individual tasks which need to be mapped to distributed resources to be executed. The process of

V. Samandi (✉) · D. Mukhopadhyay (✉)
Department of Information Technology, Maharashtra Institute of Technology, Pune, India
e-mail: vahabsamandi@gmail.com

D. Mukhopadhyay
e-mail: debajyoti.mukhopadhyay@gmail.com

H.R. Vishwakarma and S. Akashe (eds.), *Computing and Network Sustainability*, Lecture Notes in Networks and Systems 12, DOI 10.1007/978-981-10-3935-5_30

mapping the tasks to resources is performed by workflow management system. The workflow management system describes, manages, and executes the cloud jobs which are mostly in the form of workflows [1]. In this paper, we first propose a workflow model and mathematical representation of workflow which is in the form of directed acyclic graph (DAG). Then, we propose a general architecture for workflow management system, followed by a model for resource monitoring tool which performs higher-level analyses of workflow execution.

2 Related Work

Several workflow management systems have been developed, for example, Pegasus [2], Triana [3], Taverna [4], and Kepler [5], and the focus of these WMSs are mostly on converting abstract workflow into an executable form by handling data and control-flow dependencies among tasks, and also generating task clusters. For other required functionalities such as resource monitoring, scheduling, and resource provisioning, even these systems handle some of the functionalities, but they require higher-level monitoring and analysis capabilities; hence, they integrate different tools to fulfill these requirements. As an example, Pegasus integrate Condor for the scheduling and resource management, and along with Triana, they integrate Stampede monitoring tool that provides new analysis capabilities to these WMSs [6].

Cloud providers like Amazon EC2 [7], Salesforce [8], Microsoft Azure [9], etc. three major cloud services, as IaaS, PaaS, and SaaS, but for scheduling and resource provisioning, not all of them allow users to access infrastructure. For example, Salesforce provides SaaS cloud and a part of its PaaS cloud that allows users to develop their applications, but Amazon EC2 allows users to access and provision the resources.

3 Workflow in Cloud

A scientific workflow is a set of computational tasks that are usually dependent on each other. The dependencies among tasks are data or control-flow dependencies [10]. The complex workflow applications such as gravitational waves physics, astronomy, and bioinformatics, which require a tremendous amount of processing power, utilize cloud resources to process a large number of data sets. Cloud jobs are in the form of workflows, and the workflows are mostly represented in the form of directed acyclic graph (DAG). Workflow management system takes abstract workflow as an input and converts it into executable form.

3.1 Workflow Model

We have proposed a workflow model illustrated in Fig. 1. In the represented graph, vertices are corresponding to cloud jobs or tasks, and arcs are corresponding to data dependencies.

Following is the mathematical representation of a workflow:

Workflow (DAG) = < T, E, D_T, W>

T = {Set of t_i | t_i is a ith task, where 1 < i < n, n = |T| is total number of tasks}

E = {Set of d_{ij} | d_{ij} is a directed edge from ith *task to* jth task and i < j}

W = {Set of wl_i | wl_i is the amount of workload (computation) assigned to ith task}

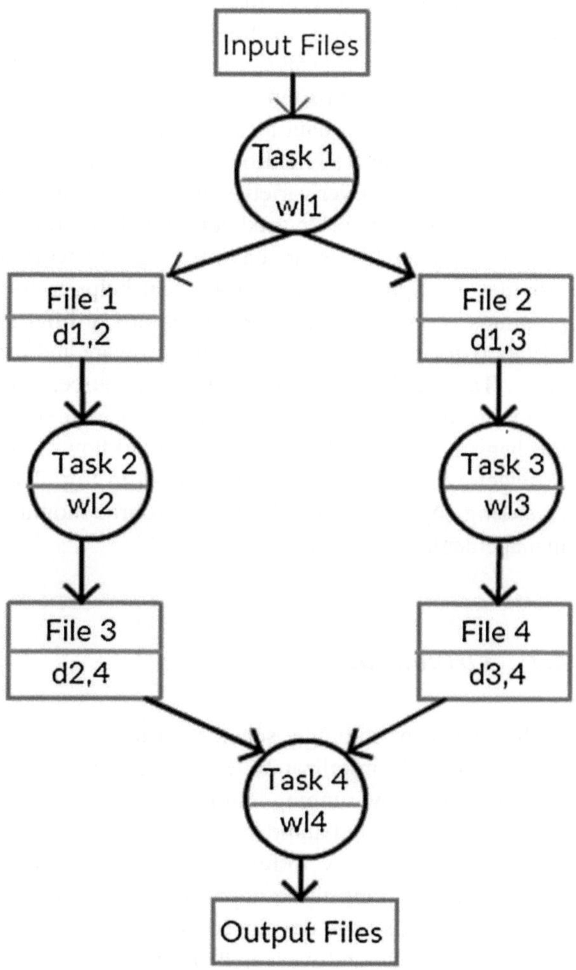

Fig. 1 A workflow model

$D_T = \{t_i \rightarrow t_j \mid t_i$ is the start point of data transmission and t_j is the end point and it represents the task t_j is data dependent on task $t_i\}$

Each vertex t_i represents task or job with an amount of workload (computation) assigned to it, and each edge d_{ij} represents data or control dependencies which indicate task t_j can start execution after task t_i is completed, and then the generated results of completion of task t_i transfer to task t_j.

The workflow model which is shown in Fig. 1 consists of four tasks, each with a certain amount of workload, that are connected by six files. The intermediary files are the outputs produced by some tasks that are inputs to other tasks.

4 Workflow Management System Architecture

Several scientific workflow management systems are developed to help scientists, analysts, and developers in different scientific domains to create and execute scientific workflows and analyses across broad areas of scientific communities. Kepler [5] is a free and open-source workflow management system which operates in a variety of formats on data locally and globally. By using Kepler's GUI, users are able to create, execute scientific workflows. Pegasus [2] is a workflow management system which runs over varieties of hardware including a laptop, a campus cluster, a grid, or a commercial or academic cloud environment such as Amazon EC2 and Nimbus. Triana [3] is an environment for workflow and data analysis, which provides a graphical user interface that helps users to develop and run their own programs. Triana has been developed at Cardiff University, initiating as a part of GEO600 gravitational wave detector software and more recently in a wider range of users. Taverna [4] is a powerful, open-source, and domain-independent tool for designing and executing workflows. It uses textual language SCUFL which is a mechanism for specifying Taverna workflows. We have proposed a workflow management system architecture shown in Fig. 2 which consists of workflow and clustering engines, a workflow and resource monitoring tools, a scheduler and data management.

4.1 Virtualization

In the context of computer science, virtualization has different aspects; at the outset development of virtualization technology, it was the idea of dividing the resources of a server to allow running multiple processes simultaneously [11]. Another aspect of virtualization is to handle legacy applications, software applications developed at the different time, by different people through different technologies, and the heterogeneity among these applications is challenging due to which environment their applications have developed. Virtualization provides an environment to adopt these applications and weaves them into a single coherent application [11].

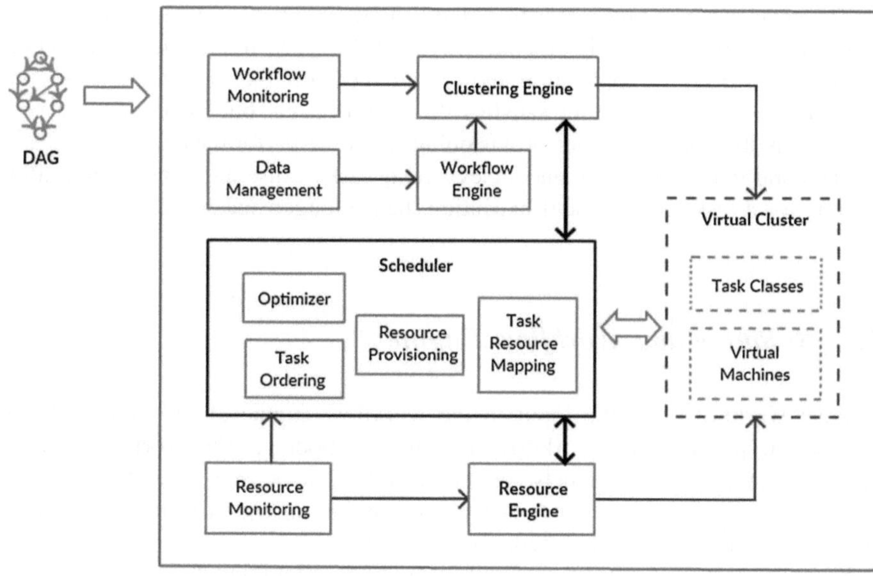

Fig. 2 Workflow management system architecture

A virtual machine (VM) is a software implementation of a computer that acts as a computing unit. On one physical node (computer), number of virtual machine instances can be created, and each of them can run a different operating system.

A large quantity of computing resources is required to execute workflows in cloud environments [12]. These computing resources are provided in the form of virtual machines. Virtual machines are processing or computational units which provide compute cycles to process tasks.

4.2 Scheduler

Scheduling is a process of mapping tasks on heterogeneous and distributed resources over time. Scheduler discovers distributed resources and assigns the tasks to the relevant resources based on user's specified parameters and key factors. During scheduling data dependencies and data transfer between tasks is handled by data management module. Tasks or activities are applications with an amount of computations assigned to them, and the unit of computation measurement is usually taken as million instructions (MI) [13]. Scheduling of scientific workflows in cloud computing introduces the following challenges:

1. Mapping task classes into virtual resources generates a large makespan, and the difficulty is to find a minimum set of optimal schedules with the highest performance and based on user-defined QoS, such as cost and speed.

2. The provisioning of resources in a cloud environment is handled by user-control scheduler [10], and the issue is to figure out the types and amount of resources that the workflow application requires to perform the computations, as the resource over-provisioning satisfies the performance, but increasing the cost, whereas the resource under-provisioning hurts the performance.
3. Data and control-flow dependencies among tasks cause an increase of waiting time for a task ready to start execution, hence longer makespan.

4.3 Resource and Workflow Engine

Abstract workflows are the input to the workflow engine along with workflow information provided by workflow monitoring module. The workflow engine handles data dependencies between tasks and releases free tasks to clustering engine. Clustering engine creates task classes by joining up small tasks into larger jobs which lead to reducing scheduling overhead.

Resource engine is a heart of cloud computing system which is responsible for resource management [14], and the information related to storage and computing resources is collected and maintained by resource monitoring tool which will be handed over to resource engine for the creation of virtual clusters.

4.4 Cloud Monitoring System

The existence of a monitoring system is crucial in workflow management system. As the workflow involves many sub-workflows and many tasks, a resource monitoring is essential to track resources and network availability and to provide real-time information about resources and workflow execution to users. Users need to be notified whenever there is a problem such as a resource or network failure, or a software bug that hampers the system performance.

Figure 3 illustrates a cloud monitoring system, and the log files that will be loaded in monitoring tool include a variety of information such as, input abstract workflow, executable workflow, and the status of each task. These log files are stored in a database, and the query interface module extracts data from the data store and forwards these data to error handling, analysis, and reporting, and dashboard tools.

During execution of workflows, uncertain problems take place that can be tracked and reported by analysis and reporting tool. Debugging of workflows is handled by error handling tool, and the users are notified of the occurrence of an error. The dashboard provides a statistical view of all details about a successful or failure of workflow execution and completion, and it displays all the information in a summary page.

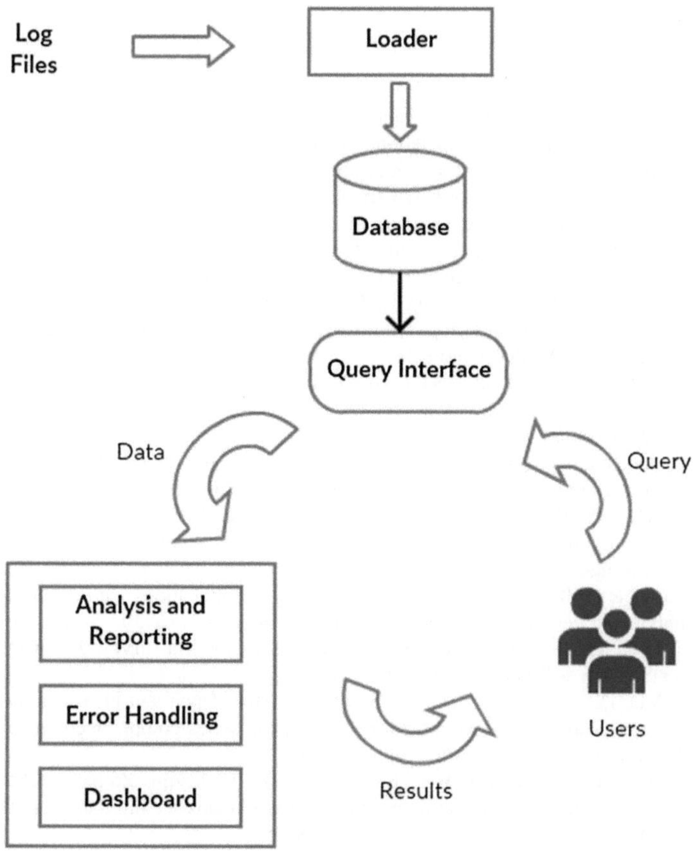

Fig. 3 A cloud monitoring system

5 Conclusion and Future Work

Many cloud tasks are in the form of workflows; in this paper, we have introduced a graphical and mathematical representation of a workflow which is in the form of a directed acyclic graph (DAG), followed by introducing a general architecture for workflow management systems and a model for monitoring tool in a cloud. Then, we survey each component of workflow management system in detail. In future, we will focus more on details of workflows' structure and dependencies between tasks and how to segregate dependent tasks so that they can execute in parallel. We also propose an optimization scheduling algorithm in order to reduce scheduling overhead.

References

1. Yu J, Buyya R, Ramamohanarao K (2008) Workflow scheduling algorithms for grid computing. In: Metaheuristics for scheduling in distributed computing environments, volume 146 of the series studies in computational intelligence, pp 173–214
2. Pegasus workflow management system. https://pegasus.isi.edu/
3. Triana scientific workflow. http://www.trianacode.org/
4. Apache Taverna. http://www.taverna.org.uk/
5. Kepler scientific workflow system. https://kepler-project.org/
6. Vahi K, Harvey I, Samak T, Gunter D, Evans K, Rogers D, Taylor I, Goode M, Silva F, Al-Shakarchi E, Mehta G, Deelman E, Jones A (2013) A case study into using common real-time workflow monitoring infrastructure for scientific workflows. Springer Science + Business Media, Dordrecht
7. Amazon web services. https://aws.amazon.com/
8. Salesforce cloud. http://www.salesforce.org/
9. Microsoft azure cloud. https://azure.microsoft.com/
10. Juve G, Deelman E (2011) Scientific workflows in the cloud. in: grids, clouds and virtualization. Comput Commun Netw 71–89 (Springer, London Limited)
11. Jin H, Ibrahim S, Bell T, Qi L, Cao H, Wu S, Shi X (2010) Tools and technologies for building clouds. In: Cloud computing part of the series computer communications and networks, Springer, London Limited, pp 3–20
12. Yu J, Buyya R (2006) Scheduling scientific workflow applications with deadline and budget constraints using genetic algorithms. Sci Program 14(3–4):217–230
13. Wu F, Wu Q, Tan Y (2015) Workflow scheduling in cloud: a survey. J Supercomput 1–46 (Springer Science + Business Media, New York)
14. Cunsolo VD, Distefano S, Puliafito A, Scarpa M (2010) Open and interoperable clouds: the Cloud@Home Way. In: Cloud computing part of the series computer communications and networks. Springer, London Limited, pp 93–111
15. Zhang Fan, Cao Junwei, Tan Wei, Khan Samee U, Li Keqin, Zomaya Albert Y (2014) Evolutionary scheduling of dynamic multitasking workloads for big-data analytics in elastic cloud. IEEE Trans Emerg Top Comput 2(3):338–351
16. Li X, Song J, Huang B (2015) A scientific workflow management system architecture and its scheduling based on cloud service platform for manufacturing big data analytics. Int J Adv Manuf Technol 1–13 (Springer, London)
17. Zhang F, Cao J, Hwang K, Wu C (2011) Ordinal optimized scheduling of scientific workflows in elastic compute clouds. In: Third IEEE international conference on cloud computing technology and science, pp 9–17
18. Varalakshmi P, Ramaswamy A, Balasubramanian A, Vijaykumar P (2011) An optimal workflow based scheduling and resource allocation in cloud. In: Advances in computing and communications volume 190 of the series communications in computer and information science, Springer, Berlin, Heidelberg, pp 411–420
19. Ho Y-C, Zhao Q-C, Jia Q-S (2007) Ordinal optimization, soft optimization for hard problems. Springer Science + Business Media, New York, USA
20. Cafaro M, Aloisio G (2011) Grids, clouds and virtualization, computer communications and networks. Springer, London Limited, pp 71–91
21. Prodan R, Wieczorek M (2010) Bi-criteria scheduling of scientific grid workflows. IEEE Trans Autom Sci Eng 7(2):364–376
22. Chen C, Liu J, Wen Y, Chen J (2015) Research on workflow scheduling algorithms in the cloud. In: Process-aware systems, volume 495 of the series communications in computer and information science. Springer, Berlin, Heidelberg, pp 35–48
23. Benoit A, Marchal L, Pineau J-F, Robert Y, Vivien F (2009) Resource-aware allocation strategies for divisible loads on large-scale systems. In: Proceedings of IEEE international parallel and distributed processing symposium (IPDPS), Rome, Italy, pp 1–4

24. Zhang F, Cao J, Hwang K, Li K, Khan SU (2015) Adaptive workflow scheduling on cloud computing platforms with iterative ordinal optimization. IEEE Trans Cloud Comput 3 (2):156–168
25. Rodriguez MA, Buyya R (2014) Deadline based resource provisioning and scheduling algorithm for scientific workflows on clouds. IEEE Trans Cloud Comput 2(2):222–234
26. Emeakaroha VC, Maurer M, Stern P, Labaj PP, Brandic I, Kreil DP (2013) Managing and optimizing bioinformatics workflows for data analysis in clouds. J Grid Comput 407–427 (Springer Science + Business Media Dordrecht)
27. Calheiros RN, Masoumi E, Ranjan R, Buyya R (2014) Workload prediction using ARIMA model and its impact on cloud applications' QoS. IEEE Trans Cloud Comput 1–11
28. Freund RF et al (1998) Scheduling resources in multi-user, heterogeneous, computing environments with smartnet. In: Proceedings of 7th heterogeneous computing workshop (HCW), Washington, DC, USA, pp 184–199
29. Li H, Buyya R (2007) Model-driven simulation of grid scheduling strategies. In: 3rd IEEE international conference on e-science and grid computing 287–294
30. Smith J, Siegel HJ, Maciejewski AA (2008) A stochastic model for robust resource allocation in heterogeneous parallel and distributed computing systems. In: Proceedings of IEEE international parallel and distributed processing symposium, Miami, FL, USA, pp 1–5
31. Lin C, Lu S, Fei X, Chebotko A, Pai D, Lai Z, Fotouhi F, Hua J (2009) A reference architecture for scientific workflow management systems and the VIEW SOA solution. IEEE Trans Serv Comput 2(1):79–92
32. Macías M, Guitart J (2012) Client classification policies for SLA negotiation and allocation in shared cloud datacenters. In: Economics of grids, clouds, systems, and services, volume 7150 of the series Lecture Notes in Computer Science pp 90–104
33. Moses J, Iyer R, Illikkal R, Srinivasan S, Aisopos K (2011) Shared resource monitoring and throughput optimization in cloud-computing datacenters. In: IEEE international parallel and distributed processing symposium, pp 1024–1033
34. Ge J, Zhang B, Fang Y (2010) Research on the resource monitoring model under cloud computing environment. In: Web information systems and mining, volume 6318 of the series Lecture Notes in Computer Science, pp 111–118
35. Wang J, Korambath P, Altintas I, Davis J, Crawl D (2014) Workflow as a service in the cloud: architecture and scheduling algorithms. In: 14th international conference on computational science, vol 29. Elsevier, pp 546–556

Hand Gesture-Based Control of Electronic Appliances Using Internet of Things

Ritima Paul and Bhanu Prakash Joshi

Abstract In this paper, the author has developed a system of wirelessly controlling remote units using hand gesture and Internet. A lot of research has already made communication between silent people and the general audience using flex sensor. Here, we will control remote devices with different hand gestures depending upon the measured flexity of the flex sensor. Using IoT, anything across the globe can be monitored and controlled from any place. The microcontrollers used as well as the remote units that are being controlled are connected to Internet either via LAN or Wi-fi module. This system will help those who generally forget small things such as switching the power off when not in use!

Keywords IOT · Flex sensor · Tilt sensor · Microcontroller · Wi-fi module

1 Introduction

We are living in a world where we have less resources but high demands due to growing population. It is an urgent requirement to judiciously utilise non-renewable resources such as water, electricity, and fuels. as over a period of time nothing will be left behind. Otherwise, everything will go with this twenty-first-century generation!

Here, we have done a bit to preserve electricity at our routine places by wirelessly turning them off with just a movement of hand. The system explained in this paper used Wi-fi modules to communicate with the electronic systems that need to be controlled. The specific hand gesture will decide what action is to be performed. Hand glove has been used previously for gesture recognition to provide aid to dumb people by making them to communicate to the general world [1, 2]. Some of the emotions described by hand gesture can be seen in Fig. 1.

R. Paul (✉) · B.P. Joshi
Amity School of Engineering & Technology, Amity University, Noida, India
e-mail: ritima21@gmail.com

B.P. Joshi
e-mail: bhanuprakashj7@gmail.com

H.R. Vishwakarma and S. Akashe (eds.), *Computing and Network Sustainability*,
Lecture Notes in Networks and Systems 12, DOI 10.1007/978-981-10-3935-5_31

Fig. 1 Hand gestures for various emotions

The main aim of this project was to design an entire wirelessly controlled system by readily translating the hand gestures into predefined commands. Here, hand glove has been used by us in this project, which consists of flex sensor, tilt sensor, ARM microcontroller LPC2148, and Wi-fi module ESP8266, which will control remote units.

We hope that the practical use of this system will definitely help the world to save power and will prove out to be the best solution for those who forget to turn off the electric appliances before leaving their place.

The main sequences that set apart the designing of this project from others are as follows: (i) selecting and fixing the hand gestures for specified task, (ii) sending of the data from the integrated sensor system over Internet and (iii) experimenting with the results for controlling devices.

2 Previous Works

Home automation is also based on the same line but it has its own drawback of high set-up cost, management problems, security issues and low adaptability.

IoT-based home automation system has already designed but it lacks in a way that there the automatic control of electronic appliances has been done based on the

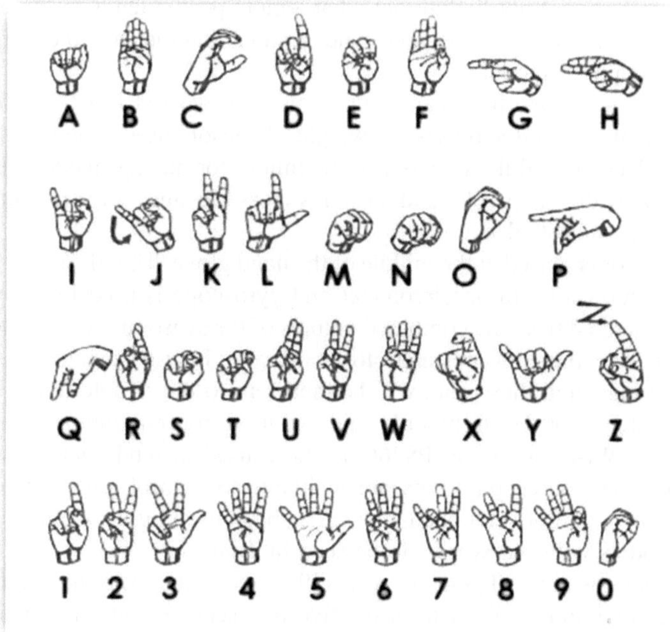

Fig. 2 American sign language hand gestures

temperature, gas sensor values. There is no manual control of status of devices when no hazardous situation is detected [3].

Also, previous uses of flex sensor include hand gesture recognition for sign language using a couple of additional sensors, Fig. 2 [4, 5]. A hand gesture-to-speech conversion has also been modelled [6]. In these models, the recognition of alphabets has been done and not the emotions such as sorry, thank you, yes or no.

There is no such comparison with our work as we have controlled the devices with different hand gestures and not recognised what the mute person wishes to say.

3 Proposed Methodology

The sign language generally used by people has a hand gesture for a complete word, whereas, individual alphabets in a word can also be represented using hand gestures. American sign language comprised of 26 different English alphabets hand gestures.

A lot of work has already been done where the hand gestures control the thing like television for switching channels or changing volume or a hand gesture-controlled robotic arm, etc. [7, 8]. But nothing like controlling things without having any direct physical communication medium has been done previously. This is what our work does.

In our work, we are not finding out what a mute person tends to say, but we are controlling remote device using any hand gesture. Internet has the major role to play in our proposed model.

A normal-sized glove has been used by us with desired flexibility for various hand movements. Also it supports the weight of sensors used. The flex sensors are made to be placed on all the fingers and the thumb for having maximum accuracy, as it made a wide range of hand gestures possible and correspondingly more devices can be controlled.

The tilt sensor is placed at the middle of the hand glove. The tilt sensor MPU6050 which is a combination of accelerometer and gyroscope is used in our work. The results are generated from the combined outputs of the sensor integrated glove (SIG).

The data from the SIG are made to travel over Internet using the concept of Internet of things, and these data will be then sent to the remote devices.

Ethernet shield or Wi-fi module can be used for communication with the Internet. Here, Wi-fi module ESP8266 has been used to send data to Internet.

A lot of open-source platforms are available to control and monitor devices easily using IoT, one of such open source has been used by us. Comparison of the proposed models with our work can be seen in Table 1.

A particular gesture will generate a specific fixed value from the combination of sensors. The individual values from the five flex sensors and one tilt sensor, and their combined value, both will contribute to the decision of the action performed. This has been done to add to the accuracy of the system.

Table 1 Comparison with existing models

Reference	Sensors used	Action performed	Remote devices control	Communication medium
[3]	Temperature, humidity, motion, light level	Light on/off fan on/off	Automatic depending on sensor, not manual	Wi-fi, IoT
[7] and [8]	Flex [7, 8], accelerometer [7], electronic compass [7], ultrasonic [7]	Robotic arm replicating human arm	No control	Zigbee module
[1, 4] and [6]	Flex, accelerometer	Hand gesture recognition, audio conversion	No control	Bluetooth (3) wired (4, 8)
[5] and [2]	Accelerometer [2, 5], flex [2, 5], contact sensor [5]	Hand gesture recognition	No control	Wired
My work	Flex, accelerometer, tilt sensor	Hand gesture recognition	Wireless control of remote devices	Wi-fi, IoT

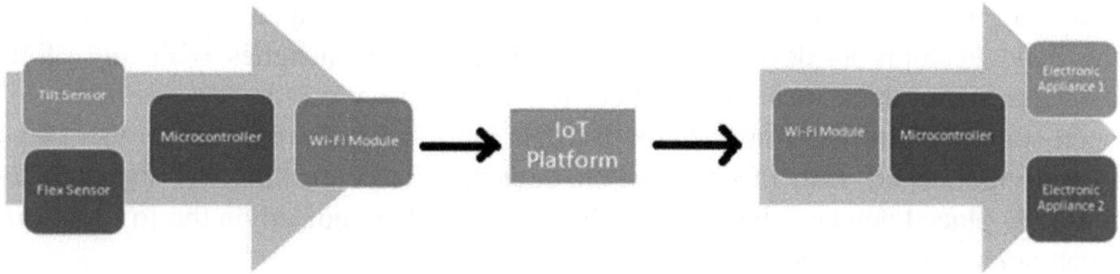

Fig. 3 Flow of our proposed methodology

Figure 3 shows the flow of our proposed method, right from the sensing of data to the controlling of remote devices.

The analog values from the flex sensors and tilt sensor are converted to digital ones using one of the two 10-bit ADC port of microcontroller. Thus, making it simple to determine whether the hand gesture position is stationary or in motion. Stationary hand gesture makes use of only flex sensor, whereas moving gesture uses the combination of both sensors.

4 Implementation Setup

The flow of the work can be seen from Fig. 4. After the start of the system, the Wi-fi needs to be configured with the existing Internet either from the router or mobile Internet portable hotspot. Then, the data from the sensor integrated glove

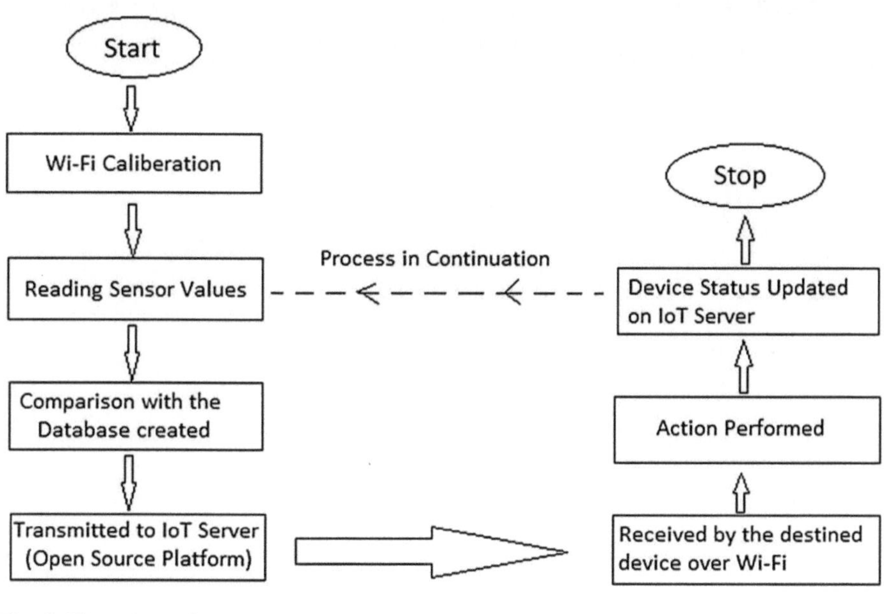

Fig. 4 Flow chart of process followed

(SIG) are measured and compared with the database we created earlier for some of the hand gestures as described in Fig. 2. If the value matches with any of the database data, then it is transmitted for the destination device over the IoT open-source platform from where the data will be gathered via another Wi-fi module associated with the destination device, thus controlling the status of the remotely placed device. The status of the device is also updated on the IoT platform as an acknowledgement for the action performed.

5 Results

We have successfully controlled five devices using five different hand gestures, which are based on ASL alphabets A, G, L, O and V. The controlling of devices is done in way that whatever be the current status of devices it is reversed when these gestures are detected. The status of the devices can be observed on the open-source IoT platform cloud, which is also updated after the action has been performed. It takes 15 s for IoT open source to update the values, so a delay of 15–16 s has to be maintained between each transmission.

In Figs. 5 and 6, we can see the values from the combined system that has been sent over IoT platform are plotted against date and IST, respectively. Depending upon these values, the status of devices is changed; each device has a predefined range of values according to which only those devices will respond. For this value

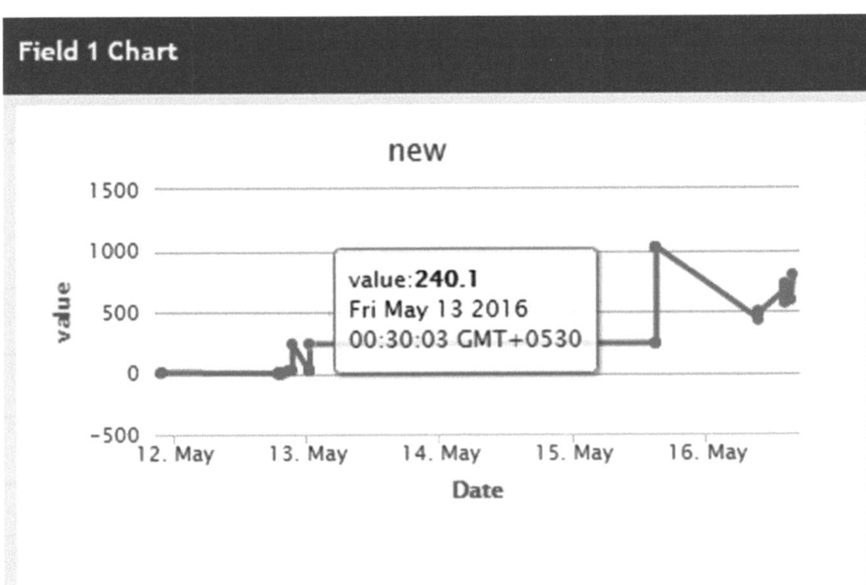

Fig. 5 Display of *values* from sensors sent over IoT platform

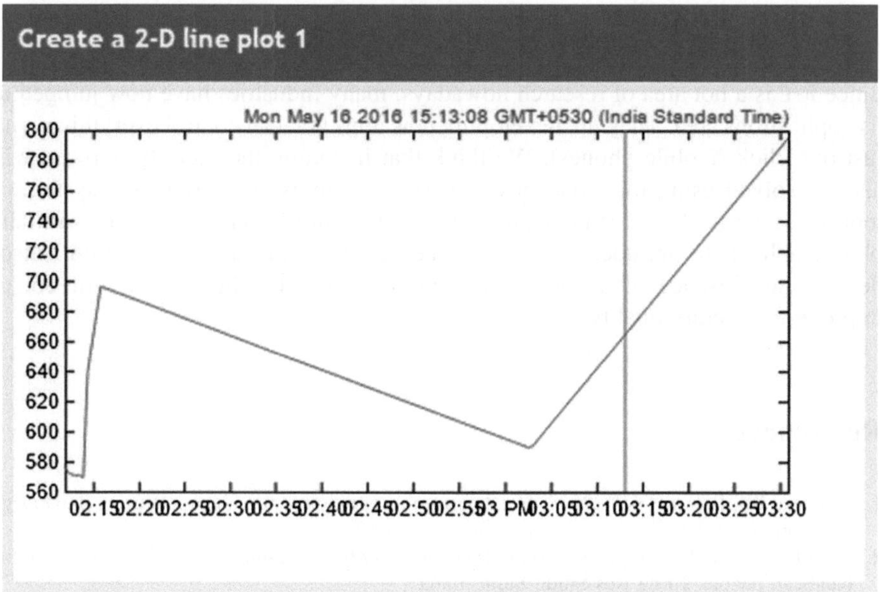

Fig. 6 2-D line plot between *values* sent over IoT and *IST* on 16 *May 2016*

of 240 in Fig. 5, the status of tube light is reversed according to our work. It corresponds to ASL alphabet A.

Certain alphabets showed conflicting results, which means some gestures showed the same values from the SIG, for example, the ASL alphabet O and numeral 0, ASL alphabets M and N and ASL alphabets G and H. That is why we have chosen those alphabets that show completely different results to achieve 100% accuracy in our work.

6 Conclusions

The system gives full flexibility to control remote devices using hand gestures with limited cost and power issues. It helps to save power and energy, which otherwise would have been wasted since there was no means to control the distant appliances. Since the world now is shifting completely towards digitalisation and every bit of it is finding its place on Internet, thus our presented work will also add value to it. This fact proved advantageous for our project.

7 Future Prospects

Since IoT is a hot area of research nowadays, many industries have now jumped to its applications area. They are developing new apps that can control everything with just one click (mobile phones). We think that in future, the security purpose will also be solved using this concept of Internet of things. The combined system of motion sensor, CCTV camera and tube light control can provide us with the photograph of the intruder which is clicked by the camera, when motion sensor detects the presence of a person, and automatically the lights will turn on, to improve the picture quality.

References

1. Gaikwad PB, Bairagi DVK (2014) Hand gesture recognition for dumb people using indian sign language. Int J Adv Res Comput Sci Softw Eng 4(12)
2. Patel B, Shah V, Kshirsagar R (2011) Microcontroller based gesture recognition system for the handicap people. J Eng Res Studi, Surat, India
3. Vinay sagar KN, Kusuma SM (2015) Home automation using internet of things. Int Res J Eng Technol (IRJET), 02(03)
4. Shriharipriya KC, Arthy K (2013) Flex sensor based hand gesture recognition system. Int J Innovative Res Stud (IJIRS), Vellore, India
5. Pathak V, Mongia S, Chitranshi G (2015) A framework for hand gesture recognition based on fusion of flex, contact and accelerometer sensor. In: 3rd IEEE international conference on image information processing (ICIIP). IEEE press, India, pp 312–319
6. Raut A, Singh V, Rajput V, Mahale R (2012) Hand sign interpreter. Int J Eng Sci (IJES), Pune, India
7. Dixit DSK, Shingi NS (2012) Implementation of flex sensor and electronic compass for hand gesture based wireless automation of material handling robot. Int J Sci Res Publ 2(12)
8. Doshi MA, Parekh SJ, Dr. Bhowmick M (2015) Wireless robotic hand using flex sensors. Int J Sci Eng Res 6(3)

VTrack: Emergency Services for Vehicular Networks with Enhanced Security Using Raspberry Pi

Pradnya Shidhaye, Pooja Pawar, Hitesh Jha and Jeril Kuriakose

Abstract A lot of literature is found in the area of vehicular networks. There are several studies that focus on the security issue. Security and privacy become major issues since the public disclosure of identity and location of vehicle is possible. Not only security but also vehicular networking lacks in applications such as emergency services. This paper presents anonymity of driver or passenger in vehicle, the privacy issue by authenticating the vehicle based on a time-dependent secret. The vehicle communicates with the remote server over a reliable and secure medium. For the purpose of simulation, Raspberry Pi, a Linux-based minicomputer, acts as a vehicle interfaced with a GPS module which gives the position of the vehicle. In the case of emergency, the driver pushes the emergency button on his vehicle and its most recent position is recorded at the server. Server sends this location to ambulance or other emergency services which in turn provides the service at the accident-prone zone with minimum delay. Security of entire communication is enhanced by using AES algorithm along with RSA key exchange technique. The entire implementation is a wireless system with added security and can aid to provide quick emergency services in the accident-prone area. This can save the lives of many people.

Keywords Beacon · Security · VANET · Anonymity

1 Introduction

The number of vehicles on the road is increasing continuously which gives rise for the need of efficient traffic management. Accidents or other roadside emergency situations arise frequently where emergency help from hospital and police station may not reach in time. This has resulted in loss of human life. The growing traffic

P. Shidhaye · P. Pawar · H. Jha · J. Kuriakose (✉)
St. John College of Engineering and Technology, Palghar, India
e-mail: jeril@muj.manipal.edu

H.R. Vishwakarma and S. Akashe (eds.), *Computing and Network Sustainability*,
Lecture Notes in Networks and Systems 12, DOI 10.1007/978-981-10-3935-5_32

problems can be solved by implementing vehicular networks. The field of vehicular communication is growing rapidly. This paper deals with the problems arising out of lack of management of traffic- and safety-related applications.

A wireless Ad-hoc network (WANET) is composed of mobile nodes that are connected via wireless links. A mobile ad hoc network (MANET) is a continuously self-configuring, infrastructure-less network of mobile devices, i.e., vehicle. Each vehicle in network is free to move independently in any direction.

Vehicular ad hoc network (VANET) is a specific type of mobile ad hoc network (MANET) that provides communication between vehicles and nearby roadside units or with infrastructure. Vehicular communications can be V2V (vehicle-to-vehicle), V2R (vehicle-to-roadside units), or V2I (vehicle-to-infrastructure). V2V or inter-vehicle communications are emerging as a new class of wireless networks enabling mobile users in their vehicles to communicate to the roadside and to each other.

The application of the system depends on the nature of the nodes. All of the above applications require a system of mobile nodes that can be controlled by a central entity. Although such a system is easily accomplishable through a mobile ad hoc network (MANET), several challenges need to be tackled.

The existing systems developed so far do not ensure secure positioning since the data obtained over wireless channel are vulnerable to packet sniffing, Sybil attack, sniffing of data obtained over GPS, and many more. The users' privacy, i.e., anonymity, is not maintained. Computational costs of security are also high.

The paper [1] analyzes the problems that may arise from falsified position data and proposes detection mechanisms that are capable of recognizing nodes cheating about their locations in position beacons. Evaluation based on simulations shows that this position verification system successfully reveals nodes circulating false positions and hence prevents attacks. It does not rely on external infrastructure. But it is less secure and there is a delay in response.

Objective in literature [2] is to present a novel protocol that verify a vehicle's announced location using a multihop cooperative approach in a Nonline-Of-Sight (NOS) condition whenever direct verification and communication are not possible. With such a solution, a vehicle's awareness of its neighbors increases, theoretically improving the reliability and availability of many safety, travel, and traffic management applications and services while maintaining its confidentiality. The problems arise since it relies on external infrastructure as well as on its protocol limitations.

Literature [3] deals with MANET, where location is obtained through node-to-node communication. Nodes correctly establish their own location as well as verify the positions of their neighbors. Also, it does not rely on infrastructure. The limitation of this paper lies in its dynamic topology and energy since mobile devices rely on battery charge.

The security issues have been well understood from paper [4] where a new approach Anonymous Verification and Inference of Positions (A-VIP) is proposed. In this, anonymous beaconing is done for sharing secret information among users.

Apart from the studies done so far, we can say there is a need for a reliable and efficient service in case of emergency which can be achieved in models of vehicular networks. This model will help provide service in case of accidents, disasters, systematic car parking, automated toll booth, and traffic management and help forces of law in chasing criminals and many more. VTrack is the solution to all above problems where it becomes the reliable as well as efficient system which can be implemented with low cost and time. The upcoming sections describe the basic model where the security is acknowledged along with working of application of emergency service.

2 System Model

VTrack presents an approach for circulating the emergency messages and reporting the event to hospitals, police stations, fire stations, and paramedics, using the existing infrastructures of vehicular ad hoc networks (VANETs) with minimum notification delay.

VTrack server is scripted using PHP and MySql languages. Python is used for implementing some algorithms. Google Maps API is used at the server side for mapping the location of vehicle. Server communicates with the vehicles using wireless medium typically Internet. The circulation of messages to emergency services is achieved using a gateway Vianett or Twilio. The security issue is handled using AES encryption algorithm and RSA key exchange algorithm. Additional security is provided against theft of car by using authentication provided by RFID module. Figure 1 shows the device used in our work.

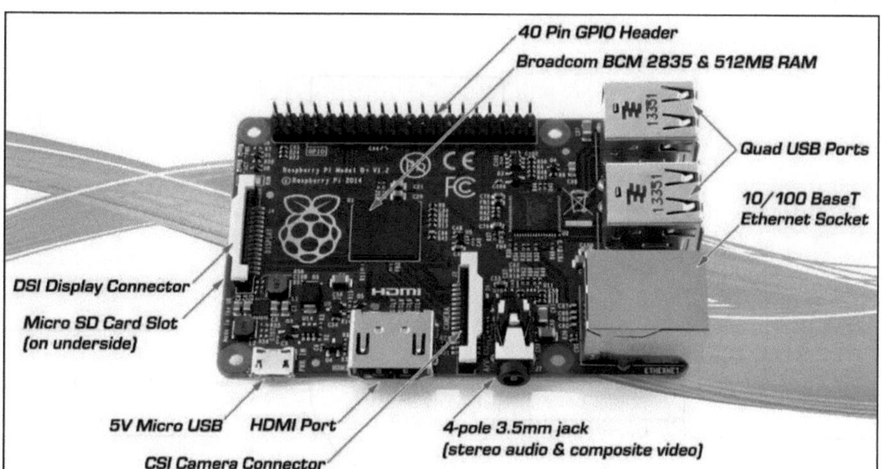

Fig. 1 Raspberry Pi 2 model B+

2.1 Using AES

The Advanced Encryption Standard (AES) is a popular and widely adopted symmetric encryption algorithm published by National Institute for Standards and Technology (NIST) in December 2001. It is a replacement for data encryption standard. AES operates on a fixed block of data and outputs a block of data of same size. It allows data length of 128, 192, and 256 bits and three supporting key lengths of 128, 192, and 256 bits. It is a symmetric key algorithm which means that same key is used for encryption and decryption. The latitude and longitude values of position of vehicle need to be encrypted before sending to server. AES algorithm is used for this purpose.

2.2 Using RSA

It is an asymmetric cryptographic algorithm used widely for key exchange. RSA stands for Ron Rivest, Adi Shamir, and Leonard Adleman, who described it in the year 1978 [5]. In this algorithm, two large prime numbers are considered and their product is used to obtain values of public key and private key. The prime factors must be kept secret. Anyone can use the public key for encrypting message. Decryption is done using private key which is kept secret. RSA is adopted widely because of the fact that it is very simple to multiply two large prime numbers but very difficult and time-consuming to factorize them back. VTrack makes use of RSA to encrypt AES key. Figure 2 shows the typical working of our work. The server encrypts AES key with RSA public key and sends to vehicle. Vehicle decrypts it and gets back its AES key which is used for encrypting latitude and longitude values.

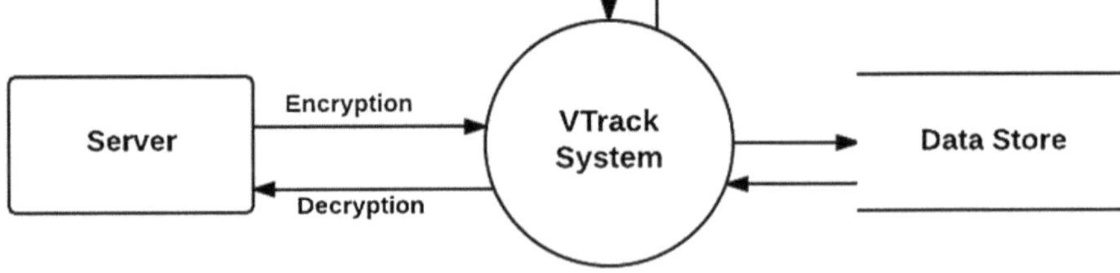

Fig. 2 Typical working of our system

2.3 Twilio

Twilio is a SMS gateway which helps to send or receive SMS to set of registered mobile numbers. Twilio offers wide range of services such as calls, MMS, voice-mail, and play music for caller. Twilio achieves this by sending the data over HTTP or its REST API. The phone numbers of ambulance, police stations, and other services are registered with a Twilio account. In case of any emergency, SMS is sent to these registered numbers informing about the location of area where emergency help is needed. Other gateways such as ICO and Vianett are available which can work easily with VTrack.

2.4 RFID and GPS Modules

The Global Positioning System (GPS) is a satellite-based navigation and positioning system developed by US department of defense. GPS is widely used for navigation and tracking location, surveying of vehicles, aircrafts, and ships. We are using Antenna GPS 3 V Magnetic Mount SMA GPS module for locating position of vehicle. This module is magnet-mount antenna operating at 3 V which is designed for use with automobiles. The magnet will hold the antenna in place at speed greater than 100 mph. It has 5-m cable terminated with standard male SMA connector with a gain of 26 dB.

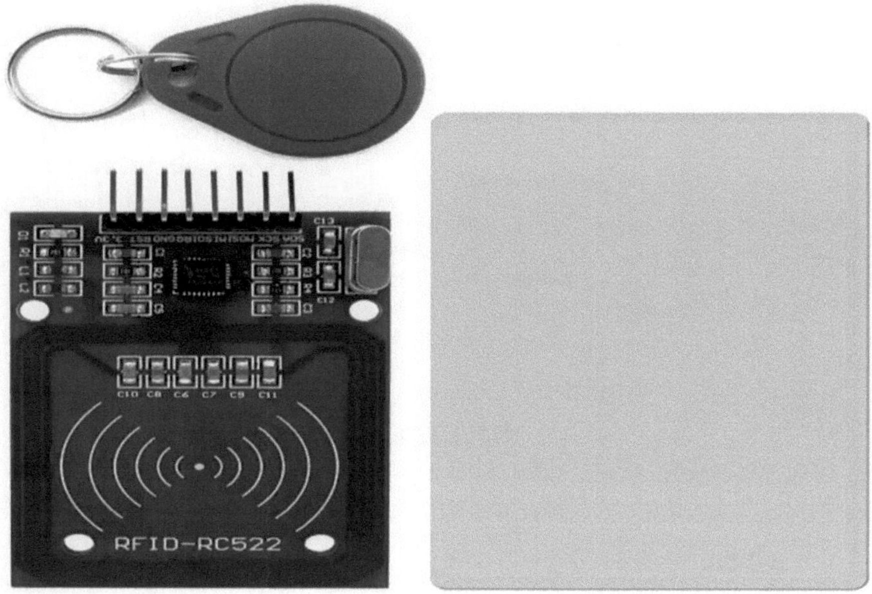

Fig. 3 RFID MFRC522

RFID is a tracking technology used to identify and authenticate tags in order to identify objects and people [6, 7]. An RFID reader continuously emits signal. When a tag comes in the proximity of the reader, the tag responds with the data on its memory. The reader then authenticates that tag based on the information it provided. Tags can be passive or battery-operated. We are using MFRC522 which is compatible with Raspberry Pi and Arduino boards. Figure 3 shows the RFID used in our work.

3 VTrack Working

The RFID tag defines the user identity, and RFID reader saves songs' wish list of that user. The wish list is the context related to that particular user. When RFID reader senses the RFID tag, it authenticates and plays its songs from wish list. This demonstrates context-aware computing. This area can be future implemented using cloud services of providing audios or videos giving entertainment.

The vehicles authenticate using RFID tags and get access to start the engine of vehicle. This prevents unauthorized access to vehicle controls. The start of engine is shown by using a push button [8]. As soon as the vehicle engine starts, it gets registered at the server. Now, a secure medium is set between server and vehicle. The position coordinates are fetched by the magnet-mount GPS module and given to the server in an encrypted form. Server continuously records the location.

Figure 4 shows the typical block diagram of our project. When there is an emergency condition, the user or the driver pushes the emergency button in the

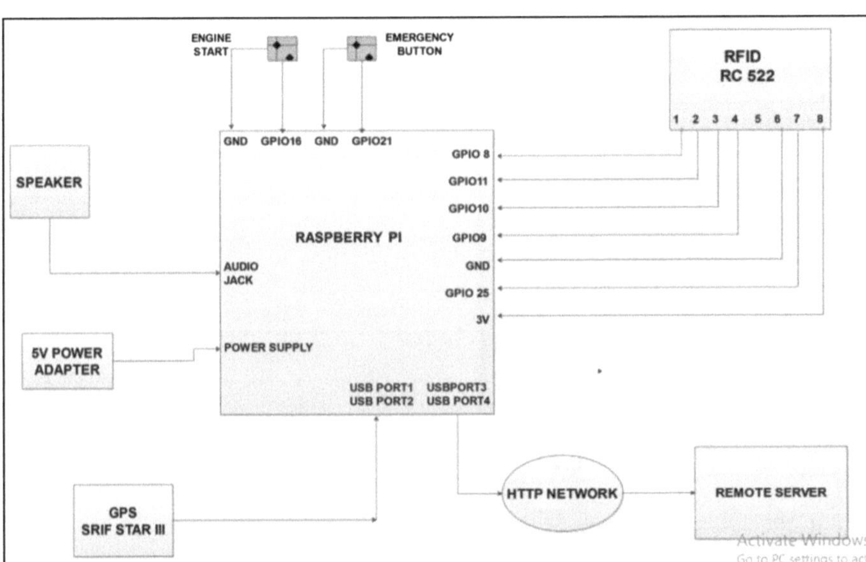

Fig. 4 Block diagram for VTrack

vehicle. The push of emergency button is notified at the server end, and it records the latest position coordinates. This position is sent to the emergency service via SMS.

Anonymity and security of position packet being sent is the most important and challenging fundamental in vehicular Networks. It is attained by using encryption techniques. Figure 5 shows the actual working of our system. AES key is generated by server randomly and sent to vehicle. This key is used by vehicle to encrypt location coordinates. Location coordinates are in the form of latitude and longitude which are Boolean values. AES key is exchanged using RSA key exchange algorithm. A vehicle is registered at server based on its AES key and not unique identification number or any personal details. As soon as the vehicle is registered, its time is recorded at both server and vehicle side. A time span is set suppose 30 min after which the AES key expires and registration is discarded. If the vehicle is still moving, the process of registration restarts. Now, a new AES key is used for encryption of location coordinates. This helps to maintain the anonymity of vehicle.

The server maintains a database wherein it sores the latitude and longitude values. Google Maps API is used to map the coordinates to human understandable location. For emergency need, the SMS is sent to mobile numbers that are registered. Any number of users can be registered. These users can be fire stations, ambulance and hospital services, police stations, etc. To send this message, SMS gateway Vianett or Twilio can be used.

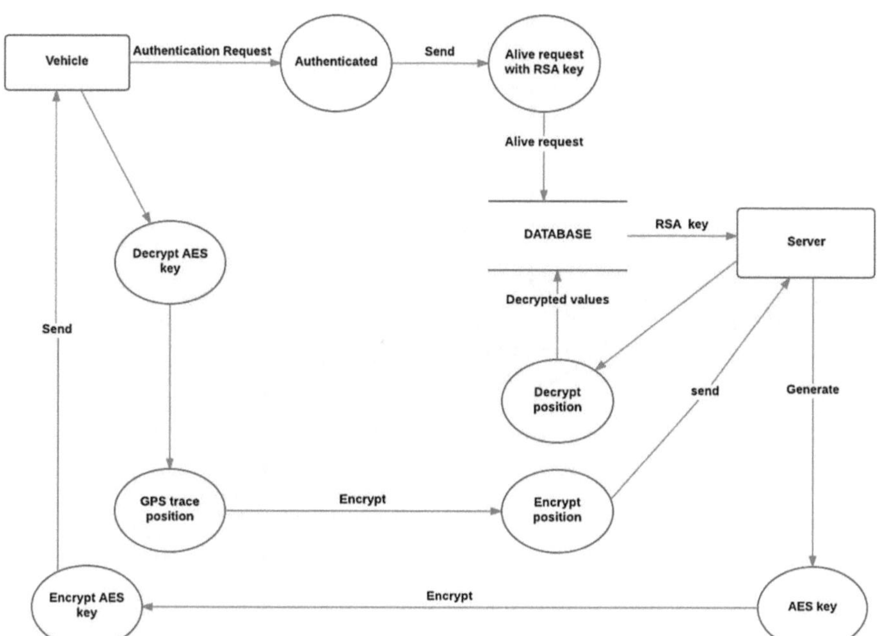

Fig. 5 Actual working of VTrack

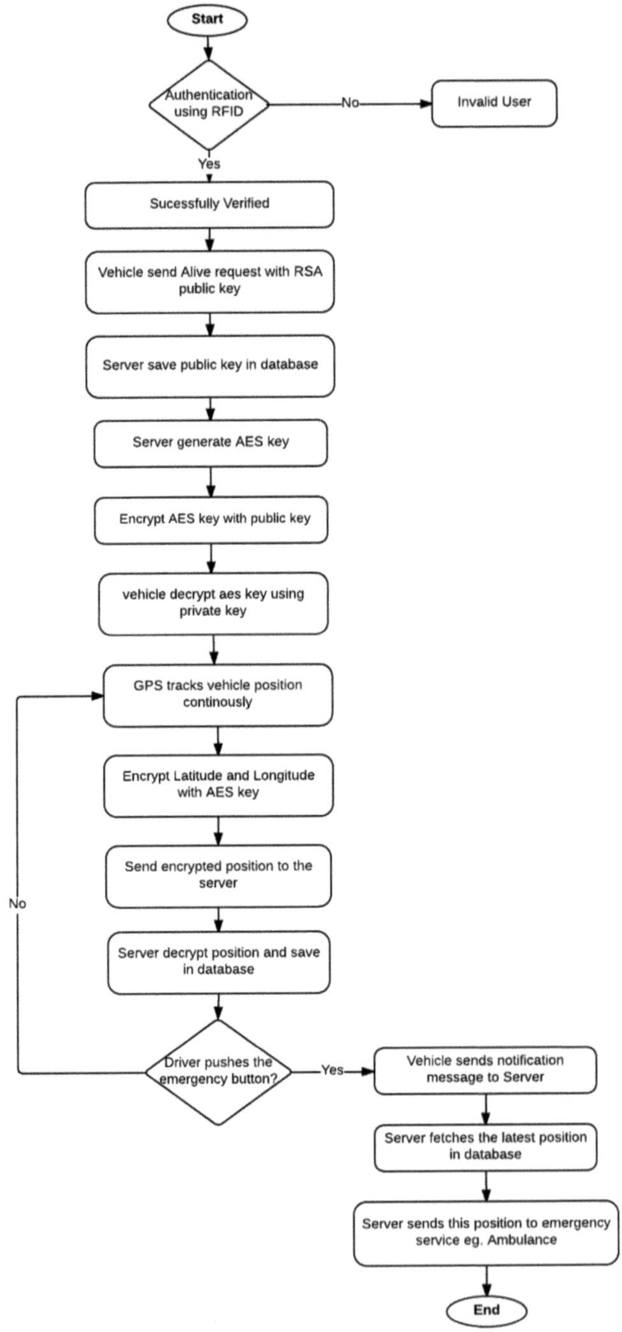

Fig. 6 Flowchart of VTrack

4 Results

As a cooperative approach, vehicular networks can be more effective in avoiding accidents and traffic congestions than each vehicle trying to solve these problems individually. Paper shows communication of vehicles to a server which collects position information of vehicles. Location is sensitive data so it should be sent on secure medium. At same time, privacy of identity of driver is of utmost importance. Traffic monitoring and messaging applications need a system which is secure and reliable. This paper tries to solve these issues (Fig. 6).

The computational time required for AES encryption is 0.0150001049042 ms and that for RSA is 0.0744615412046 ms. Figure 7 shows the time taken for AES and RSA. The time delay between the report is that emergency situation has arisen and its notification to the registered number is very less approx 1–2 s. This shows minimum notification delay and serves the purpose of emergency service which is the primary aim of this paper. Figure 8 shows the localization error faced during the working of our system.

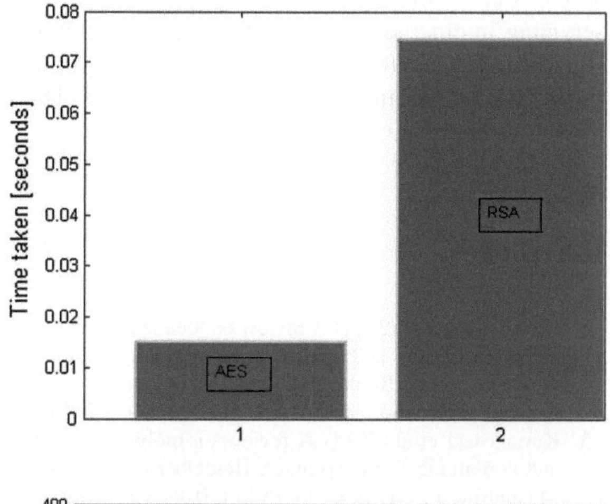

Fig. 7 Time taken for simulation

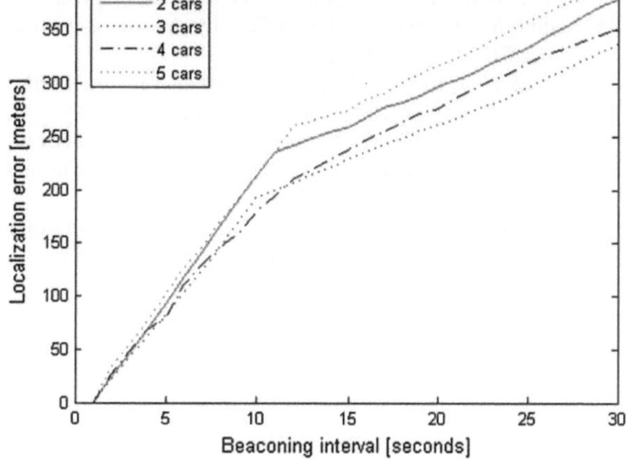

Fig. 8 Localization error

5 Discussions and Future Works

Future of VTrack is quite huge. VTrack can be used for conducting rescue operations where it is humanly impossible or difficult to reach in time. The concept of VTrack can be future extended for traffic monitoring, e-tolling, automatic car parking systems, smart car or brake messaging, and so on. As more automated navigation systems are being developed, the range of applications of VTrack increases, and hence, we can ensure its commercialization also. Thus, VTrack can be used in day-to-day working ensuring well-being of human race.

6 Conclusion

Finally, we can say that major concern for systems that implement vehicular networks is security and anonymity. If not, then it can motivate unlawful conduct that in fact lowers the advantages of its deployment. Hence, we presented a model that includes anonymity, authentication, and secure data transmission. It also provides servicing in emergency situation. Thus, its functionality is monitored under human supervision, henceforth being beneficial toward public help and safety applications. It can be used for automatic car driving system. The system can be further enhanced for future applications.

References

1. Kuriakose J et al (2014) A review on localization in wireless sensor networks. In: Advances in signal processing and intelligent recognition systems. Springer, pp 599–610
2. Abumansoor O, Boukerche A (2012) A secure cooperative approach for non line-of-sight location verification in VANET. IEEE Trans Veh Technol 61(1):275–285
3. Kuriakose J et al (2014) A review on mobile sensor localization. In: Security in computing and communications. Springer, Heidelberg, pp 30–44
4. Malandrino F, Casetti C, Fiore M (2014) Verification and inference of positions in vehicular networks through anonymous beaconing. IEEE Trans Mobile Comput 13(10)
5. Diffie W, Hellman M (1979) Privacy and authentication: an introduction to cryptography. Proc IEEE 67(3):397–427
6. Kuriakose J, Joshi S (2015) A comparative review of moveable sensor location identification. Int J Rob Appl Technol (IJRAT) 3(2):20–37
7. Amruth V et al (2015) Attacks that downturn the performance of wireless networks. In: 2015 International conference on computing communication control and automation (ICCUBEA). IEEE
8. Kuriakose J, Amruth V, Raju RV (2015) Secure multipoint relay node selection in mobile ad hoc networks. In: Security in computing and communications. Springer, pp 402–411

9. Kuriakose J, Amruth V, Nandhini NS (2014) A survey on localization of wireless sensor nodes. In: 2014 International conference on information communication and embedded systems (ICICES). IEEE
10. Kuriakose J et al (2016) Assessing the severity of attacks in wireless networks. In: Proceedings of the international conference on recent cognizance in wireless communication and image processing. Springer, India

Pre-processing Algorithm for Rule Set Optimization Throughout Packet Classification in Network Systems

V. Anand Prem Kumar and N. Ramasubramanian

Abstract With recent advancement in various networking technology, many field packet classifications have evolved from traditional classification so as to classify large rule sets. Most of the previous algorithms provide excellent performance when rule set was small. As rule sets grew in size, performance degraded due to lack of memory and do not have enough processing capabilities to route incoming packet at such a high rate. Packet pre-processing is one of the most important aspects of classification, as it will increase the throughput as well as improve the search performance. The proposed method mainly focuses on pre-processing of pre-defined rule set used during classification. In proposed approach, double hashing technique is to optimized memory usage for high throughput. Proposed algorithm implemented on Xilinx ISE design suite 14.2 with 10000–50000 rules. Simulation results shows that the memory consumption is only three fourth compared to existing approaches.

Keywords Packet pre-processing · TCAM: ternary content addressable memory · Rule set · Memory

1 Introduction

In order to enhance advanced network services like quality of services, flow routing, security and network measurement in network systems, packet classification is a key factor [1, 2]. Packet classification is a process of categorizing the packet using pre-defined rules shown in Table 1. Packet will be forwarded only if incoming packet matches the rule, otherwise deny. Figure 1 shows the flow of packet matching in network systems. It has been studied broadly in the past, however increasing number of rule-set size encourage to study packet classification [3]. Rule sets consists

V.A.P. Kumar (✉) · N. Ramasubramanian
Department of Computer Science and Engineering, National Institute of Technology, Tiruchirappalli 620015, Tamil Nadu, India
e-mail: 306113001@nitt.edu

N. Ramasubramanian
e-mail: nrs@nitt.edu

H.R. Vishwakarma and S. Akashe (eds.), *Computing and Network Sustainability*, Lecture Notes in Networks and Systems 12, DOI 10.1007/978-981-10-3935-5_33

Table 1 Filter set with five fields

Sl. no	IP (8 byte)		Port (4 byte)		Protocol (1 byte)
	SA (32 bits)	DA (32 bits)	SP (16 bits)	DP (16 bits)	
1	209.237.201.208	250.13.215.160	88:88	53:53	0x01/0xFF
2	240.178.169.176	250.222.86.16	123:123	22: 22	0x2f/0xFF
3	69.0.206.0	0.0.0.0/0	0:65535	0: 65535	0x00/0x00
4	63.99.78.32	55.186.163.16	750:750	22:22	0x00/0x00
5	209.67.92.32	159.102.36.48	69:69	21:21	0x06/0xFF

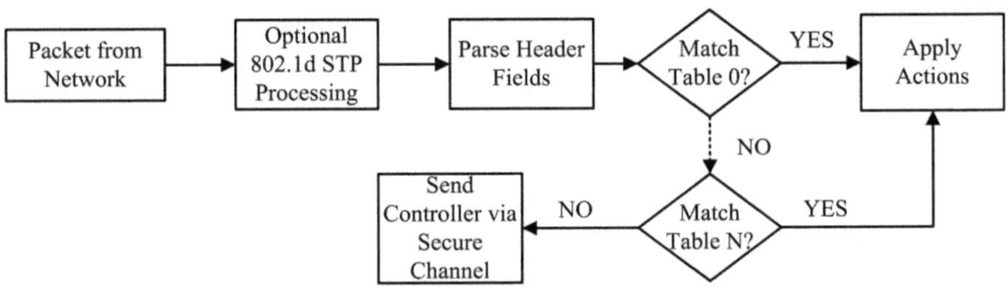

Fig. 1 Flow of packet matching in network systems

of a number of rules, each consisting of several fields and corresponding policies. Traditional classifier have five fields: first two fields are source Address (SA) and destination address (DA), next two source port (SP) and destination port (DP) and last field related to protocol [4].

In case if packet get multiple hits, it returns highest priority rule of most accurate classifier and return matched rules for multi match classification problem [5]. Generally packet classification is divided into following types: Hardware based approach and Software based approach. To achieve high processing speed, hardware devices such as ASIC, FPGA [6] and Ternary content addressable memory (TCAM) are adopted. But it is difficult to update and supports the new network applications. For example TCAM [7] supports only small rule sets due to limited memory space as well as hardware devices consume too much power and area. FPGA provides high frequency operating environment, logic density and it is also being much faster than general purpose processor [8] because of its parallel processing nature.

2 Problem Background

Surveying wide choice of existing hardware and software techniques that try to accomplish the packet classification needs, here are the list of challenges and issues in the implementation of these algorithm.

1. **Pre-processing time**: To fetch rules from rule set in an efficient way, we need to build the appropriate data structure. Time taken for construction of this data structure (pre-processing time) should not be high; otherwise it will have an adverse effect on the overall speed of the system.
2. **Classification speed**: The main objective of classification technique is to aggregate the packet at high speed. Generally time-lapse measured by processor clock cycle is dependent on the number of memory requests. We found one memory access takes more than 150 clock cycles. So memory accesses should be reduced to get better speed.
3. **Memory utilization**: The memory utilization plays an important role in packet classification. For small rule set, less memory and less number of memory accesses required. When rule sets are large, it is expecting more memory for building data structure as well as memory access.
4. **Incremental revise**: Future we may add or delete rules in rule sets according to current network applications. Most of the current solutions are unable to do incremental updates. Therefore pre-process phase has to run again to build reconstructing data structure and it will take additional time during classification.
5. **Number of fields**: It is important to report that how many fields will be able to handle future algorithm. Traditional algorithms support only five fields but current applications need more than 13-fields like Software Defined Network (SDN). The proposed methodology must be able to support maximum fields in Rule set.
6. **Rule size support**: When rule sets are small like 5K–10K most of the algorithms provide attractive results, In larger rule sets expected throughput are under estimation due to the number of rules, processing time and memory requirement increase exponentially. So researcher take care of future algorithms should support large number of rules.

3 Related Works

In software based approach many algorithms are proposed. They are commonly two parts: Decision tree and Decomposition technique. Decision tree based algorithms are most popular technique [9–12]. Hicut algorithm decides to cut single dimension at every internal node for this appraisal request in all nodes [9]. Hypercut is the conservatory of Hicut technique. Here multiple cut allowed simultaneously reducing tree height [10]. For reducing rule replication, Hypersplit algorithm make equal sized cuts, but still unable to eliminate all replication [11]. To overcome this problem, Efficuts algorithm implemented, reducing rule replication separate independent trees were build all overlapping rules [12]. However, in terms of memory utilization in decision tree approach degrades as the number of rule increases and it is not scalable.

In Decomposition based method, rules are decomposed and reconstructed for easy classification such as Bit vector technique. It performs parallel search on every individual field, bit is set as '1' if a particular field match otherwise set as '0' for non matching field. Finally Bitwise AND operation tells us whether the rule match the given packet or not. Bit vector can give high lookup but consume extra memory for vectors. Baboescu and Varghese came up with Aggregate Bit Vector (ABV) technique to enhance the Bit vector approach by statistical analysis of rule set. ABV algorithm essentially partitions the N bit vector into chunks. However the final stage of aggregation part consumes more memory [13]. Fong and Wang came up with Range point conversion. Here, all Rules represented as points and they group similar rules together using a cluster algorithm. Preprocessing time is much higher in this approach [14]. Tuple space search is a multiple match fields technique; tuple define how many bits representing the particular field, for example source address and destination address having four bit and source port and destination port having 2 bits and protocol having one bit, the tuple will be (4, 4, 2, 2, 1). For matching rule set, probe this tuple using exact match hashing but the problem is their assumption in no collision [15]. To reduce memory requirement saran and song, came up with bloom filter techniques. It gives efficient performance due to O(1) search and sometime it gives false positive which is in a set search value not available but it may give positive results [16].

4 Proposed Method

The proposed method consists of hash optimization using the double hashing procedure. This is a two step procedure where in first step appropriate hash function is used. In case of collision free key, there is no need to do hashing on the value rule set. If there is a collision in the key values, then the next hash function is used till the collision is resolved. Figure 2 explains how rule sets are processed using our approach. From original rule set, we split the entire field separately and then we map every field to the hash table using hash function. Finally once again we update the RAM memory. Since major fraction of values are repeated in rule set, the small hash table is sufficient to accommodate all hash values without any loss of information. The procedure is explained in Algorithm 1. In the above procedure chaining is purposefully avoided in order to maintain a constant search time. For repeated values, the hash value is dropped which makes the procedure even more efficient for rule set optimization. The classification is a complex procedure which is done at particular core of the network processor. The small hash table is easy to store and swap in the lowest level of memory and make the procedure faster. We have three main benchmark rule sets -ACL, FW and IPC given in Table 2 with number of rules and its traces. Our proposal mainly focused for large rule sets, but benchmark rule sets are having less number of rules. So we created synthetic rule sets with large number of rules and the result is analyzed only in terms of reduction in size of rule set for classification.

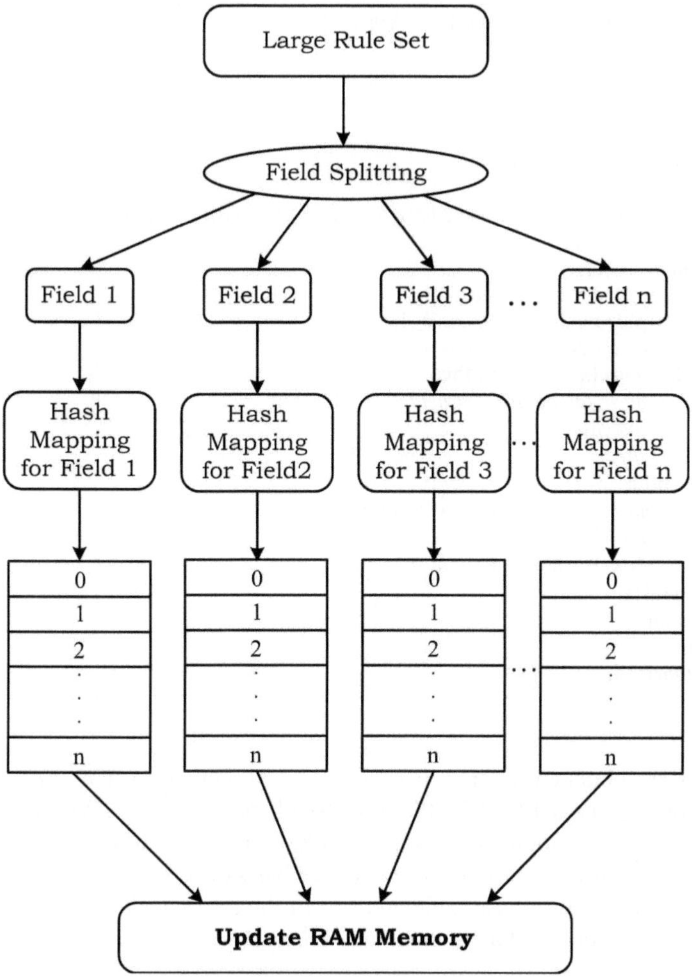

Fig. 2 Memory optimization using hash technique

Table 2 The range of standard rules and test packets

Bench mark (→)	ACL		FW		IPC	
Range(↓)	Rule count	Test packets	Rule count	Test packets	Rule count	Test packets
0.1 K	98	1000	92	920	99	990
1 K	916	9380	791	8050	938	9380
5 K	4415	45600	4653	46700	4460	44790
10 K	9603	97000	9311	93250	9037	90640

Algorithm 1 Rule set optimization Algorithm

```
Assumptions:
hash_table is a vector;
match is a vector;
Begin Algorithm
Initialize hash_modulus_1
Initialize hash_modulus_2
Where hash_modulus_2! = 0 and hash_modulus_2! = hash_modulus_1;
vector hash_table = 0;
vector match = 0;
for (each field in rule set ) do
    hash1=field mod hash_modulus_1;
    for (r=1 to sizeof(hash_table)) do
        if (match[hash1]==0) then
            hash_table[hash1] = field ;
            match[hash1] = 1 ;
            break;
        end if
        if (hash_table[hash1]!=field ) then
            hash2 = hash_modulus_2 - (field mod hash_modulus_2);
            hash1 = hash1 + r * hash2;
        end if
    end for
end for
end Algorithm
```

Results are obtained by applying the proposed method on FW 9000, Synthetic ACL-15000, Synthetic FW IPC 30000 and Synthetic FIC50000 . We used hash function 16000 for destination Address field, 14000 for source Address field and 10000 for port fields. Hash function is selected on the basis of previous studies about extent of repletion of values in hash table and size of the rule set. For example, if rule set is smaller we can choose hash function as 100 or 200.

5 Result and Analysis

The proposed method is implemented on Xilinx ISE Design suite 14.2 for purpose of evaluation. The standard rule set FW 9000, Synthetic ACL-15000, Synthetic FW IPC 30000 and Synthetic FIC50000 were tested for optimization in the size of all five fields. The hash table evaluated is of size 16000 for destination Address field, 14000 for source Address field and 10000 for port fields. Table 3 shows the significant reduction in all the five fields, the table does not occupy more than 74.64% and it occupies only 5.76% of protocol fields because of high redundancy. The firewall rule set is well known for the high extent of repeated values due to which it appears to be the most optimized one in the results.

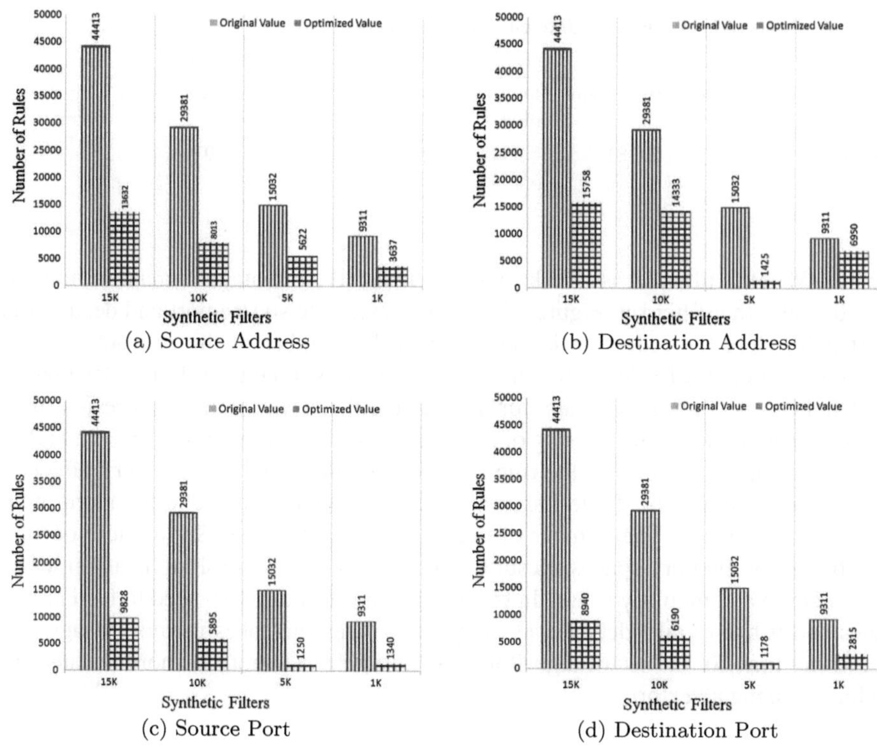

Fig. 3 Memory utilization

Table 3 ACL, FW & IPC rule set memory usage (in %) for HASH MOD 50

No. of rules(→)	9311	15032	29381	44413
Types(↓)	FW	Synthetic_ACL	Synthetic_FW_IPC	Synthetic_all
Source address	39.061325	37.400212	27.272727	30.693715
Destination address	74.642895	9.479776	48.783227	35.480602
Source port range	14.391579	8.315593	20.063986	22.128656
Destination port range	30.233057	7.836615	21.068037	20.129241
Protocol	5.369992	4.626609	2.382492	5.761151

Figure 3a represents the change in size with respect to source address where the 'x' axis denotes the different types of synthetic filter set and the 'y' axis denotes the number of rules. The Graph shows reduction in the number of rules when adapting our optimized solution. Figure 3b shows memory reduction for destination address. Due to excessive redundant values in destination address field, we achieve better

Table 4 Protocol field values for different synthetic filters

15 K		10 K		5 K		1 K	
Real count	Optimized count	Real count	Optimized count	Real count	Optimized count	Real count	Optimized count
44413	7	29381	7	15032	4	9311	5

results. When a filter set is a 15 k memory, reduced percentage is maximum compared to all other filter set. Figure 3c and d represent the source port and destination port field. Here the reduction is higher when the number of the rule set is large. For lower number of rule set it is not as effective in source port. In the destination port field, it gets optimal results for all the rule set sizes. Storage space is a well known issue in the pre-processing of rules at the core of the processor. So the higher extend of repeated values in the rule set make the proposed method very effective for pre-processing task. Table 4 shows the protocol field original value along with the optimized value. The number of unique values is too less. So the field occupy few blocks of memory spaces. Results acquired that double hashing beats in terms of searching and memory usage. In tree based data structure [9], need to feed all the rules and it has to take decision at every node. This increases memory usage and searching time. In bit vector algorithms [13] usage of memory is more because all fields are using a vector.

6 Conclusion

The proposed pre-processing method exploits the repeated values in the rule sets to reduce the requirement of the storage space. The pre-processing task is done at the embedded network processors having very limited storage space. High reduction of size of the field in a rule set makes pre-processing very effective, especially when the new rules sets are having values in the range of 10000 to 40000. The method is tested with different hash function to show the generosity of the proposed method. In the future with the development of standard benchmarks, the proposed method can be evaluated for the performance in the terms of processing speed.

References

1. Chao HJ (2002) Next generation routers. Proc IEEE 90(9):1518–1558
2. Specification, O.S. (2009) version 1.0.0 (wire protocol 0x01)
3. Ma Y, Banerjee S (2012) A smart pre-classifier to reduce power consumption of TCAMS for multi-dimensional packet classification. In: Proceedings of the ACM SIGCOMM 2012 conference on applications, technologies, architectures, and protocols for computer communication. ACM, pp 335–346

4. Gupta P, McKeown N (2001) Algorithms for packet classification. IEEE Netw 15(2):24–32
5. Qu YR, Zhang HH, Zhou S, Prasanna VK (2015) Optimizing many-field packet classification on FPGA, multi-core general purpose processor, and GPU. In: Proceedings of the Eleventh ACM/IEEE symposium on architectures for networking and communications systems. IEEE Computer Society, pp 87–98
6. Cho YH, Mangione-Smith WH (2004) Deep packet filter with dedicated logic and read only memories. In: 12th Annual IEEE Symposium on Field-programmable custom computing machines, FCCM 2004. IEEE, pp 125–134
7. Yu F, Katz RH, Lakshman T (2005) Efficient multimatch packet classification and lookup with tcam. IEEE Micro 25(1):50–59
8. Brebner GJ (2011) Reconfigurable computing for high performance networking applications. In: ARC, p 1
9. Gupta P, McKeown N (1999) Packet classification using hierarchical intelligent cuttings. In: Hot interconnects VII, pp 34–41
10. Singh S, Baboescu F, Varghese G, Wang J (2003) Packet classification using multidimensional cutting. In: Proceedings of the 2003 conference on applications, technologies, architectures, and protocols for computer communications. ACM, pp 213–224
11. Qi Y, Xu L, Yang B, Xue Y, Li J (2009) Packet classification algorithms: from theory to practice. In: IEEE INFOCOM 2009. IEEE, pp 648–656
12. Qi Y, Fong J, Jiang W, Xu B, Li J, Prasanna V (2010) Multi-dimensional packet classification on FPGA: 100 gbps and beyond. In: 2010 international conference on field-programmable technology (FPT). IEEE, pp 241–248
13. Baboescu F, Varghese G (2001) Scalable packet classification. ACM SIGCOMM Comput Commun Rev 31(4):199–210
14. Qu YR, Zhou S, Prasanna VK (2013) Scalable many-field packet classification on multi-core processors. In: 2013 25th international symposium on computer architecture and high performance computing (SBAC-PAD). IEEE, pp 33–40
15. Srinivasan V, Suri S, Varghese G (1999) Packet classification using tuple space search. In: ACM SIGCOMM Comput Commun Rev 29. ACM, pp 135–146
16. Dharmapurikar S, Song H, Turner J, Lockwood J (2006) Fast packet classification using bloom filters. In: ACM/IEEE symposium on architecture for networking and communications systems, ANCS 2006. IEEE, pp 61–70

Simulation and Comparison of AODV Variants Under Different Mobility Models in MANETs

Shiwani Garg and Anil Kumar Verma

Abstract AODV (ad hoc on-demand distance vector) is a reactive routing protocol for MANETs, and it enables multihop routing within the mobile nodes taking part in initiating and preserving an ad hoc network. In multihop routing, route is requested only when it is required and it does not keep track of routes to the destination node. The primary idea of this paper was to assess the AODV variants underneath acquainted mobility models. We have considered three mobility conditions, i.e., Group mobility model, Random Waypoint, and Manhattan model. In our paper, performance metrics such as PDR (packet delivery ratio) and throughput are used to evaluate the performance of AODV variants. The performance metrics are examined with varying node density, and Ns-2 is used as a network simulator to carry out these simulations.

Keywords AODV · AOMDV · MAODV · MANET · Mobility models · PDR · Throughput

1 Introduction

MANET (mobile ad hoc network) [1] is a cluster of mobile nodes which are wirelessly connected. It builds a network which is temporary, and the network is neither built using any access point nor through centralized administration. Its topology changes dynamically due to which routing is very challenging. MANETs can communicate with different networks that are not ad hoc. MANETs comes with a feature of multihop routing which states that once one node likes to send data to the other node but the node is out of its range, then the packet is forwarded in the network through one or more intermediate nodes. In this paper, survey of some

S. Garg (✉) · A.K. Verma
CSED, Thapar University, Patiala, Punjab, India
e-mail: shiwani92@gmail.com

A.K. Verma
e-mail: akverma@thapar.edu

H.R. Vishwakarma and S. Akashe (eds.), *Computing and Network Sustainability*, Lecture Notes in Networks and Systems 12, DOI 10.1007/978-981-10-3935-5_34

mobility models is presented and to simulate an ad hoc network, these models are used.

The objective of this paper was to analyze the AODV variants performance under different mobility conditions using varying node density. RandomWaypoint is used as reference model in previous studies. MANETs in another era are supposed to be used with various node configurations and topographies. Therefore, one ought to grasp the influence of varied mobility conditions on protocol performance along which they should develop a thorough understanding of these models.

The paper is structured as follows: Sect. 2 provides description of AODV variants. Section 3 gives an overview of mobility models used. Section 4 describes about the simulation environment. Section 5 shows the results. Conclusion and future scope are given in Sect. 6.

2 AODV Variants

2.1 AODV (Ad Hoc On-Demand Distance Vector)

AODV [2] is a reactive routing protocol. As the definition of AODV says, it is an on-demand routing protocol, neither the nodes in the chosen path should maintain the route nor they should participate in the exchange of tables. It is pure-on-demand procurement system. In AODV, route discovery process is used to discover the routes through which messages can be delivered. It always make efforts to discover the loop-free routes which are shortest. If it finds any error, it always maintains the routes by creating new routes and this process is called as route maintenance.

A broadcast Route Discovery Mechanism

RREQ (Route Request Packet) is broadcasted to find a route.

RREP (Route Reply Packet) is used to set up forward path.

AODV routing table entries

Destination Address—It provides IP address for the destination.

Hop Count—How many hops are required to send data to the destination.

Destination Sequence Number—It specifies the freshness of the information received at the destination.

Next-hop Address—The address of the adjacent node which is appointed to forward packets to the destination for the purpose of current route entry.

Life-time—How long this path can be used.

2.2 AOMDV (Ad Hoc On-Demand Multipath Distance Vector Routing)

AOMDV [3] is a supplement of the eminent AODV routing protocol. In every route discovery process, it locates multiple routes within the destination node and the source node. It is used to determine various link disjoint and loop-free paths. These multiple paths can be used for loop spreading and when main route fails, these paths serve as backup routes. In this routing protocol, RREQ packet moves from the source node toward the destination node and multiple number of reverse routes have been established at intermediaries and at the destination. Several RREPs move across these reverse routes in backward direction, so that multiple number of forward routes can be created from the source and intermediaries to the destination.

2.3 MAODV (Multicast Ad Hoc On-Demand Distance Vector Routing)

Many of the ad hoc applications are efficiently supported by multicasting that represents a feature of a close degree of association. Due to the increase in the dynamic nature of mobile nodes, MAODV evolves. It is used to provide an effective multicasting service. MAODV extends AODV to support multicasting. When needed, it creates multicast trees to connect the members of a group. MAODV protocol is to "multicast" a packet to several destinations or a group of destinations. Route discovery in MAODV follows a RREQ/RREP discovery cycle. It is a reactive routing protocol which discovers routes when demanded. A multicast tree consisting of group members is created when nodes join a group.

3 Mobility Models

Mobility models [4, 5] can be differentiated with respect to random models, geographic restriction models and spatial dependency models as shown in Fig. 1.

3.1 RandomWayPoint

RandomWaypoint [4, 5] is a widely used mobility model. It became a standard model to assess and analyze the routing protocols of MANET. Every moment, a node moves toward the destination chosen randomly with a desired uniform velocity (Fig. 2).

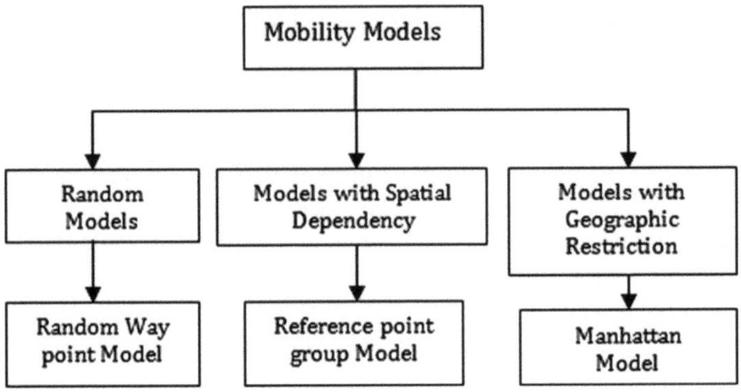

Fig. 1 Mobility models

Fig. 2 RandomWaypoint mobility model

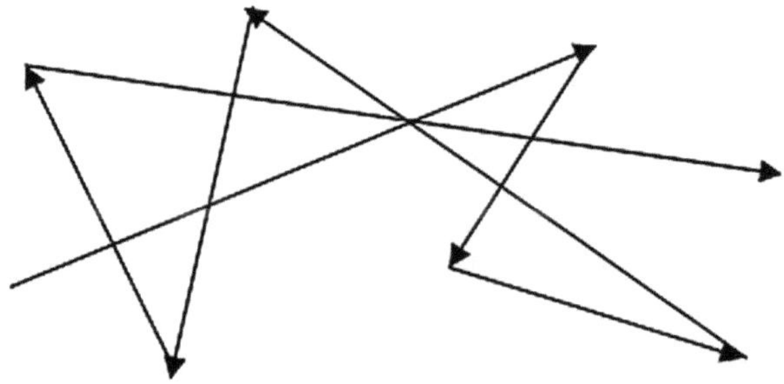

It includes pause times whenever the node changes its direction or velocity till it approaches its destination. This action is constantly repeated till the end of the duration of the simulation.

3.2 RPMG (Reference Point Group Mobility)

Each group of this mobility model consists of either a *group leader node* or a *logical center*. The nodes in a group [6] of this model are randomly placed in adjacent to its reference points. Afterward due to the scheme of reference point which allows every node to have its own speed and direction. In addition to group motion, each node has an independent motion, i.e., it deviates from its group leader randomly (Fig. 3).

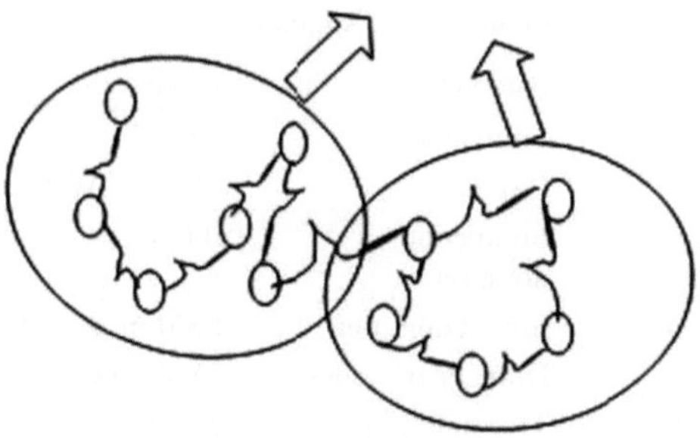

Fig. 3 RPMG (reference point group mobility model)

3.3 Manhattan Mobility Model

Mobile nodes in this model imitate the movement scheme of locomotives on roads as stated by maps. In this model, the mobile nodes moves in pseudorandom manner on already defined pathways in the simulation area (Fig. 4).

4 Simulation Environment

4.1 Simulation Scenario

Operating System: Ubuntu 14.04
Simulator: NS2
NS2 version: ns-allinone-2.34

The implementation environment is of "OPEN SOURCE." The network simulator NS-2 is used to carry out simulation. It is the software with a feature of discrete event simulation and is used for network simulations. NS is primarily useful for local and wide area networks.

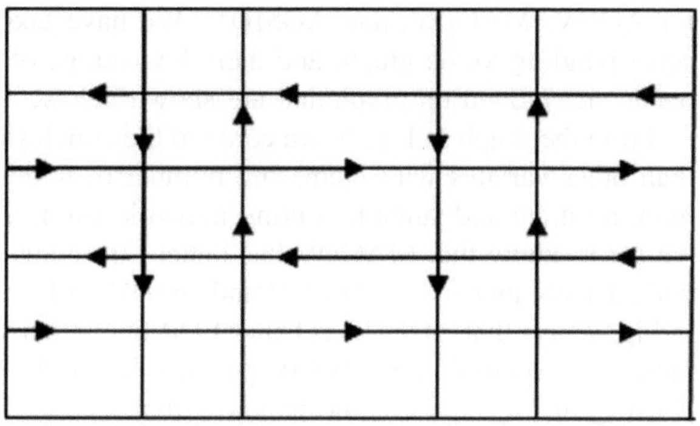

Fig. 4 Manhattan mobility model

Table 1 Parameters of simulation

S. no	Parameters	Values
1.	Simulator	NS-2.34
2.	Protocols	AODV, AOMDV, MAODV
3.	Simulation duration	200 s
4.	Simulation area	1000 m × 1000 m
5.	Number of nodes	30, 40, 60, 80, 100
6.	Transmission range	250 m
7.	Mobility models	RandomWayPoint reference point group model, Manhattan model
8.	MAC layer protocol	IEEE 802.11
9.	Pause time	100 s
10.	Max speed	20 m/s
11.	Packet rate	4 packets/second
12.	Traffic type	CBR (UDP)
13.	Data payload	512 bytes/packet

4.2 Simulation Parameters

The performance of AODV variants is assessed by varying node density. Simulation parameters to assess AODV variants are described in Table 1.

5 Results and Analysis

As already declared, we have taken three variants of AODV and three mobility models. We ran the simulation environment for 200 s with varying node density form 30 to 100 nodes. Throughput and packet delivery ratio (PDR) are calculated for AODV, MAODV, and AOMDV. We have taken 10 readings for each value corresponding to the graph, and then the average of these readings is plotted. The results are plotted on graph and are shown below.

From the graph in Fig. 5, we come to the conclusion that AODV has better PDR than other variants with increasing number of nodes. The better PDR implies the more accurate and suitable routing network. From the analysis of graph in Fig. 6, we get to know that MAODV has better throughput than other variants when the node density increases. AOMDV and AODV perform in similar manner. The graph in Fig. 7 says that AODV has better PDR than other variants when the node density increases. AOMDV and AODV perform in similar manner with increasing node density. In Fig. 8, we come to know that AODV has exceeded throughput than

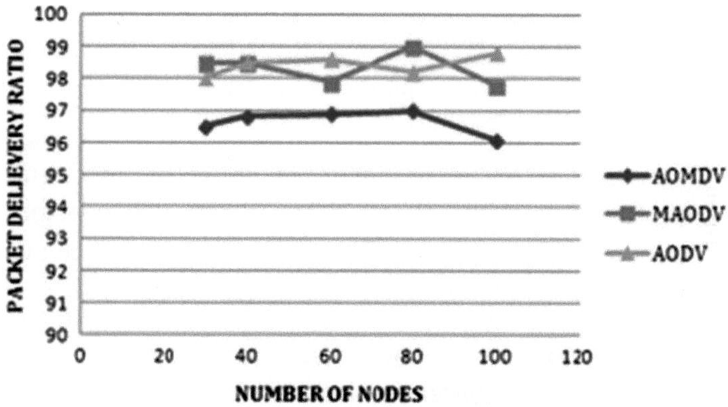

Fig. 5 Average PDR (packet delivery ratio) versus number of nodes in RandomWaypoint mobility model

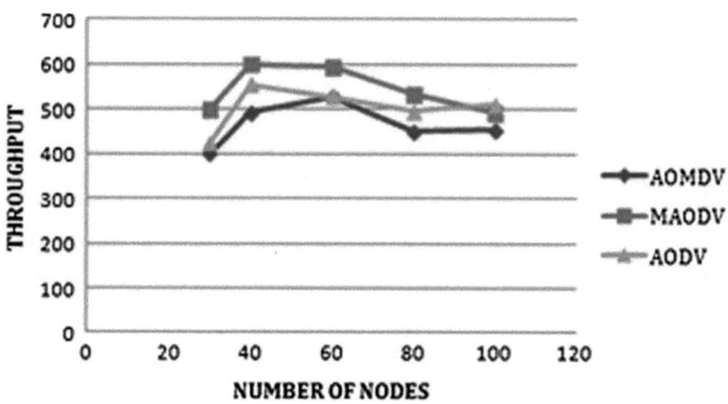

Fig. 6 Average throughput (kbps) versus number of nodes in RandomWaypoint mobility model

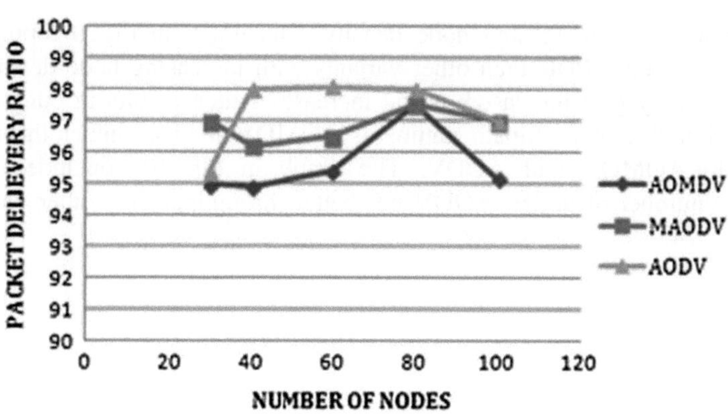

Fig. 7 Average packet delivery ratio versus number of nodes in RPMG (reference point group mobility model)

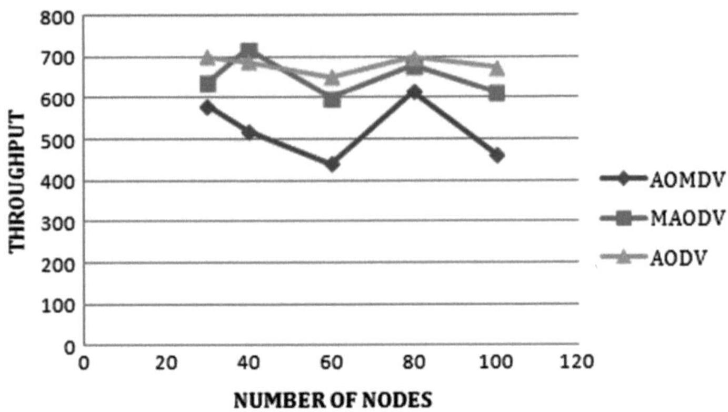

Fig. 8 Average throughput (kbps) versus number of nodes in reference point group mobility model

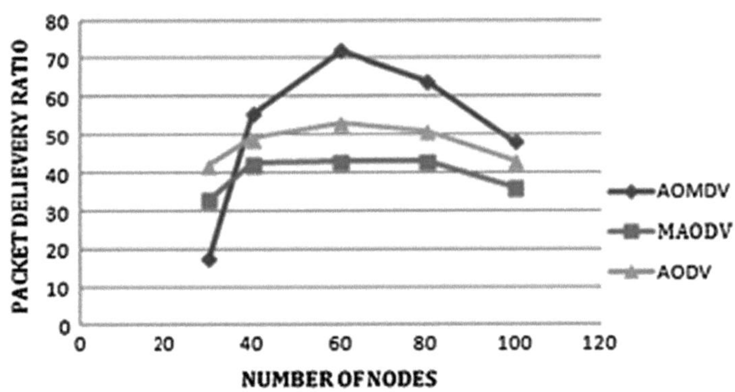

Fig. 9 Average PDR (packet delivery ratio) versus number of nodes in Manhattan mobility model

other variants with increasing node density. The graph in Fig. 9 represents that AOMDV has better PDR than other variants with increasing node density. When the node density gets increased, traffic increases which results in congestion and data loss but due to multipath nature of AOMDV it gives better throughput in contrast to AOMDV and AODV. The graph in Fig. 10 concludes that with increasing number of nodes, AODV has better throughput than other variants.

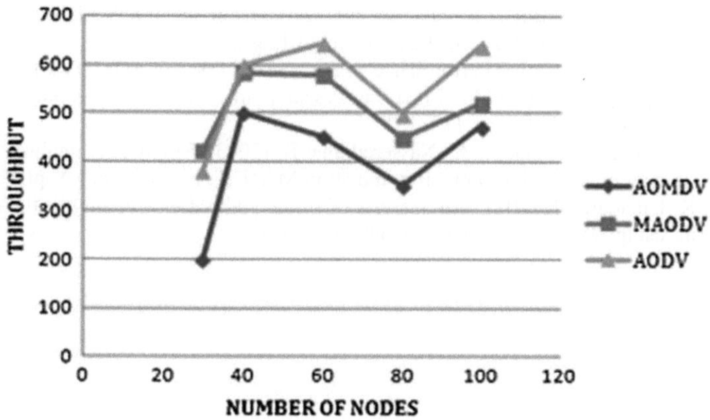

Fig. 10 Average throughput (kbps) versus number of nodes in Manhattan mobility model

6 Conclusion and Future Scope

In this paper, a short description about the AODV variants and various mobility conditions has been given. It represents an overview of performance of AODV, AOMDV, and MAODV under different mobility conditions. We have observed that performance degrades with increasing node density.

In future, we are able to study different routing protocols using other mobility models to figure out the selection of the required protocol optimally.

References

1. Aarti, Tyagi SS (2013) Study of MANET: characteristics, challenges, application and security attacks. Int J Adv Res Comput Sci Softw Eng 3(5):252–257
2. Chakeres ID, Belding-Royer EM (2004) AODV routing protocol implementation design. In: Proceedings of 24th international conference distributed computing systems workshops, 2004. IEEE, pp 698–703
3. Marina MK, Das SR (2006) Ad hoc on demand multipath distance vector routing. Wireless Commun Mobile Comput 6(7):969–988
4. Bai F, Sadagopan N, Helmy A (2003) Important: a framework to systematically analyze the impact of mobility on performance of routing protocols for ad hoc networks. In: Proceedings of IEEE information communications conference. San Francisco
5. Bai F, Helmy A (2006) A survey of mobility models. Wireless adhoc networks, vol 206. University of Southern California, USA
6. Hong X, Gerla M, Pei G, Chiang CC (1999) A group mobility model for ad hoc wireless networks. In: Proceedings of the 2nd ACM international workshop on modeling, analysis and simulation of wireless and mobile systems, vol 5360
7. Broch J, Maltz DA, Johnson DB, Hu Y-C, Jetcheva J (1998) A performance comparison of multi-hop wireless ad hoc network routing protocols. In: Proceedings of the 4th annual ACM/IEEE international conference on mobile computing and networking, pp 85–97

8. Chadha MS, Joon R (2012) Simulation and comparison of AODV, DSR and AOMDV routing protocols in MANETs. Int J Soft Comput Eng 2(3):745–749
9. Rekha B, Ashoka DV (2014) Performance analysis of AODV and AOMDV routing protocols on scalability for MANETs. In: Emerging research in electronics, computer science and technology. Springer, India, pp 173–181
10. Saadi Y, Kafhali SE, Haqiq A, Nassereddine B (2013) Simulation analysis of routing protocols using manhattan grid mobility model in MANET. Int J Comput Appl 45(23):24–30
11. Vanaja K, Umarani DR (2011) An analysis of single path AODV vs multipath AOMDV on link break using NS-2. Int J Electron Comput Sci Eng

A Novel Trust Mechanism for Collaborative Recommendation Systems

Manjeet Kaur and Shalini Batra

Abstract Collaborative filtering is one of the successful techniques in generating personalized recommendations. This paper provides a novel trust mechanism between nodes which is inspired from dynamic trust relation between crime inspectors and their secret informers. In the proposed technique, collaborative filtering has been merged with K-means clustering which improves the overall efficiency and speed of the recommendations. The trust value changes dynamically based on the confidence and similarity between users, which increase the confidence of user in recommendations and overcomes the issues such as shilling attacks. Performance of trust based recommender system is evaluated on Movielens dataset and compared with traditional collaborative filtering and K-means clustering without trust.

Keywords Recommender system · Collaborative filtering · K-means clustering · Movielens

1 Introduction

In our real life whenever one come across choices or options and one is in dilemma, he or she mostly seek for the most appropriate person to recommend. The characteristics of this person provide us idea of an intelligent recommender system. For example, one may choose a person who has similar preferences or features as ours, etc.

Physical delivery systems have very limited resources but online stores make most of existing things available to the customers. This difference between the online world and physical world is called as long tail phenomenon due to the shape

M. Kaur (✉) · S. Batra
CSED, Thapar University, Patiala, Punjab, India
e-mail: manjeet.kamboj13@gmail.com

S. Batra
e-mail: sbatra@thapar.edu

H.R. Vishwakarma and S. Akashe (eds.), *Computing and Network Sustainability*, Lecture Notes in Networks and Systems 12, DOI 10.1007/978-981-10-3935-5_35

Fig. 1 Long tail

of graph plotted between the number of times item chosen (vertical) to items ordered on the basis of popularity (horizontal). The long tail phenomenon makes recommendation system [1] really necessary. It is not possible to show all in stock items of user in online store the way physical institutions can do, so recommendations are needed in online stores. Figure 1 shows the long tail that is composed of small number of popular items, and the all left out items in heavy tail, i.e., all those items which are not selling frequently.

Various techniques of recommender system have been proposed which include collaborative filtering, knowledge-based, demographic-based, utility-based, content-based techniques [2]. Collaborative filtering relies on the behavior, preferences, and opinions of a large community of other users, and hence named community-based technique. Content-based technique takes into account additional information about items and user preferences and does not depend on rating history. In demographic approach, depending on sex, age, marital status, items which are relevant to the user at that point are recommended. Hybrid approach is a combination of two or more of the above-mentioned approaches. Collaborative filtering [3] mainly consists of three following steps.

(i) Building of user-item rating matrix
(ii) Selecting the nearest neighbors.
(iii) Generating recommendations from neighbors

In this paper, collaborative filtering has been used in combination with K-means clustering to improve the speed of recommendation process, but it has been found that such hybrid approach is prone to attacks and sparseness of the matrix.

One of the major issues is recommender system is that how trust mechanism [4] should be introduced in the users using collaborative filtering so that it increases the confidence of users on the recommendations provided. Further shilling attack which is caused when many fake profiles give faux rating to disturb true recommendations by rating high a less popular item or rating low a popular item is another important issue which needs consideration. Our approach increases the confidence of the users on the recommendations provided and checks issues such as shilling attack by introducing trust mechanism.

Collaborative filtering works by building a user item database of preferences of items by users. An active user is matched against the database to produce neighbors. In the proposed approach, clustering algorithm is applied in the initial stage. Clustering the user item rating database divides the entire set into K clusters, to one of which the new customer will belong. Now recommendations are found on much smaller part of entire user database, i.e., cluster leading to increased processing speed, and hence performance is enhanced. Trust being time-dependent, information can be analyzed using the changed trust level of crime investigators on their secret informers. Few police officers are well known for having relied on informers more than their own team.

The trust on informers is dynamic in nature, and it varies as follows:

Informers that come up with more useful information in terms of the material it possess toward solving the case are updated with higher trust, and informers with less or no clues are less relied. Hence, recommendations are produced by first constructing the clusters to find closest neighbors, and then trust is calculated that is updated each time recommendations are looked for by the active user. In our proposed approach, speed of classic collaborative filtering method is improved using clustering and attack such as shilling attack is eliminated by introducing trust mechanism.

2 New Collaborative Filtering Algorithm Based on Clustering and Trust

2.1 Clustering of Users

Initially, K-mean clustering [5] algorithm has been used to organize the dataset into meaningful groups, assuming that each object belongs to only one cluster and overlapping of cluster does not happen. These clusters should have high intra-cluster similarity and low inter-cluster similarity.

Input: Dataset user-item rating matrix and number of clusters k. Output: Set of cluster centers C.

1. Set the number of clusters. Say k.
2. Pick any centroids of k clusters.
3. Compute Euclidean distance of each user in the dataset from each of centroids.
4. Allocate each user to the cluster it is similar to on the basis of Euclidean distance calculated above.
5. Compute centroids for these clusters by calculating average of column value of users in each cluster.
6. If cluster membership is changed go to Step 3, else stop (Fig. 2).

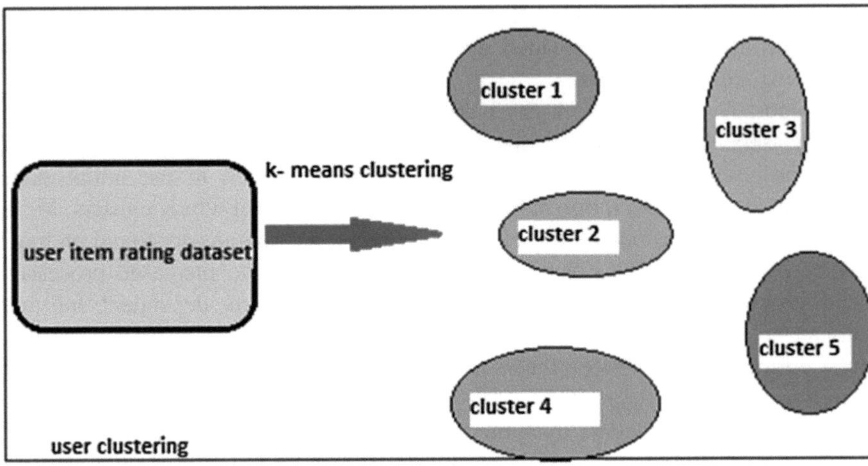

Fig. 2 Clustering of users

2.2 *Placing Active User*

Active user will reside in one of the clusters based on its closeness to the centroids of clusters. Euclidean distance of active user from centroids is calculated, and user is placed in the cluster with the least Euclidean distance. Euclidean distance between two users is calculated as:

$$D = |X1 - X2| + |Y1 - Y2|$$

It shows that active user will get its place in the cluster that has users having preferences similar to that of the active user.

2.3 *Creating Trust Database*

Once cluster is found, trust on neighbors is calculated. Trust depends on factors [6] such as similarity between profile partners and confidence of one user on another.

Similarity—This value shows how similar two users are, depending on how similar their ratings are on items they prefer. Consider x and y as two users and sim (x, y) is similarity between them. Pearson's coefficient has been used to calculate similarity among x and y in following way.

$$sim(x, y) = \frac{\sum (r_{x,i} - \overline{r_x})(r_{y,i} - \overline{r_y})}{\sigma_x \sigma_y}$$

If sim(x, y) comes out to be negative, then it is taken as 0, where $r_{x,\,i}$ is the rating of user x for item i, is average rating of user x, and σ_x and σ_y are standard deviations of rating of users x and y, respectively.

Confidence—confidence of one user on another user is high, if both have considerable number of items that both have rated.

$$conf(x, y) = \frac{no.\ of\ items\ both\ x\ and\ y}{no.\ of\ items\ rated\ by\ x}$$

As stated above, both the factors will influence trust; therefore, the final value of trust is

$$\Gamma_{x,y} = \frac{2 * sim(x, y) * conf(x, y)}{sim(x, y) + conf(x, y)}$$

2.4 Calculating No. of Items to be Fetched from Neighbors

Once neighbors are selected and trust is calculated, next step is to find how much number of items is to be fetched from these neighbors. Number of items user y can offer to active user x is given by:

$$\mathrm{Num}_{x,y} = (\Gamma_{x,y} * X) / \Gamma_{max}$$

Here, X is total number of recommendation we decide to be generated for active user x, and Γ_{max} is maximum trust of active user "x".

2.5 Store Selected Items in Another Database and Score Them

All the items in neighborhood are evaluated on the basis of score value. This score depend upon rating and trust among neighbors. Score of an item is:

$$\mathrm{Score}[i] = \mathrm{rating}[i] * \Gamma$$

Score_i is score of item "i", rating_i is rating for item "i", and Γ is trust of active user on neighbor. This score is to decide which item is best to recommend. This score of each item is stored in database.

2.6 Recommending Top T Items to Active User

This database is sorted, and top T items of the database are recommended to active user. These are actually items with relatively high score.

2.7 Updating Trust in Trust Database

Trust can be contradictory; sometimes trust has different meanings in various subjects. It depends on mind, experience, and circumstances. Trust is not symmetric in nature, and its value is unidirectional as well.

Let Γ (old) = value of trust of active user before getting recommendations.

Γ (new) = value of trust of active after getting recommendations

Count = No. of recommendations generated by particular user for active user.

Here, π is small constant to update trust value in graph

Secret informer with useful information: If informer has some useful information and trustworthy past records, he can be trusted in future too. Considering this scenario in the recommender system, if active user not only visits the recommended items but rate them as well, trust on neighbors with such items is updated as:

$$\Gamma(\text{new}) = (1 + \lambda * \text{count}) * \Gamma(\text{old}) + \underline{D}$$

where $\underline{D} = \Gamma$ (old) * (R-avg. rating)/max. rating

Secret informer with no useful information: If informer does not have any useful matter in the information that could provide lead to solve the case but has trustworthy past, trust on them may change a little bit but may not get reduced drastically. Similarly, a situation may be where the active user may only visit the recommended item but do no rate them.

$$\Gamma(\text{new}) = (1 + \lambda * \text{count}) + \Gamma\,(\text{old})$$

Secret informer with no information: Sometimes informer does not get any information for the present case and further no trustworthy past is associated with this him, thus it can be said that the trust is more influenced. Considering the same scenario in recommender system, an active user may neither have rated the recommended item nor have visited the item.

$$\Gamma(\text{new}) = \{1 - (\lambda / \text{count})\} * \Gamma\,(\text{old})$$

Here, even though no item is actually recommended to active user by this neighbor, count is considered as 1.

By above formulae, trust on neighbors who have more potential of recommendation is increased and for those who do not provide good recommendations is decreased.

3 Experimental Studies

3.1 *Metric for Evaluation*

Experiment is performed on Movielens dataset. Parameters considered for experimental evaluations are precision and recall. For this purpose, top T items that have user's rating are predicted and evaluated for corrections. The dataset is divided into training set and test set. Recall is the ratio of size of hit dataset to size of test dataset, and precision is ratio of size of hit dataset to size of top T item set.

$$\text{Recall} = |\text{test} \cap \text{top T}| \,/\, |\text{test}|$$
$$\text{Precision} = |\text{test} \cap \text{top T}| \,/\, |\text{top T}|$$

Recall increases with an increase in the number of recommendations produced, and at the same time precision is decreased. Both precision and recall are necessary for the evaluation of efficiency and performance of recommender system. Therefore, the combination of these two parameters is needed, which can be formulated as follows:

$$F = 2 * \text{Precision} * \text{Recall} \,/\, (\text{Precision} + \text{Recall})$$

3.2 *Experimental Results*

The experiments are performed in R 3.2.4, and the operating system used is Microsoft Windows. Dataset being used is Movielens which is provided on http://grouplens.org/datasets/movielens/. Different recommender system approaches such as KNN have been compared with the proposed approach, and experimental analysis has been performed with and without trust update. Values considered for experiments are as follows: Top T = 10, test set consisting of 15 items and recommendations are generated with K = 5. Trust is updated on every response of active user on recommended item; this response may increase or decrease the trust depending on the nature of feedback. Overall performance of the system increases as trusted users are now participating in recommendation process, as they are more likely to give recommendations of active user's interest.

Result in Table 1 indicates that proposed approach outperforms the traditional approaches and the similar approaches with static trust values

Proposed approach overcomes many issues [7] related to recommender system which include the following:

(i) Speed: This approach uses clustering leading to increase in the speed of the system.

Table 1 Experimental analysis of output from traditional and proposed approaches

	KNN	Trust-based approach without updating	Proposed approach with updating trust
Precision	40.58	44.25	47.38
Recall	31.82	30.44	32.05
F	35.67	36.07	38.24

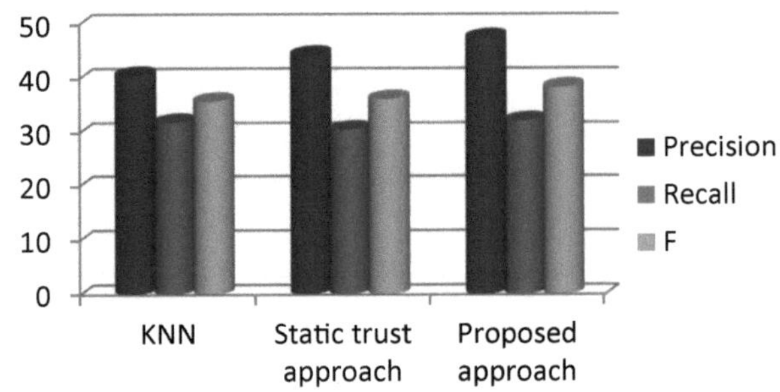

Fig. 3 Bar graph of comparison among various approaches on various evaluation parameters

(ii) Shilling Attack: Along with fundamental collaborative clustering, the use of trust mechanism eliminates the attack by fake users. Users with fake profile can disturb the system in two ways: one is by decreasing the value of popular items by rating them low and similarly increasing popularity of items of their interest. This disturbance in recommender system is known as shilling attack.

(iii) Sparseness of data in the form of user item matrix is reduced by the combination of sim(x, y) and conf(x, y) at initial stage.

Figure 3 presents the graphical comparison of results of different recommender system approaches on the various evaluation parameters.

4 Conclusion

One of the issues that recommender system suffers is being attacked by fake profiles. Our proposed approach eliminates this problem that is endured by conventional collaborative filtering technique and many other techniques as well. Shilling attack problem due to fakeness involved in the system is rectified as we included trust into the system. A static nature of trust can also lead to wrong predictions of choices as the time passes; therefore, we gave trust a dynamic characteristic. This approach gives the predictions quite easily and efficiently. Also the dynamic implementation of trust increases its prowess. The effect of sparseness of data is low for our approach, and this leads to the conclusion that even if initial information is sparse in terms of rating provided by users, the system will work efficiently without any considerable influence. All these features help in increasing the customer satisfaction as the system is fast and trustworthy, which make them decide easily.

5 Future Scope

Although trust is calculated on the basis of similarity between the users and confidence of user on others, yet it is a calculated prediction which is quite close to actual trust. Instead, we can also provide a provision to ask user the trust value between some predefined ranges, widely known as called explicit trust. Accuracy of explicit trust is much more than implicit trust value. Therefore, we can introduce in our system trust which can be explicitly. Moreover, we only consider items which user has rated and do not consider items which user has bought and liked it as well, though may not have rated it. Considering such items may lead to yet better recommendations.

References

1. Schafer JB, Konstan J, Riedl J (1999) Recommender systems in e-commerce. In: Proceedings of the 1st ACM conference on Electronic commerce. ACM, pp 158–166
2. Burke R (2007) Hybrid web recommender systems. Springer, Heidelberg, pp 377–408
3. Su X, Khoshgoftaar TM (2009) A survey of collaborative filtering techniques. Adv Artif Intell 4
4. Massa P, Avesani P (2007) Trust-aware recommender systems. In: Proceedings of the 2007 ACM conference on Recommender systems. pp 17–24
5. Hartigan JA, Wong MA (1979) Algorithm AS 136: a k-means clustering algorithm. J Roy Stat Soc Ser C (Appl Stat) 28(1):100–108
6. Massa P, Avesani P (2007) Trust metrics on controversial users: Balancing between tyranny of the majority. Int J Seman Web and Inf Syst (IJSWIS) 3(1):39–64
7. Chirita PA, Nejdl W, Zamfir C (2005) Preventing shilling attacks in online recommender systems. In: Proceedings of the 7th annual ACM international workshop on Web information and data management. ACM, pp 67–74
8. Goldberg D, Nichols D, Oki BM, Terry D (1992) Using collaborative filtering to weave an information tapestry. Commun ACM 35(12):61–70
9. Ricci F, Rokach L, Shapira B (2011) Introduction to recommender systems handbook, Springer, US, pp 1–35
10. Golbeck J (2006) Computing with Trust: Definition, Properties, and Algorithms. In: Securcomm and workshops. IEEE, pp 1–7
11. O'Donovan J, Smyth B (2005) Trust in recommender systems. In: Proceedings of the 10th international conference on Intelligent user interfaces. ACM, pp 167–174
12. Ricci F, Rokach L, Shapira B, Kantor PB (2010) Recommender systems handbook. Springer
13. Guo G, Zhang J, Thalmann D (2012) A simple but effective method to incorporate trusted neighbors in recommender systems. In: User modeling, adaptation, and personalization. Springer, Heidelberg, pp 114–125

Comprehensive Data Hiding Technique for Discrete Wavelet Transform-Based Image Steganography Using Advance Encryption Standard

Vijay Kumar Sharma and Devesh Kumar Srivastava

Abstract The steganography is the discipline of trouncing data into innocuous media in a manner that reality of the hidden data remains invisible to an antagonist. Varieties of methods have been developed over time in the area of image steganography. Robust steganography technique increases data protection from an adversary even if an attacker clutches the knowledge about the embedding process. Modern information security system combines both cryptography and steganography techniques. This paper proposes an image steganography method, at the initial stage secret image is encrypted using advance encryption standard (AES) and afterwards hide the results of AES into the picture (i.e. cover media or cover image) with the assistance of Haar Discrete Wavelet Transform and alpha blending. These efforts ensure that the proposed data hiding mechanism gives higher imperceptibility and trustworthiness which is the essential requirements of any steganography technique. All practical implementation perform on MATLAB.

Keywords AES · Steganography · Steganalysis · Haar DWT · Haar IDWT · Alpha blending · PSNR · MSE · NCC

1 Introduction

Information system security (Cryptography, Watermarking and Steganography) is a discipline that protects the confidentiality and accessibility of information. From earlier days, there were many ways to keep confidential information secure when they are communicated. The steganography is a technique that is used for secure communication. These vital aims separate it from supplementary practices as cryptography watermarking [1]. Sometimes, sending encrypted information may

V.K. Sharma (✉) · D.K. Srivastava
SCIT Manipal University, Jaipur, Rajasthan, India
e-mail: vijaymayankmudgal2008@gmail.com

D.K. Srivastava
e-mail: devesh988@yahoo.com

H.R. Vishwakarma and S. Akashe (eds.), *Computing and Network Sustainability*,
Lecture Notes in Networks and Systems 12, DOI 10.1007/978-981-10-3935-5_36

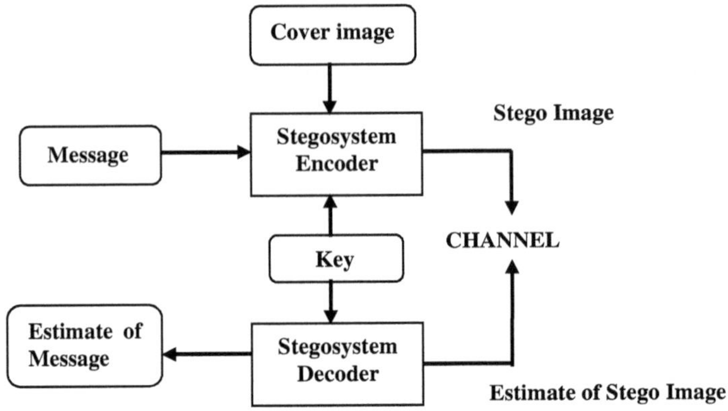

Fig. 1 Basic steganography model (*Source* [1])

draw the attention of an observer, while imperceptible information will not. In cryptography, feature is always visual, whereas in steganography, the feature is always invisible. Watermarking is used for copyright protection. The main aim here is to place the identification of author or artist. It can be either visible or invisible.

For secure communication, steganography is the most demanding technique; it is applicable to all data objects that contain redundancy, in this paper, images (i.e. JPEG and TIFF) are considered. People frequently broadcast digital pictures over Internet or other communication like e-mail, most common image format is JPEG. The use of JPEG picture format in steganography systems seems more interesting because system operate in transformed domain (like DWT, DCT) and not affected by existing visual attack

Figure 1 shows the basic image steganography model. At the initial stage (sender end) of model, the secret data is inserted inside the cover or spread media (Key is optional) that gives the stego picture. At the collector end, the reverse procedure is applied to get mystery information.

Today internet plays the great role as communication media. Each and all thing goes in digital so data protection is the main concern for secure communication. So there is a need to develop the more robust data hiding technique.

2 Proposed Steganography Technique Using AES

This paper presented AES-based secret image encryption, afterwards host image (cover image) added to the secret image for Haar DWT coefficients [2–5]. The alpha blending or mixing is utilized to attach wavelet coefficients of both cover image and mystery image. Later than the mixing operation, obtained stego image by

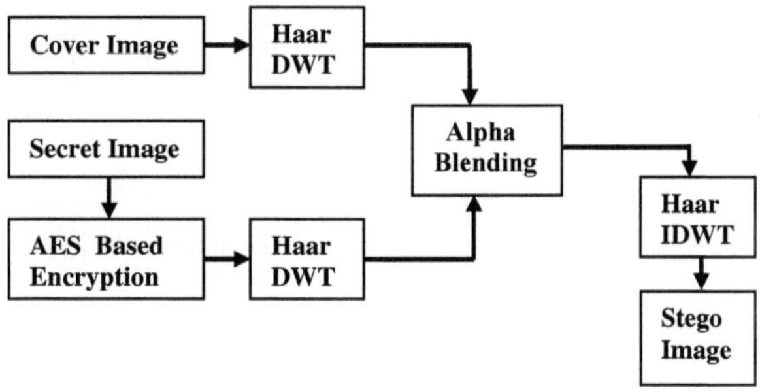

Fig. 2 Data concealing process of proposed steganography technique

taking Inverse Haar Discrete Wavelet Transform (Haar IDWT). Figure 2 illustrates the broad representation of the projected steganography Algorithm.

2.1 Alpha Blending

Alpha blending is the blending function which basic meaning is to mix up two images and gives final single image. The following equation gives the mixed image.

$$\text{Final Image} = \text{Image}_1 + \alpha * \text{Image}_2 \tag{1}$$

Here, the value of α lies in between $0 < \alpha > 1$, Image_1 and Image_2 is the cover image and covert image, respectively.

Alpha builds the inserting quality so it is known as strength factor [2].

2.2 Encoding Process

The proposed technique contains the following algorithmic steps that were implemented using MATLAB [2, 6].

Step 1: Input both cover and secret image.
Step 2: Use advance encryption standard (AES) on secret image and get encrypted image
Step 3: Use a two-level 2-D Haar DWT on the covert image.
Step 4: Use a one-level 2-D Haar DWT on the cover picture or image.
Step 5: Pull out the estimated coefficients of matrix LA and feature coefficients matrices LH, LV and LD of 1-level 2-D Haar DWT of the cover picture or image.
Step 6: Pull out the approximation coefficient of matrix named LA1 and feature coefficients of the matrices named as LH1, LD1 and LV1 of one-level 2-D Haar DWT of secret picture or image.

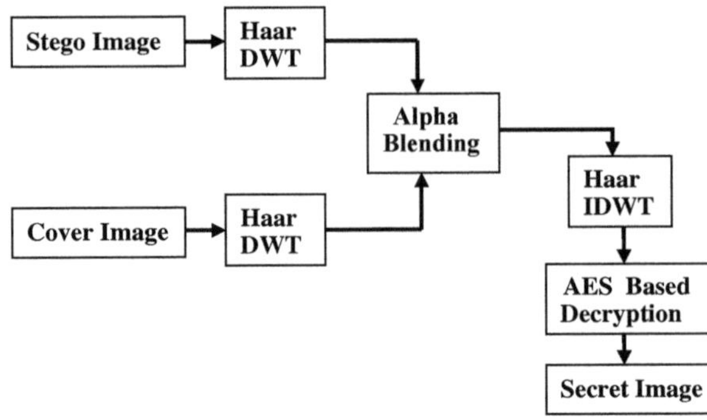

Fig. 3 Decoding process for proposed technique

Step 7: Mixed both cover and secret image (which is obtained from step 5 and step 6) by using alpha blending process

Step 8: Finally, stego image is obtained by performing 2-D Haar IDWT on results obtained from alpha operation.

2.3 Decoding Process

The reverse process of encoding is known as decoding process. In the earlier step of decoding, use Haar DWT on both stego image and well-known cover image pursued by alpha operation. Next, Haar IDWT is used to reconstruct encrypted secret or covert image. This encrypted image is passed through AES-based decryption to acquire original covert image. Figure 3 illustrate the steps used in decoding

Step 1: Perform single level 2-D Haar DWT on both stego image and well-known cover image or original image.

Step 2: Perform the alpha operation on Haar DWT transferred images obtained from step1.

Step 3: Perform Haar IDWT on resultant image from step 2, to get an encrypted secret image.

Step 4: Apply AES algorithm-based decryption to get secret image.

3 Materials and Tool

3.1 MATLAB Software Tool

For implementing the proposed algorithm for steganography, MATLAB software is used. By taking pirate .tiff as the cover image and vijay .jpeg as the secret image, corresponding MATLAB results are as shown below in the Figs. 4 and 5.

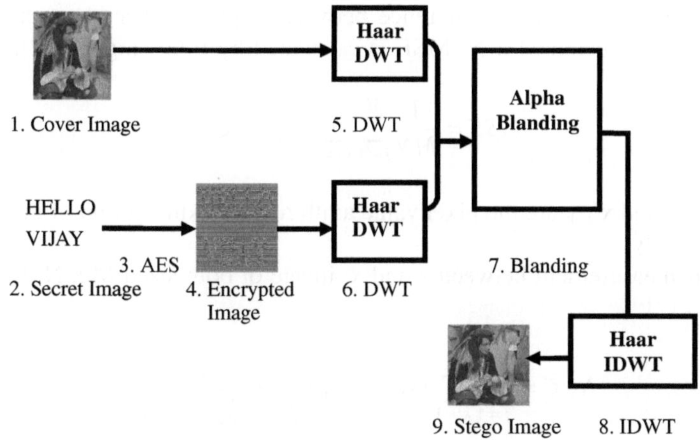

Fig. 4 Experiment results of the encoding process of proposed algorithmic Steps

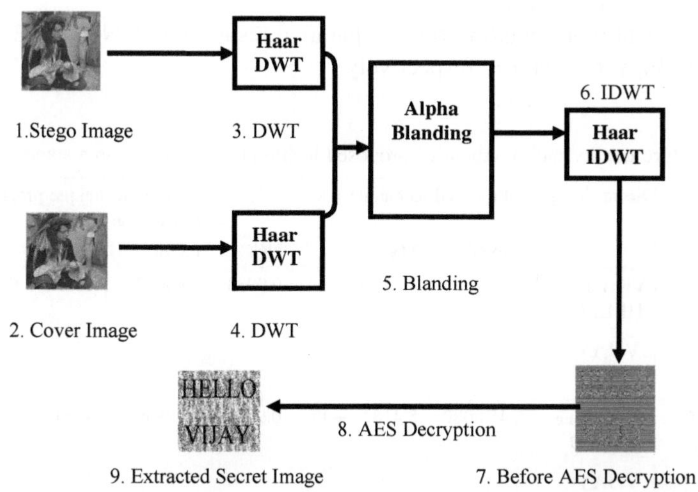

Fig. 5 Results obtained after decoding process of proposed algorithmic steps

The strength of proposed work is tested by comparing the peak signal-to-noise ratio (PSNR), mean square error (MSE) and normalized cross correlation (NCC) of the cover image and the stego image. PSNR evaluate by following equation [2, 7].

$$PSNR = 10\log\left(\frac{L^2}{MSE}\right)\text{dB} \tag{2}$$

where L is the dynamic variety of pixel value.

MSE is measured as the difference between original cover image x and stego image x′ of dimension M x N. MSE is calculated by following Eq. [2].

$$MSE = \frac{1}{MN} \sum_{j=1}^{M} \sum_{k=1}^{N} (x_{j,k} - x'_{j,k})^2 \quad (3)$$

Here $x_{j,k}$ and $x'_{j,k}$ are the pixel value at jth row and kth column of images x and x′, respectively.

NCC is measurement between x and x′ image of both sized M x N. It is defined as follows [2]:

$$NCC = \sum_{j=1}^{M} \sum_{k=1}^{N} (x_{j,k} - x'_{j,k}) \frac{1}{\sum_{j=1}^{M} \sum_{k=1}^{N} (x_{j,k})^2} \quad (4)$$

Larger PSNR value indicates the better visual quality. The proposed method is tested for the different cover image and same secret image. There corresponding results shown in Table 1 as follow.

The graphical representation for comparative results for PSNR, MSE and NCC is shows in Figs. 6, 7 and 8, respectively.

Table 1 Picture quality measurement of proposed technique, tested on some standard image

Cover image	Secret image	Proposed technique			Existing technique proposed by Prabakaran. G		
		PSNR	MSE	NCC	PSNR	MSE	NCC
Pirate.Jpeg 	**Vijay.Jpeg** HELLO VIJAY	58.7501	4.6464e + 04	0.9983	49.3077	5.1215e + 04	0.9951
Pepper.Tiff 	**Vijay.Jpeg** HELLO VIJAY	45.8089	8.4182e + 04	0.9934	45.7859	8.4384e + 04	0.9926
Flower.Jpeg 	**Vijay.Jpeg** HELLO VIJAY	47.9688	5.6795e + 04	0.9942	46.8089	6.8182e + 04	0.9934

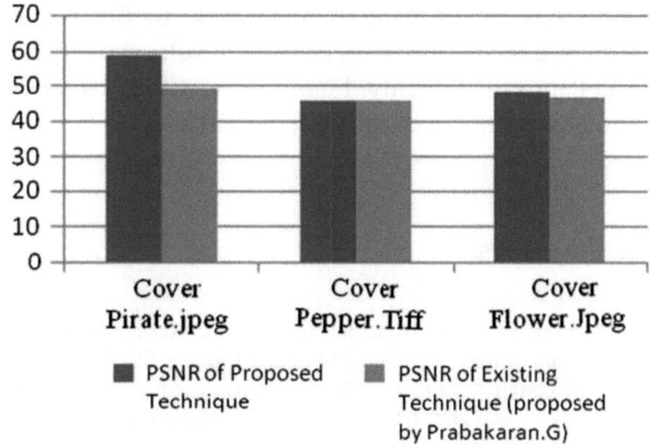

Fig. 6 Comparative results of PSNR for same secret image based on Table 1

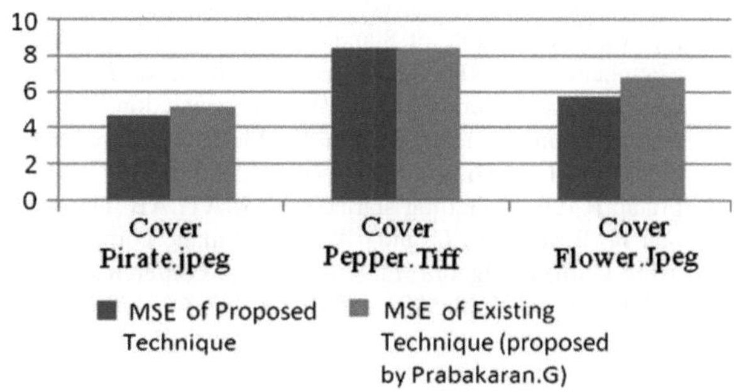

Fig. 7 Comparative results of MSE for same secret image based on Table 1

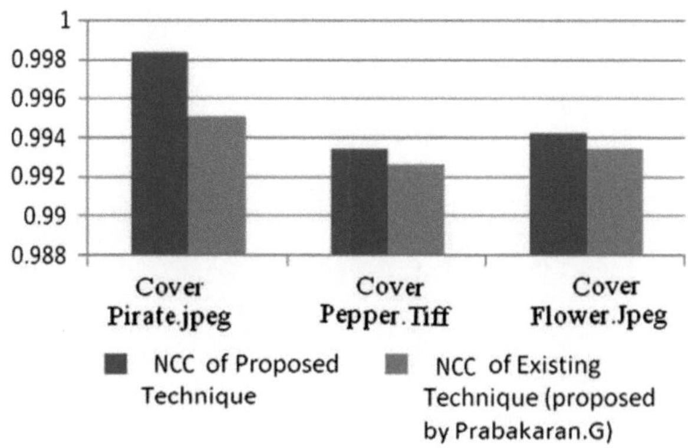

Fig. 8 Comparative results of NCC for same secret image based on Table 1

4 Conclusions

In this paper, it is observed on the experiment basis that the technique fulfils the objective by embedding and extracting secret image. The technique increases the picture quality of stego image so nobody can identify the difference between stego and cover image without nuke eyes. In Fig. 8, graphs illustrate the strength of proposed technique. Experimental results show that extracted secret image also has excellent visual quality. Future work will focus the on blind detection, so that it can eliminates the cover image requirements.

References

1. Marvel Lisa M, Boncelet Charles G, Retter Charles T (1999) Spread Spectrum Image Steganography. IEEE Trans Image Process 8:1075–1108
2. Prabakaran G, Bhavani R (2012) A modified secure digital image steganography based on discrete wavelet transform. In: International conference on computing, electronics and electrical technologies. pp 1096–1100
3. Bassham L (2000) Efficiency testing of ANSI C implementations of round 2 candidate algorithms for the advanced encryption standard. In: The Third AES candidate conference, by the National Institute of Standards and Technology Gaithersburg. 136–148
4. Manoharan S (2008) An empirical analysis of RS steganalysis. In: The third international conference on internet monitoring and protection. IEEE, pp 172–177
5. Foster, Wang S, Yang B, Niu X (2010) A secure steganography method based on genetic algorithm. J Inf Hiding Multimed Signal Process. 1:2073-4212
6. Pratap R (2013) Getting started with MATLAB. Oxford University Press
7. Sau K, Basak RK, Chanda A (2013) Image compression based on block truncation coding using Clifford Algebra. In: International conference on computational intelligence: modeling techniques and applications (CIMTA). pp 699–706

Comparison and Analysis of RDF Data Using SPARQL, HIVE, PIG in Hadoop

Anshul Chandel and Deepak Garg

Abstract In the modern generation of "Semantic Web Data", cloud data services play a important role. These services are based on the MapReduce Programming Model. Hadoop is an open source implementation of MapReduce. Hadoop based extensions such as PIG and HIVE are query languages which provide high level data flow. Although SPARQL is considered as the backbone of the semantic web based applications but in this paper we introduce HIVE and PIG for querying RDF data. The goal of the paper is to compare the results of SPARQL, HIVE and PIG and analyze the retrieval time for a query in RDF data.

Keywords Hadoop · SPARQL · HIVE · PIG · MapReduce · RDF · PIG-LATIN · HDFS · HIVE-QL

1 Introduction

With the advent of semantic web data new challenges regarding the query evaluation rose. Semantic Web data represented as Resource Description Framework (RDF) is growing rapidly. RDF is a core technology of the semantic web for representing data in machine-readable formats. RDF is very useful in semantic web and data integration. It gives a universal model of data which is a milestone in data integration. SPARQL is the standardized query language specifically designed for RDF, just as SQL is the standardized query language for the relational type of databases. A single SPARQL query comprises of specific set of triplets where the subject, the predicate and/or object constitute the variables. The idea behind this format is to match triples in the SPARQL query with the already existing RDF triples and consequently find the solutions of the variables. It is a challenging task

A. Chandel (✉) · D. Garg
CSED, Thapar University, Patiala, Punjab, India
e-mail: rbanshul1143@gmail.com

D. Garg
e-mail: dgarg@thapar.edu

H.R. Vishwakarma and S. Akashe (eds.), *Computing and Network Sustainability*, Lecture Notes in Networks and Systems 12, DOI 10.1007/978-981-10-3935-5_37

to query an RDF dataset but with the use of Hadoop MapReduce infrastructures, we can spread the processing across the clusters. One of the primary concerns of the paper is to study the big data challenges faced by developers due to its volume, variety and velocity. The paper on hands presents the arms of the Big Data such as Hadoop, SPARQL, HIVE and PIG which are used to provide innovative solutions for all the big data challenges. The rest of the paper has been organized as follows: Section 2 provides some of the related work. Section 3 gives an overview of query languages. Section 4 describes about the work environment. Section 5 gives an overview of the proposed architecture. Section 6 shows the test cases, which highlights the implementation and results. Conclusion along with the future scope is given in Sect. 7.

2 Related Work

In this section, review of literature related to the paper and motivation for our work is described. At present Google and other search engines are best example of big data and semantic web. Resource Description Framework (RDF) is used to represent the semantic data and SPARQL is used for semantic web analysis. Generating data in RDF/XML configuration by the usage of the LUBM data creator and transforming the designed data to N-Triples form using N-Triple Converter was proposed in [1].This shows that a query time rate is less than rate in data size. In last few years linked data cloud has expanded rapidly to hold billions of triples. Because of this large amount of data is required for a scalable and dynamic approach. MapReduce computing model has become an essential component of current large scale refining solutions. Google and search engine uses linked data [2] and MapReduce technique to obtain huge amount of results. How Hadoop and MapReduce is used to obtain huge amount of semantic data was proposed in [3]. Later an algorithm was discussed to convert SPARQL into MapReduce [3]. A structure placed on Hadoop titled "HadoopSparql" was proposed in [4] to hold queries together based on servicing multi-way join operator. How to load data from Hadoop to RDF is given in [5]. One drawback of the MapReduce model is that users should convert their works into refined map and reduce code. Apache HIVE and PIG solved this issue by giving dataflow languages in Hadoop. There are not many approaches that consider the comparison of SPARQL, PIG and HIVE together. There are some papers published that address the HIVE techniques to retrieve data. Our system depends on translating the SPARQL query into HIVE-QL and PIG and reduces the query processing time in SPARQL. However our work needs an optimization for transforming of SPARQL to PIG and SPARQL to HIVE and also demands to filter its data model.

3 Query Languages and RDF Dataset

RDF (The resource description framework) data model has been around a decade. RDF being a data model having the property of scheme-free structured information has gained momentum in context of the semantic web based data. RDF is supported by data repositories as an import/export format and also for the construction of information mash-ups. There are four existing interface types for navigating RDF data

- Keyword Search-Information lookup.
- Explicit Queries-Requires Schema Knowledge.
- Graph Visualization-RDF data is stored in native graph form.
- Faceted Browsing-is a technique to access information from datasets which is based on the facet theory.

SPARQL (SPARQL Protocol and RDF Query Language) [6] is an query language for RDF. It is officially recommended by W3C.SPARQL is quite similar to SQL. It is a key semantic web technology. It is used to get information from RDF graphs and is also used to express queries based on RDF data. Today, most business data is stored in RDBMS because RDBMS have problem with synchronization and replication. In order to prevent synchronization problem, in many cases we require direct access to data without copying it into RDF.

PIG is an open-source scripting platform to analyze huge dataset using high level data flow language PIG-LATIN. PIG is made up of two parts: the first is called PIG-LATIN that is language itself and another is a runtime environment where PIG-LATIN scripts are executed. PIG can work with structured, semi-structured or unstructured data. PIG translates PIG-LATIN script into MapReduce [3] and is extended with User Defined Functions (UDFs).

HIVE is a Petabyte scale data-warehousing package/infrastructure built on top of Hadoop and uses HIVE Query Language (HIVE-QL) for querying Hadoop Clusters. It summarize query and analyze the data, although does not support transactions and is best suitable for batch processing data like log processing, text mining etc. HIVE language (HIVE-QL) has a basic data analyzing method for valuable data. HIVE-QL is open source given by Apache software foundation. HIVE allows users to collect, analyze, and store their data on the cloud. HIVE-QL is one of the languages that is supported by Hadoop.

4 Work Environment

In this paper, all the tasks are performed on single node with following configuration: AMD octa core 2.0 GHZ processor

- 8 GB RAM
- 1000 GB hard disk
- Operating system-red hat and windows 7

- Java version-1.7.0
- Hadoop version-1.0.3
- Eclipse-java-luna

We have used apache jena fuseki server for implementing RDF and SPARQL. We have also used HIVE and PIG package on Hadoop for implementing HIVE-QL and PIG-LATIN [7].

5 Architecture Analysis

5.1 Proposed Architecture

When we are working with huge size of semantic data then we are using HIVE-QL or PIG-LATIN instead of SPARQL because we want to reduce our query processing time. In our work as shown in Fig. 1, we are converting our dataset into CSV file instead of using triple-store. Our architecture consists of following four stages:

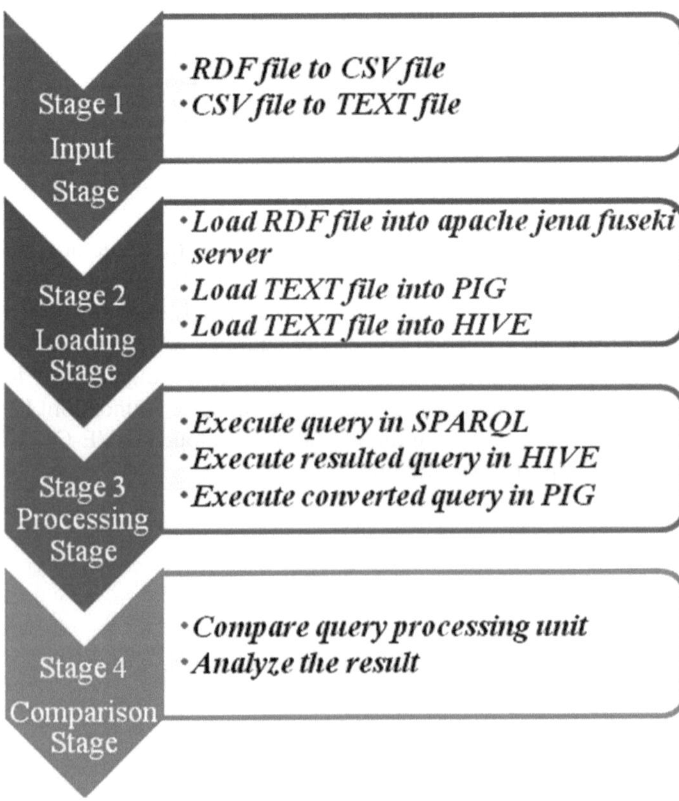

Fig. 1 Phases of a proposed architecture

Data Input Stage: In this stage, we require two types of files. First file is for SPARQL and second file is for HIVE-QL and PIG-LATIN. Here we have used RDF file for SPARQL and then we are converting this RDF file into CSV file and later into text file so that we can use it in HIVE-QL and PIG-LATIN.

Loading Stage: In this stage we will load two files. Firstly we will load RDF file into apache jena fuseki server and then we will load the text file into HIVE and PIG.

Processing Stage: In this stage we will execute our query in respective query languages. Firstly we will execute our query in SPARQL then we will execute the converted query in HIVE-QL. At last we will execute the resulted query in PIG. After the execution, output of all the queries is similar but query processing time is different.

Comparison Stage: In this stage, we compare the resulting query processing time of each Query language. Finally we conclude which one is better.

6 Results and Analysis

In this part, we have shown two test cases and have applied each of them separately on SPARQL, PIG and HIVE. This part will confirm that which query language is best for our dataset. We know that SPARQL is best language for semantic web, but when the data set get larger in size it lacks with its time (SPARQL take more time for query execution) hence we move on to Hadoop for reducing the query processing time. So in this section we have applied our test cases on SPARQL, PIG and HIVE and show the comparison with respect to time. Here we are using a dataset with 5000 triples or rows.

First Case: Here the first query is to find out the email-id of a person named 'craig'. So we execute the query in SPARQL, PIG and HIVE as shown in Figs. 2, 4, and 5 respectively and results are same in all of them as shown in Fig. 3 but query processing time is different for each query language. The query has taken 5.34 s in SPARQL, 1.1 s in HIVE and 0.57 s in PIG in our environment.

```
PREFIX ab: <http://learningsparql.com/ns/addressbook#>

select ?craigmail
where
{ab:craig ab:email ?craigEmail}
```

Fig. 2 SPARQL query

QUERY RESULTS

Raw Response | Table

	craigEmail
1	"craigellis@yahoo.com"
2	"c.ellis@usairwaysgroup.com"

Fig. 3 SPARQL query result

```
emp = load '/tmp/cloudera/mydata.txt' using pigstorage(',')
as (name:chararray, info:chararray, comm:chararray);
data = FILTER emp by name == 'craig' and info == 'mail';
dump data;
```

Fig. 4 PIG query

```
CREATE TABLE IF NOT EXISTS employee(name string, info string, comm string )
ROW FORMAT DELIMITED
FIELDS TERMINATED BY '\t'
LINES TERMINATED BY '\n'
STORED AS TEXTFILE;
LOAD DATA LOCAL INPATH '/tmp/cloudera/mydata.txt'
overwrite into table employee;
select comm from employee;
where name = 'craig' and info= 'mail';
```

Fig. 5 HIVE query

```
PREFIX ab: <http://learningsparql.com/ns/addressbook#>

select (count(?craigEmail) as ?mail)
where
{ab:craig ab:email ?craigEmail}
group by ?craigEmail
```

Fig. 6 SPARQL query

Second Case: In this query the goal is to obtain the count of record given by previous query. Here we used group by clause in all of above language to observe which one is more efficient for that clause. We have used group by which is more often used in semantic web.

Fig. 7 SPARQL query result

```
emp = load '/tmp/cloudera/mydata.txt' using pigstorage(',')
as (name:chararray, info:chararray, comm:chararray);
emp1 = group emp by comm;
data = FILTER emp1 by name == 'craig' and info == 'mail';
dump data;
```

Fig. 8 PIG query

```
CREATE TABLE IF NOT EXITS employee(name string, info string, comm string)
ROW FORMAT DELIMITED
FIELDS TERMINATED BY '\t'
LINES TERMINATED BY '\n'
STORED AS TEXTFILE;
LOAD DATA LOCAL INPATH '/tmp/cloudera/mydata.txt'
OVERWRITE INTO TABLE EMPLOYEE;
select count(comm) from employee
where name = 'craig' and info = 'mail'
group by comm;
```

Fig. 9 HIVE query

When we execute this query in SPARQL has taken 7.23 s in our single node as shown in Fig. 6 and it has taken 1.4 s and 0.75 s in HIVE and PIG respectively as shown in Figs. 8 and 9 respectively (Fig. 7).

The results on the two cases tested here in SPARQL, PIG and HIVE-QL prove that PIG-LATIN takes less retrieval time and performs better, while every query language is efficient in querying semantic datasets, but we prefere PIG-LATIN for large datasets (Fig. 10).

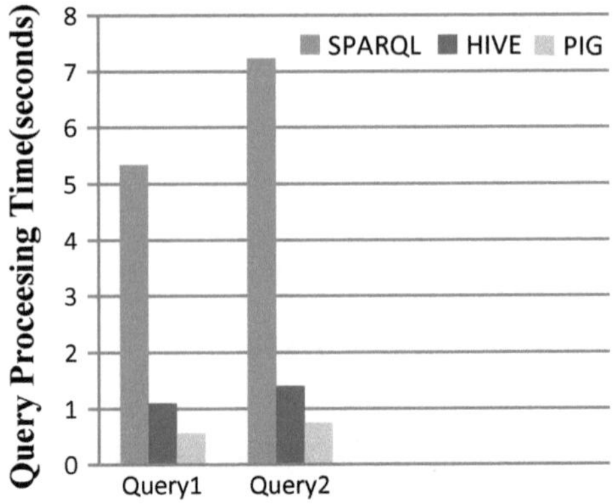

Fig. 10 Comparison analysis of SPARQL, PIG and HIVE

7 Conclusion and Future Scope

This work is done to propose a new method that is capable of efficiently querying huge amounts of semantic data content on Hadoop environment based on HIVE or PIG. This paper discusses our proposed method and shows the implementation of this method based on SPARQL, HIVE-QL query and PIG-LATIN. Results shows query processing time in SPARQL, HIVE and PIG. In the result we have executed the query in SPARQL, PIG and HIVE. We find that SPAQRL has taken more time than PIG and HIVE, HIVE has taken more time than the PIG. According to our analysis PIG is more efficient than HIVE and SPAQRL, PIG is 35% faster than HIVE for arithmetic operation and 45% faster than HIVE for filtering operation. Also PIG is 15% faster than HIVE for filtering operation and 30% faster than HIVE for joining operation. The results show how our proposed method outperforms using SPARQL in retrieving the data for future task, use Apache Spark and Impala as an attempt to get good results to the system and to get better Efficiency and Accessibility.

References

1. Schätzle A, Przyjaciel-Zablocki M, Lausen G (2011) PigSPARQL: mapping SPARQL to pig latin. In: Proceedings of the international workshop on semantic web information management. ACM
2. Anyanwu K, Kim H, Ravindra P (2013) Algebraic optimization for processing graph pattern queries in the cloud. Internet Comput IEEE 17(2):52–61
3. Lehmann J, Isele R, Jakob M, Jentzsch A, Kontokostas D, Mendes PN, Bizer C (2015) DBpedia–a large-scale, multilingual knowledge base extracted from Wikipedia. Semant Web 6(2):167–195
4. Schätzle A, Przyjaciel-Zablocki M, Neu A, Lausen G (2014) Sempala: interactive SPARQL query processing on hadoop. In: The semantic Web– ISWC 2014. Springer International Publishing, pp 164–179

5. Thusoo A, Sarma JS, Jain N, Shao Z, Chakka P, Anthony S, Murthy R (2009) Hive: a warehousing solution over a map-reduce framework. Proc VLDB Endow 2(2):1626–1629
6. Quilitz B, Leser U (2008) Querying distributed RDF data sources with SPARQL. Springer, Heidelberg, pp 524–538
7. Olston C, Reed B, Srivastava U, Kumar R, Tomkins A (2008) Pig latin: a not-so-foreign language for data processing. In Proceedings of the 2008 ACM SIGMOD international conference on Management of data (pp. 1099–1110). ACM
8. Kim H, Ravindra P, Anyanwu K (2011) From sparql to mapreduce: the journey using a nested triplegroup algebra. Proc VLDB Endow 4(12):1426–1429
9. Ravindra P, Kim H, Anyanwu K (2011) An intermediate algebra for optimizing RDF graph pattern matching on MapReduce. In: The semantic web: research and applications. Springer, Heidelberg, pp 46–61
10. Ismail AS, Al-Feel H, Mokhtar HM (2015) Querying DBpedia using HIVE-QL. In: Proceedings of the 14th international conference on telecommunications and informatics (TELE-INFO '15) proceedings of the 2nd international conference on power. pp 102–108
11. Ravindra, P., Hong, S., Kim, H., Anyanwu, K.: Efficient processing of RDF graph pattern matching on MapReduce platforms. In: Proceedings of the second international workshop on Data intensive computing in the clouds, ACM, 13–20(2011)
12. Arias M, Fernández JD, Martínez-Prieto MA, de la Fuente P (2011) An empirical study of real-world SPARQL queries
13. Thusoo A, Sarma JS, Jain N, Shao Z, Chakka P, Zhang N, Murthy R (2010) Hive-a petabyte scale data warehouse using hadoop. In: 2010 IEEE 26th International Conference on Data Engineering (ICDE). IEEE, pp 996–1005

IoT-Enabled Integrated Intelligence System for Automatic Pothole Detection, Pollution Monitoring, Post-Accident Medical Response, and Breakdown Assistance

Nikhil Bhat, Krunal P. Bhatt, S. Prithvi Alva, B.M. Tanvi Raj and R. Sanjeetha

Abstract Some of the problems faced by commuters today are as follows: (i) accidents due to running over potholes; (ii) reporting accidents to call for medical help (especially if the accident is on isolated/remote roads); (iii) inability to identify the vehicle which was involved in the accident; (iv) unexpected vehicle breakdown; and (v) monitoring the levels of gas emission from vehicles. In our paper, we propose a solution for the above-mentioned problems using an application that works based on the concept of IoT. The application uses various sensors to collect the relevant details and send it to the (i) traffic-related authorities to fix the identified potholes; (ii) hospitals/ambulance services to help the accident victims; (iii) legal authorities to use it as evidence while investigating accidents; (iv) vehicle service centers to fix the vehicle; and (v) to monitor vehicular pollution.

Keywords Smart vehicles · Cloud computing · IoT in vehicles · Pothole detection · Pollution monitoring · Accident response · Breakdown assistance

N. Bhat (✉) · K.P. Bhatt · S. Prithvi Alva · B.M.T. Raj · R. Sanjeetha
MSR College Road, MSR Nagar, Bengaluru 560054, Karnataka, India
e-mail: nikhil.bhat93@gmail.com

K.P. Bhatt
e-mail: krunal.bhatt9@gmail.com

S. Prithvi Alva
e-mail: al.prithvi@gmail.com

B.M.T. Raj
e-mail: tanvirajbm@gmail.com

R. Sanjeetha
e-mail: sanjeetha.r@msrit.edu

H.R. Vishwakarma and S. Akashe (eds.), *Computing and Network Sustainability*, Lecture Notes in Networks and Systems 12, DOI 10.1007/978-981-10-3935-5_38

1 Introduction

An automobile, as seen from a layman's view, is just a means of transportation, to travel from place A to place B. However, from the automobile's perspective, it is a source that experiences a lot of real-time factors. The automobile experiences various hardships such as severe weather conditions, accidents, and traffic jams.

The modern-age automobile is well equipped with top-notch sensors—a rain sensor, a light sensor, a temperature sensor, a humidity sensor, an impact sensor, sensors for various functional units, along with embedded systems that log all the data collected.

This is just the data a single automobile is able to collect. Millions of such automobiles can collect huge volumes of such data in real-time and assist in solving many problems.

(i) Majority of road accidents are caused due to terrible road conditions, mainly potholes. The authorities are in a constant struggle to identify potholes and repair them as they are created, but still the number of potholes needing attention is very high and they are not able to prioritize which pothole needs attention. In this paper, we show how accelerometers can be used in automobiles to detect potholes and inform the same to the authorities. Also, higher the number of vehicles reporting a particular pothole, higher is need to fix it, thereby assigning a priority for each pothole.

(ii) In the case of accidents causing grievous injuries, it is noticed that onlookers do not generally call for medical help, leading to loss of lives due to delay in medical treatment. This also applies if the accident occurred on an isolated or remote road. Our system detects the occurrence of accidents and notifies to the nearest hospital/ambulance service immediately.

(iii) In many cases, the person guilty of causing an accident will go unpunished because of lack of evidence. Our system provides necessary evidence for such crimes by providing details such as vehicle registration number, vehicle owner details, cyber time stamp, and acceleration values recorded on which further analysis is possible.

(iv) Pollution is a big problem that authorities are desperately trying to solve. Our system tracks the levels of harmful gases in automobile exhausts and alerts the vehicle user if the emission levels cross a certain threshold. If the user does not fix it after a week, the information is reported to concerned authorities.

(v) In the event of a breakdown, a lot of time is spent on trying to find a service center. Our system notifies the user when his vehicle's engine health is poor and alerts him to visit the service center and fix the problem. In the case of a breakdown, the system notifies the nearest service center, with the vehicle's accurate location, distance, and time to reach in real-time conditions.

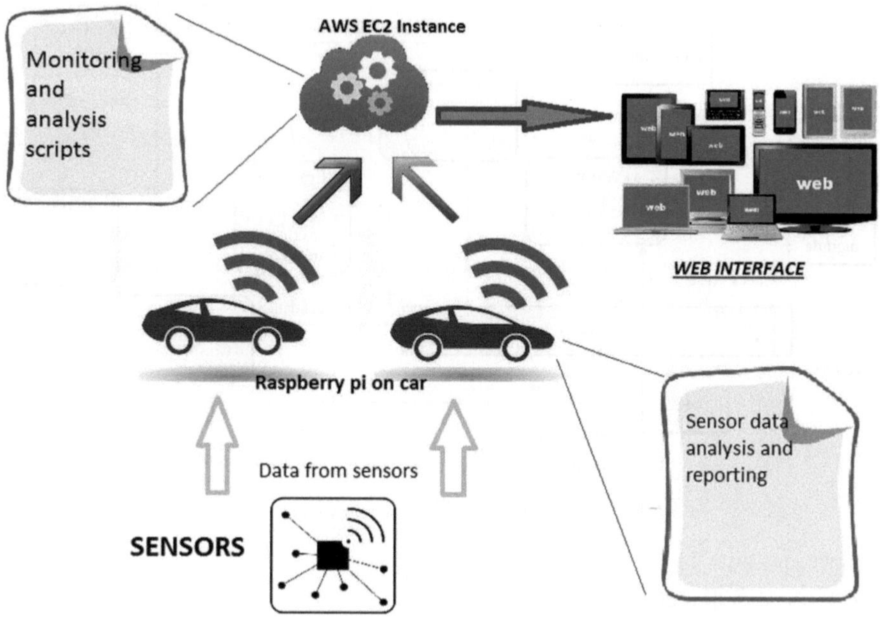

Fig. 1 Design architecture

2 Design

Figure 1 shows our system which consists of a Raspberry Pi chip with multiple sensors attached, i.e., a temperature sensor for sensing engine temperature, which will aid us in breakdown detection, an accelerometer for pothole and accident detection, MQ-x gas sensors for checking levels of pollutants in the exhaust gases. It also has a GPS module that collects location data of the vehicle.

The system uses the data from the various sensors to perform local analysis on the chip and send the values to the server. The data flow diagram of our system is shown in Fig. 2. The Web application interface developed is hosted on the server and is used by vehicle users and concerned authorities. "It achieves the goal of intelligent identifying, locating, tracking, monitoring, and managing things" [1].

3 Implementation

Sensing Modules. This module will obtain the required data from the sensors and sends it to the monitoring module. The sensing modules are an aggregation of the following individual sensor modules: accelerometer, gas sensor, temperature sensor, and GPS module.

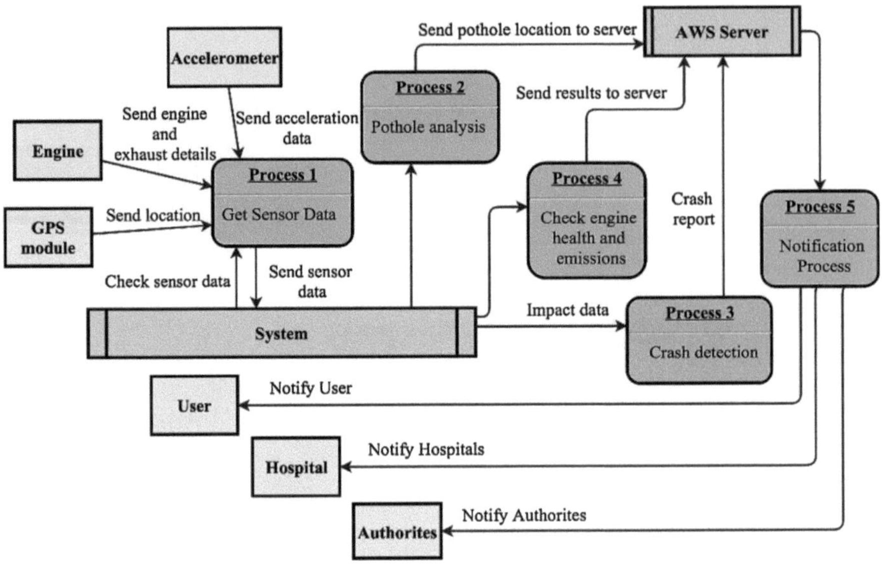

Fig. 2 The data flow diagram

Pothole Processing Module. This module reads the acceleration values from the sensor and then processes it on the Raspberry Pi. If the z-axis acceleration value crosses the lower threshold, then it categorizes this reading as a pothole and sends the location to the server. Alternatively, speed-breakers are avoided by using an upper threshold limit.

Monitoring Module. The Raspberry Pi collects all the data from the sensor modules and creates reports, i.e., engine breakdown reports, crash report, pothole report, and emission report, and sends it to the cloud application module for analytics. This module also provides the GPS location data for the reports created and time-stamps them accordingly. Similar to the pothole module, the monitoring module too has a little computational work on the data, for example, checking the value of data and deciding whether or not to forward to the other modules or take other action. For example, the accelerometer data is compared to threshold (40 G) to decide whether a severe accident has occurred [2] and whether the data needs to be uploaded to the server.

Cloud Application Module. This module is located on the server and receives the reports from all the monitoring modules of different vehicles, performs analysis, and sends related information/results to the corresponding Web interface modules and to the vehicle itself.

Web Interface Modules. This module provides the users the required information in a clean and user friendly Web interface. Figure 3 is the screenshot of the pothole Web interface.

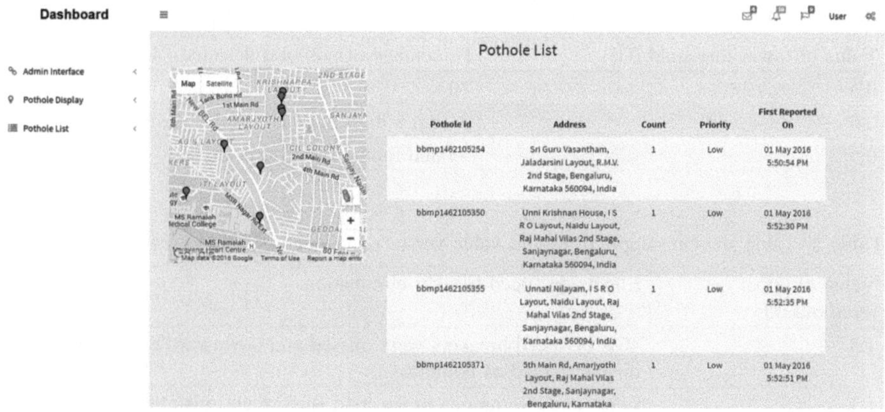

Fig. 3 Pothole web interface

4 Observations and Graphs

The readings shown in Fig. 4 were recorded on a road with irregularities, and the downward spikes corresponded to potholes.

In Table 1, we can see the lower thresholds selected during our experiments for detecting potholes and its effect on detection.

From Fig. 4, it is observed that on the occurrence of a pothole, the spike goes down immediately, but in the case of a speed-breaker, there is an initial upward spike followed by a downward spike (which could be misinterpreted as a pothole). To avoid this, we set an upper threshold to detect speed-breakers. Hence, if there is an upper spike caused by the speed-breaker followed by a downward spike, and the gap between the two spikes is less than 300 ms, then that downward spike is not considered to be a pothole.

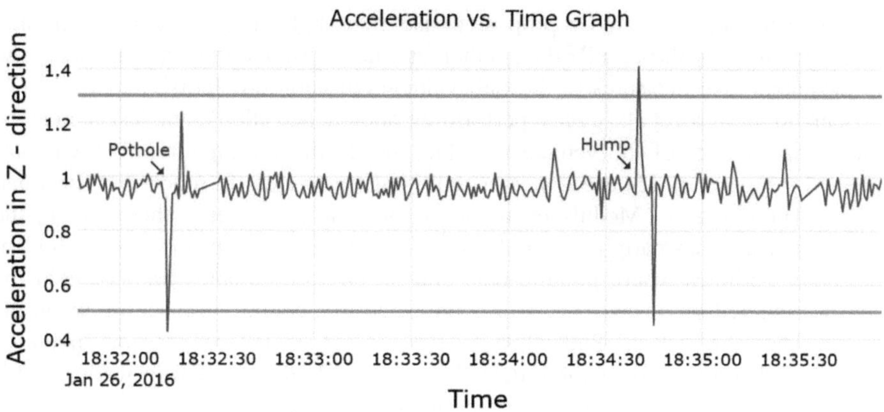

Fig. 4 Accelerometer versus time graph

Table 1 Table showing lower threshold value set versus percentage of potholes detected

Value of lower threshold (G)	Percentage of potholes detected (%)
0.3	70
0.6	97 but a high number of false positives
0.5	93 and minimum false positives

Table 2 Table showing upper threshold value versus percentage of speed-breakers eliminated

Value of upper threshold (G)	Effects on speed-breaker elimination
1.5	40% of speed-breakers were missed and downward reading were misinterpreted as potholes
1.1	Even small variations in the road surface got misinterpreted as speed-breakers and a lot of potholes were mistaken as speed-breakers
1.3	Had the maximum correct identification of speed-breakers

In Table 2, we can see the upper thresholds selected for speed-breaker elimination and its effect on detection of potholes.

An added advantage to our method is scale, the sheer number of vehicles passing over the potholes and speed-breakers correct any discrepancies, due to factors like speed of the vehicle and different types of driving. The vast number of automobiles on the streets provides us with sufficient readings for reporting and prioritizing potholes highly accurately.

5 Related Work

This section summarizes some of the other methods explored in the domain of pothole detection and emission monitoring.

Chi-Man Vong et al. in [3] propose to use traffic lights to get a reliable emission reading wirelessly through RFID and then use the information to warn the drivers to immediately take measures to fix their vehicle emission problems.

Christoph Mertz et al. suggest performing pothole detection using lasers attached under the carriage of the vehicle to detect the depth of the pothole [4]. Christopher M. Jengo et al. [5] use hyperspectral imagery to classify road conditions and identify potholes. A. Mednis et al. in [6] propose to detect potholes using the accelerometers in smartphones. G.D. De Silva et al. in [7] propose a method of detecting potholes where pothole data and its GPS coordinates are stored in the sensor mounted on buses. This data is uploaded to a central base station when the bus reaches the station. Sudarshan S. Rode et al. in [8] propose an inventive Wi-fi-based solution for pothole detection and an alerting system to warn the drivers about the potholes using access points mounted on the roads.

The method we have chosen has the following advantages over the above methods:

(i) Sensors like accelerometers are already present in most vehicles, using which we may collect data from multiple vehicles on the roads.
(ii) It is much more cost effective, than adding lasers to vehicles [4] or hyper-spectral imagery [5].
(iii) Implementing it is much easier when compared to installing the required sensors in traffic lights [3].
(iv) Our system is an integrated solution for solving many road-related problems using hardware already present in most vehicles today, in a future proof way, thereby making vehicles smart.

6 Scope and Future Work

Other parameters such as speed of the vehicle and suspension capacities of different vehicles can be used to set the threshold value dynamically so that potholes can be detected with a higher accuracy.

7 Conclusion

In this paper, we explore the concept of the Internet of Things by applying it to vehicles in a real-time situation to solve traffic-related and civic problems. The solution is feasible because most of the modern vehicles are already equipped with a range of high-end sensors for optimizing vehicle operations.

Acknowledgments We acknowledge the support and help provided by the HoD, Department of Computer Science & Engineering, Principal, and Management of M.S. Ramaiah Institute of Technology, Bengaluru (an autonomous college affiliated to VTU).

References

1. Chen S, Xu H, Liu D, Hu B, Wang H (2014) A vision of IoT: applications, challenges, and opportunities with China perspective. IEEE Internet Things J
2. Chidester A, Hinch J, Mercer TC, Schultz KS (1999) Recording automotive crash event data. In: International symposium on transportation recorders on 3–5 May 1999, Arlington, Virginia
3. Vong C-M, Wong P-K, Ip W-F (2011) Framework of vehicle emission inspection and control through RFID and traffic lights. In: Proceedings 2011 international conference on system science and engineering
4. Mertz C (2011) Continuous road damage detection using regular service vehicles. In: Proceedings of the ITS world congress, Oct 2011

5. Jengo CM, Hughes D, LaVeigne JD, Curtis I (2005) Pothole detection and road condition assessment using hyperspectral imagery
6. Mednis A, Strazdins G, Zviedris R, Kanonirs G, Selavo L (2011) Real time pothole detection using Android smartphones with accelerometers. In: 2011 International conference on distributed computing in sensor systems and workshops (DCOSS), 27–29, June 2011, pp 1–6. doi:10.1109/DCOSS.2011.5982206
7. De Silva GD, Perera RS, Laxaman NM, Thilakarathna KM, Keppitiyagama C, de Zoysa K (2008) Automated pothole detection system. In: International IT conference (IITC 08), Colombo, Sri Lanka, Oct, 2008
8. Rode, SS, Vijay S, Goyal P, Kulkarni P (2009) Pothole detection and warning system: infrastructure support and system design. In: 2009 International Conference on electronic computer technology, 20–22 Feb 2009. doi:10.1109/ICECT.2009.152

Railway Security Through Novel Machine-to-Machine Network Implementation

Chitra Suman, Lokesh Tharani and Saurabh Maheshwari

Abstract In railways, if all the systems work synchronously and accurately then only security can be ensured. To avoid manual errors, we are hereby proposing a machine-to-machine communication network which will connect all critical machinery and subsystems. It will be a data aggregation and decision-making system during the railway operations. Whole critical machinery will be having their local error detection and correction mechanisms through sensors and transducers. They also have the proposed communication capability with the sink. All threats will be collected at a common node outside the train at a global sink which forwards the train required information for next 10-km propagation. A risk score is calculated which when higher than threshold then the train is signaled to stop immediately. All the threats have their priority score. Collected threats' priority score is added to get final risk score. This is first ever implementation of machine-to-machine network for railway security. The implementation of work is done using MATLAB v14 which is used to simulate the proposed algorithm. The simulation results are compared with the previous approaches on the basis of average transmission time. Priority score-based analysis also is done in the simulation.

Keywords Railway security · Machine-to-machine network · Data aggregation · Security · Clustering · Critical systems

---

C. Suman (✉)
Department of Computer Science and Engineering,
University College of Engineering, RTU Kota, Kota, India
e-mail: sumanchitra.suman3@gmail.com

L. Tharani
Department of Electronics & Communication Engineering,
University College of Engineering, RTU Kota, Kota, India
e-mail: tharani123@gmail.com

S. Maheshwari
Department of Computer Science and Engineering,
Government Women Engineering College Ajmer, Ajmer, India
e-mail: saurabh.maheshwari.in@ieee.org

H.R. Vishwakarma and S. Akashe (eds.), *Computing and Network Sustainability*,
Lecture Notes in Networks and Systems 12, DOI 10.1007/978-981-10-3935-5_39

1 Introduction

Accidents in railways due to the poor maintenance are seriously hazardous for human health and safety [1]. Derailment is a major cause of accidents; this is basically caused due to the rail defects [2].

Machine-to-machine (M2M) communications offer ubiquitous and direct connectivity between devices. M2M devices that are organized as a network exchange information without any human interference [3]. It is a part of Internet of Things (IoT), and it contains wide range of applications. M2M devices are battery-operated and could be located at wide variety of places; thus, wide coverage and low energy consumption are the important requirements of M2M networks. We have proposed a novel M2M network architecture and multi-hop routing method to manage the flaws of approaches in literature. The motivation of this work is to benefit thousand lives with the safety are following. A completely extensible solution is required which can be extended for very large number of machine nodes and several machine networks. Since the nodes are battery-operated, TCP/IP may not be used for data packaging and routing. So a lightweight protocol sending and receiving very small but significant amount of information is required. There have been a lot of approaches to save the track and detect breakages. Some approaches are ensuring security for peripheral systems like railway crossings. But an approach is required which can collect working status of not only large systems but also small machines and combine them to identify the probable threats for train propagation. Data should be kept in simplest form with minimum preprocessing involved in it. It should be able to work with all types of status sensors. It should be able to incorporate multiple protocols as small sensors will be using ZigBee networks while long-distance transmission needs other wireless and wired protocols for that TCP/IP can be used. Working of all the sensors properly is essential, but stopping of the train propagation is only necessary and critical when some vital sensors are not working properly. To maintain this requirement and unnecessary stopping, all the machine nodes should be given some priority score. Objective of the proposed methods is proposing a machine-to-machine communication network to connect all critical machinery and subsystems in railway network. Preparing a data aggregation and decision-making system during the railway operations. To make decision by calculating risk score at train to take decision to stop the train or keep-on running.

2 Proposed Methodology

In this section, we will show the proposed architecture of the machine-to-machine communication network. The network has various entities that include local sensor nodes, which are equipped with problem detection sensors. Local sink is the collection node for all the sensing nodes (Fig. 1).

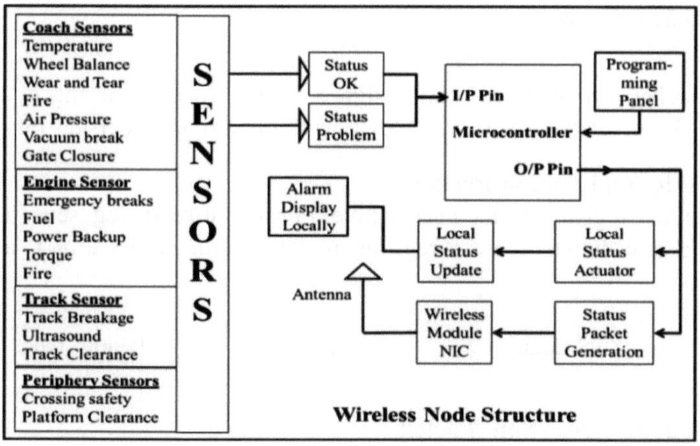

Fig. 1 Components of machine node

Collection poles are the fixed poles which collect the data from the local sinks. Global sink has the responsibility of complete network data collection and processing the data. Risk score is the total score of all threats to the movement of the train x. Unique IDs are assigned to each train, track segment for which a breakage detection sensor is attached, signaling poles, platform clearance, and for the railway crossings. Risk priority value is the priority value assigned to each risk type.

Then, architecture includes local machine-to-machine network that is the network of the sensor nodes which are connected wirelessly to the local sink. Collection pole network is where status packets that have 'problem' status are forwarded that is nearest to the train/railway crossing/track location. Data aggregation is done at a global sink where complete network data is collected from all the machine networks. Decision making is the final step taken at the train by calculating the risk priority score (Fig. 2).

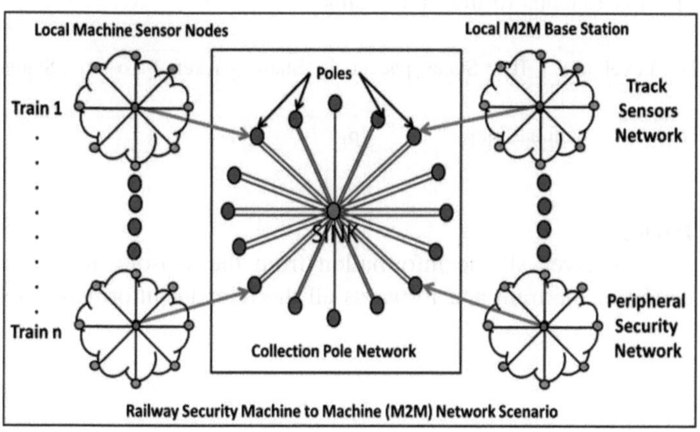

Fig. 2 Railway security M2M network scenario

2.1 *Packet Formats*

Specific packet formats have been designed for this machine-to-machine communication. These packets are used for data transmission and decision making.

Sensor Node Registration Packet

Each node first registers to the nearby pole. For this, it sends reg_sens_pole packet whose fields are shown below. It registers its ID at particular location in a particular direction.

ID	GPS Location	Type	Priority_score

Sensor Node to Local Sink Packet

This is the packet which is created by the sensor node and forwarded to the local sink of the machine network. The status is ON '1' if status is abnormal, else this line is OFF '0'. Local_sink_ID is the unique ID of the local sink.

ID	Local_sink_ID	Status

Local Sink to Level 1 Poles Packet

This packet contains status of all the registered nodes as payload and its own header.

Local_sink_ID	Level1_pole_ID	Status 1	Status 2	o	o	o	Status n

Level 1 to Higher-Level Pole Packet

Level 1 pole collects data from all the registered local sinks and then forward all of them to the higher level by embedding in its own packet. The same process keeps on till the packet reaches to the global sink.

Level_n_ID	Level_n + 1_ID	Status_packet_1	Status_packet_2	o	o	Status_packet_n

Level_n_ID	Global_sink	p_1	p_2	o	p_n

Decision Packet

Once the sink receives all the information from the sensors, then it checks the current location of the train and forwards all the relevant information required for

10 km of propagation by embedding it in decision packet. The destination is the pole to which the train is currently registered or nearest to.

Sink	Pole_ID	Total_Info_number	Info_1	Info_2	o	o	o	Info_n

Here, the total_info_number field is important for security of packet. This field tells how many info fields are following.

Data forwarding technique
Proposed method includes different data forwarding steps that first include machine-to-local sink network processing in which sensors check the status of machinery working for each cycle either 'OK' or 'Problem' and forward to the microcontroller input pin which processes the input and generates a programmed output; one output is fed to the actuator to convert the signals to physical output like alarm or blinking LED and other to the wireless module where a status packet is created with sensor ID, machine ID, and status bit, which is sent wirelessly to the local sink. Second, in local sink to pole forwarding, all the local sinks at all on and off train machine networks forward the collected problem status packets and prepare for forwarding to the poles. All the local sinks are registered to unique pole at a particular time instant. Stationary machine networks are registered to single pole statically whereas moving train keeps on registering with nearest pole while moving. Third, pole to global sink collection network processing, in which all the poles collect the instantaneous data from all registered sinks' local machine networks. Poles forward these status packets to the global sink for aggregation; then, it is finally send to train from global sink for decision making.

3 Simulation Results

For implementation of this algorithm, MATLAB has been used due to its property of allowing creation of various packets in the form of structure variables. In this section, we have shown the operation of railway security network equipped with various sensors. The figure below shows the simulation done in MATLAB v14 to implement the algorithm. There is an imaginary railway line horizontal track of 100 km long; on this track, vibration estimating sensors are equipped at each kilometer shown in small red dots. The red stars encircled by red circle represent the railway station machine nodes whereas the black stars encircled by black circle represent the machine nodes of railway crossings. The poles at different levels are situated on both sides of the track. Level 1 poles just near the track shown as small blue circles collect the data from the local sinks and registered vibration sensor (Fig. 3).

All other nodes forward their data to their local sinks. The level 1 poles forward the collected status data to next-level poles and so on till the status reaches to the

Fig. 3 Network architecture

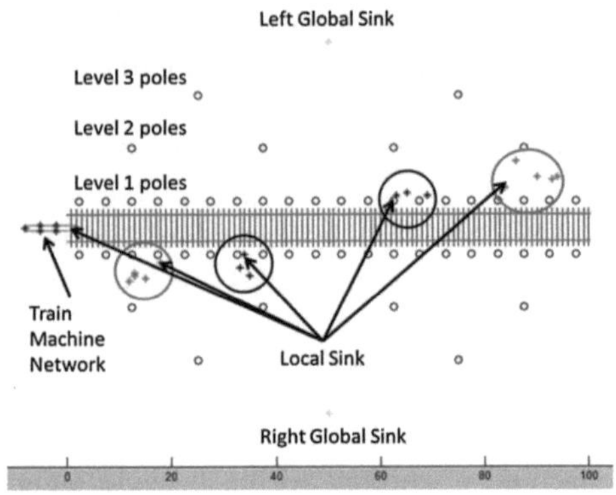

sink. The sink makes the decision by adding the risk priority values of the risk factors and calculates the resultant risk score. If this score is more than the threshold, then a signal to stop the train is generated and sent through a reply packet to the train through the shortest path as shown in figure. This shortest path keeps on changing depending on the current train location. The train keeps on registering to next poles as it moves. Only train is movable in this complete network, and all other nodes are stationery and statically registered to unique pole. For simulation, the scenario is shown where there is broken track at the railway station first from left. So the status signals generated by track breakage sensors are 'problem.' This status is forwarded to the sink as shown a chain in Fig. 4. The train is given a command to stop 2 km before the problem location.

3.1 Average Transmission Time

Average transmission time (ATT) is the time taken by the network to transfer the sensed information from the sensor nodes to the sink (TSS) and then forwarding decision from the sink to the train (TST) using shortest path. This is generally constant for a network for fixed number of levels. This brings a unique advantage to the proposed network architecture that the total time taken for decision making and forwarding is independent of the track length and the train speed. ATT can be given by the equation below:

$$ATT = \frac{Sensing\ time + T_{SS} + Decision\ making\ time + T_{ST}}{No.\ of\ cycles}$$

Here, the sensing time and the decision-making time are dependent on the hardware selection. More efficient is the hardware, and respective efficient firmware lesser is the processing time. The remaining time components TSS and TST depend only upon the number of levels and the forwarding delay at the communication node (Figs. 5 and 6).

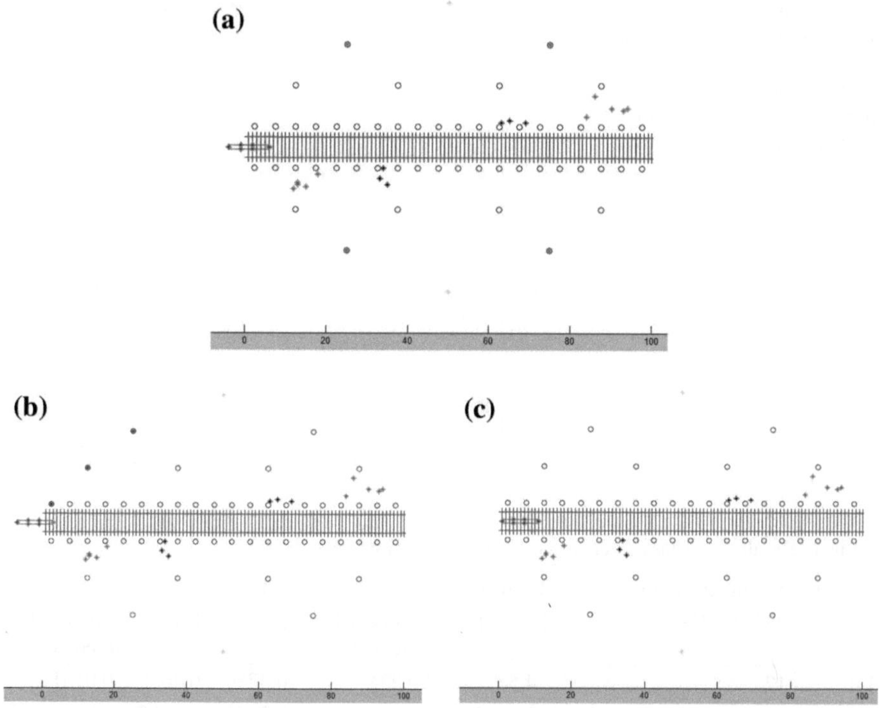

Fig. 4 **a** Collecting of all sensor status data at higher-level poles from all local sinks and forwarding to sink. **b** Sink forwarding decision to the nearest pole to the train. **c** Train stopped before entering to the railway station as risk score became more than threshold

Fig. 5 Increasing plot for the distance in KM on X-axis and total time taken for fault detection on Y-axis

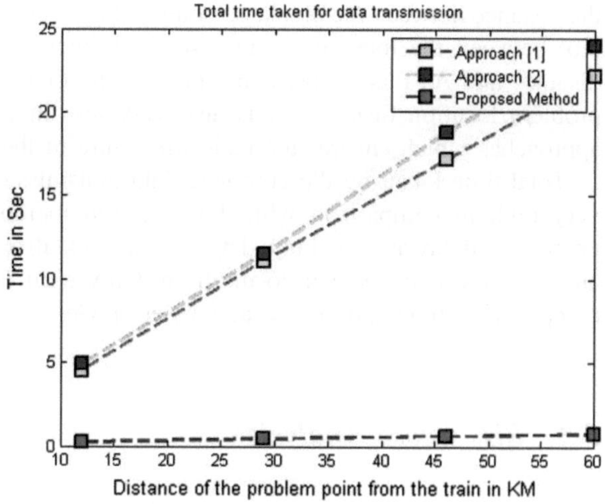

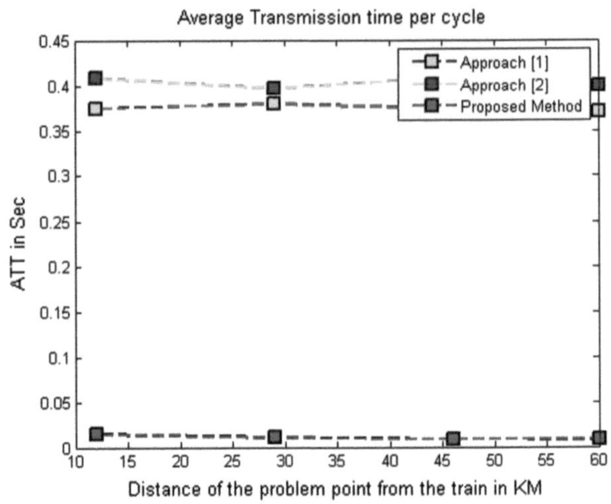

Fig. 6 Constant plot for the distance in KM on X-axis and ATT on Y-axis showing real-time nature of proposed approach and very less time taken from previous methods

Table 1 Breakage distance versus average transmission time

Sr. No.	Minimum problem distance (KM)	Total time (s)			ATT (s)		
		[5]	[6]	Proposed method	[5]	[6]	Proposed method
1	12	4.5	4.9	0.234	0.3750	0.4083	0.0151
2	29	11.1	11.5	0.453	0.3793	0.3965	0.0106
3	46	17.2	18.8	0.625	0.3739	0.4086	0.0091
4	60	22.2	24.0	0.750	0.3700	0.4000	0.0082

Another important inference is that the average time per cycle is independent of the distance between the breakage point and the train as shown in plot in figure. The plot between the total time and distance is increasing, but the plot between the distance and ATT is almost constant. It refers that the ATT is independent of the problem location distance. It is also very less in comparison with the previous approaches which ensures the real-time nature of the proposed method (Table 1).

Total time taken by the complete data transmission for previous approaches is very high in comparison with the proposed method since we do not add any processing delay and send raw data to sink. Sink also does not make any processing and send the relevant data to the train. Only train does processing which can be completed in time and also in automated mode.

3.2 *Risk Score Analysis*

In this section, we will see for constant threshold = 60 how the train stops at correct distance before the problem area using the risk priority score. Two cases are given where problem locations and sensors have been chosen randomly. The cumulative priority score is being regularly calculated by train: if this score exceeds threshold, then train stops. In case 1, the score never exceeds threshold so the train is not stopped and run for complete path.

	Node ID	Node type	Priority score	Priority score due to status-problem	Cumulative priority score
Case 1	3	'Platform clear'	20	@ 6 km = 20	20 + 10 = 30
	15	'Crossing clear'	20	@ 35 km = 20	20 + 10 = 30
	18	'Signals working'	10	Always 10	
	30	'Track broken'	20	@ 30 km = 20	20 + 30 = 50
	78	'Obstacle'	20	@ 78 km = 20	20 + 10 = 30

In case 2, the score exceeds threshold = 60 again @ 20 km due to problem location @ 30 km. The total risk score due to train system failure, signal not working, crossing not clear, and track broken is equals to 60 so the driver is given a command by the system to stop the train @ 20 km.

	Node ID	Node type	Priority score	Priority score due to status-problem	Cumulative priority score
Case 2	3	'Platform clear'	20	@ 16 km = 20	20 + 10 + 10 = 40
	15	'Crossing clear'	20	@ 34 km = 20	20 + 10 + 10 = 40
	17	'Engine proper working'	10	Always 10	
	18	'Signals working'	10	Always 10	
	30	'Track broken'	20	@ 30 km = 20	20 + 40 = 60
	78	'Obstacle'	20	@ 78 km = 20	20 + 10 + 10 = 40

3.3 *Comparative Results*

Majority of track breakage detection system requires the train to reach up to the breakage point like in the LED-LDR assembly-based track breakage detection system [4]. Here, the breakage is detected when the light from LED is transmitted to the LDR through the crack. Also, complete track is to be checked regularly, but in our case, remote monitoring can easily be done. All the decision-making power is with sink and train only so they can get all the required information and automate the process of decision making. The train can be stopped sufficiently before the problem area no need to go near the breakage point. The systems in [2] aim to find an approximate location of the problem on the track. In our case, we are able to find the exact location and sensor which is not working properly. Also, in this approach the total data overhead is very high. In our case, very less data is generated and communicated.

4 Conclusion and Future Work

Proposed algorithm is the first ever attempt to implement machine-to-machine networks for railway security. We have simulated the whole security scenario through a visual model in which the packets are being transmitted from machine node to sink and decision taken by sink to stop the train or keep on running is forwarded to the train. The train acts accordingly to the decision taken by the sink. More detailed real-time implementation is needed to check the performance for real-world situations like obstacles in the path. The results are satisfactory for the real-time requirement as proved by very less average transmission time and approach is approximately 96% faster in comparison with previous approaches. The time taken for decision making is independent of the train speed and track length. Stopping of the train using the priority risk score has been shown through various case studies. All types of systems can be connected so it is completely scalable. In future, we plan to practically implement this work using various machine-to-machine sensor nodes and verify the capability of this system on real-time railway security.

References

1. MuneedraRao Ch, Bala Jaswanth BR (2014) Crack sensing scheme in rail tracking system 4(1) 13–18
2. Daliri ZS, Shamshirband S, AmiriBesheli M (2011) Railway security through the use of wireless sensor networks based on fuzzy logic. Int J Phys Sci 6(3):448–458
3. Xiong X, Zheng K, Rongtao Xu, Xiang W, Chatzimisios P (2015) Low power wide area machine- to-machine networks: key techniques and prototype. Commun Mag IEEE 53(9): 64–71
4. Vanimiredd A, Kumari DA (2013) Automatic broken track detection using led-ldr assembly. Int J Eng Trends Technol (IJETT) 4(7)
5. Sharma K, Maheshwari S, Khanna V (2014) Railway track breakage detection using vibration estimating sensor network. In: 2014 International conference on advances in computing, communications and informatics (ICACCI). IEEE, pp 2355–2362, 24–27 Sep 2014
6. Islam MF, Maheshwari S, Kumar Y (2015) Energy efficient railway track security using vibration sensing network. In: IEEE international conference on signal processing and integrated networks (SPIN). pp 973–978, 2015

A Honeypot Scheme to Detect Selfish Vehicles in Vehicular Ad-hoc Network

Priya Patel and Rutvij Jhaveri

Abstract Vehicular ad hoc network (VANET) has been emerged as a prominent technology for intelligent transportation systems. A VANET consists of a number of vehicles equipped together for communication on road side. VANETs are used to fulfill many requirements such as drivers' safety, data transformation and traffic control, and so on. In VANET, each vehicle behaves independently and as a result, some vehicles might behave selfishly to save their resources. This issue can induce network latency, network break down, security breach, and other issues. In this paper, we address this issue by proposing a honeypot detection approach which endeavors to mitigate selfish vehicles from the network. For experimental results, proposed scheme is incorporated with AODV protocol. We present the behavior of selfish vehicles based on energy constraints. Simulation results under various network parameters depict that the proposed approach provide more robust and secured routing among the vehicles in VANETs.

Keywords Vehicular ad hoc networks · Selfishness · Bait mechanism · AODV

1 Introduction

Vehicular ad hoc network defined as number of vehicles connected together wirelessly and established the network on road route to communicate with each other for different purpose. VANET is another sort of specially appointed network that is described by its exceptionally mobile topology [1]. VANET is more preferable network because of its special features such as unconstrained deploy-

P. Patel (✉) · R. Jhaveri
Department of Computer Engineering, SVM Institute of Technology,
Bharuch 392-001, Gujarat, India
e-mail: piyubhandari11@gmail.com

R. Jhaveri
e-mail: rhj_svmit@yahoo.com

H.R. Vishwakarma and S. Akashe (eds.), *Computing and Network Sustainability*,
Lecture Notes in Networks and Systems 12, DOI 10.1007/978-981-10-3935-5_40

ment, infrastructure less network, distributed nature, self-arranging nature, and geographical independence [2]. These natures make it more utilizable in many application areas such as road safety, rout planning assistance, tolling, traffic management, and many more [3]. VANET can be categorized in three diverse types based on its communication style: (i) Vehicle to vehicle (V2V) communication, (ii) Vehicle to road infrastructure (V2I) communication, and (iii) Vehicle to broadband cloud (V2B) communication [4].

In VANET, the routing protocols are categorized in various types: Topology-based routing protocol—in which existing link used for packet forwarding, Position-based routing protocol—in which the forwarding choice by node is essentially made in view of the position of the packet's destination and node's one-hop neighbors, Cluster-based routing protocol—used for clustering network, in which routing decision taken by cluster head, Geo-cast routing protocol—these protocols are utilized for sending message to vehicle which locate in predefined geographical region, and Broadcast routing protocol—Broadcast routing is much of the time utilized as a part of VANET for sharing activity, climate and crisis, street conditions among vehicles, and conveying notices and announcements [4, 5].

There are many terms which effect network such as energy efficiency, number of vehicles, resources, and security of the data transmission over network. In VANET, each vehicles behave independently, from that some vehicles behave selfishly by dropping packets during routing or transmission to full feel their malicious purpose. The selfish vehicles use the network without pay back for the usage of network [6]. So result of these malicious activates on network QoS become low in terms of network breakage, latency occurred, low transmission rate, less security, energy constraints, etc.

To address the selfishness problem in the network, many researchers develop the different techniques and mechanisms. There are various techniques developed to solve different types of selfish vehicles such as energy based, speed based, memory based, and so on. Here in our work we proposed a solution for the energy-based selfish vehicle. Here, we developed the selfish vehicles which aim to save its energy. We designed the selfish vehicle which drop the packets when it's remaining energy less than 50% of total energy.

For prevention of the network from the selfish environment, numerous strategies were produced in past years by numerous researchers. They are as follows: Trust-based approach, watchdog model based, Reputation-based approach, and many more. These mechanisms provide security but in some scenario, some of the techniques increase overhead, and in trust-based approach more memory is required. So here we proposed honeypot mechanism-based selfish vehicle detection scheme named as honeypot selfishness detection scheme (HPSD). This HPSD uses honeypot mechanism to detect selfish vehicles when vehicle noticed as suspicious vehicles by its activities. After that, it adds into the suspicious list, bait RREQ unicast to that suspicious vehicle if it drops the bait packet then it's mentioned as selfish vehicle. And discard it from the network.

The rest of the paper is described as: Sect. 2 describes selfishness in which types of selfish vehicles and its prevention techniques mentioned, Sect. 3 presents related work of different mechanisms developed by numerous researchers in past years to solve selfishness problem, Sect. 4 represents existing and proposed approach for mitigating selfish vehicles from the network, Sect. 5 represents simulation results and finally, in Sect. 6 conclusion is mentioned.

2 Selfishness

2.1 Selfish Vehicles

In VANET, the vehicles which does not want to participate, drop the packet during routing and transmission process known as selfish vehicles. They generally expect to expand their own benefit, bringing about undesirable postponements in the message conveyance, and expansion the network inertness, which thus influences the whole execution of the network. The main goal behind these types of behavior of selfish vehicles is to save their resources such as battery power, memory, CPU utilization, and bandwidth. Another reason behind these types of behavior might be fear of getting harmful data or information from neighbor vehicles. Energy is the critical parameter for vehicular ad hoc network because of low assets, these selfish vehicles may carry on as ravenous and they take the information from other vehicles when they require, however, they abstain from sending it to other vehicles in the network [7]. Based on the vehicles' different behavior, selfish vehicles are categorized in three types [8]: Type 1: Vehicles which do not participate in routing, Type 2: No response to HELLO message, Type 3: Drop the data packet.

2.2 Selfish Vehicles in AODV-Based VANET

In Fig. 1a VANET scenario is shown, in which A vehicle acts as a source vehicle and G vehicle acts as a destination vehicle. Now, A vehicle wants to communicate with vehicle G. For that, A vehicle broadcast RREQ packet to its all neighbor vehicle to know whether G vehicle is its neighbor node or not (refer Fig. 1b).

Now in cooperative environment all vehicles check the RREQ packet and compare destination id with its own. If destination id matches, then it forwards RREP packet to source node and communicate with it directly. But if destination vehicle id does not match with its own id, then it further broadcast RREQ packet to its neighbor vehicle until it gets destination (as shown in Fig. 1c). Now, as shown in Fig. 1d, if selfish vehicle is present in network, it simply drop the packet and do not further broadcast RREQ packets to its neighbor. Here, in Fig. 1d shows that vehicle B behaves selfishly and drops the RREQ packet during routing from source vehicle A to destination vehicle G.

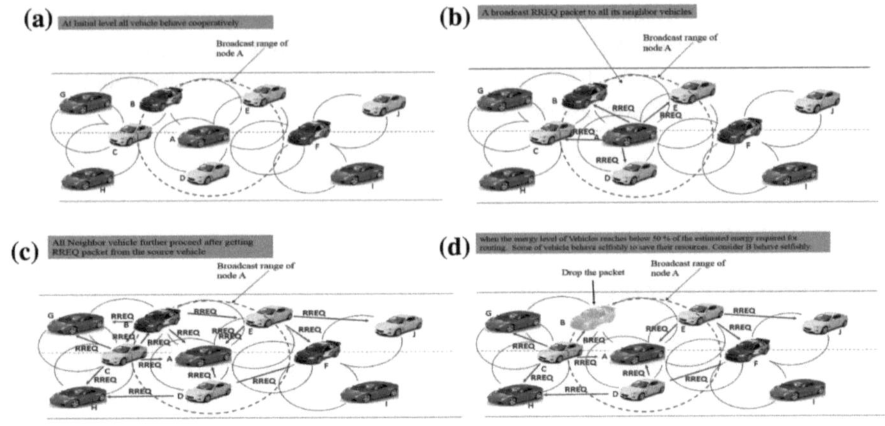

Fig. 1 Route discovery in VANET. **a** Initial scenario. **b** Route discovery. **c** Route discovery in cooperative environment. **d** Route discovery in selfish environment

2.3 *Model of Selfish Vehicles Based on Energy Constraint*

In our work, we used to implement energy-based selfish vehicles, i.e., its purpose to save its remaining energy. Here, we take Type 1 selfish vehicles, which do not want to participate in the routing process. For that it drops the RREQ packet broadcasted from its neighbor node.

As shown in Fig. 2, we implemented the selfish vehicles, which behave selfishly i.e., drop the RREQ packet when its remaining energy is less than 50% of the total energy. Here, as shown in above model, if energy level of selfish vehicles is less than 50%, then it drops the packet unless it behaves cooperatively. If vehicle is cooperative, then it first checks the destination id of the RREQ packet with its own id, if destination id matched, then it unicast RREP packet to the source or current source and communicate with its. And if destination id is not matched with its id, then it further broadcast the RREQ packets to its neighbor to reach destination.

Fig. 2 Model for selfish vehicle

Action of Selfish vehicles after receiving RREQ message

```
IF(Energy of the Vehicle less than 50% )
    Selfish Vehicle drop the packet
ELSE IF Vehicle check the Destination vehicle ID
    Destination Vehicle ID match with vehicle ID
    unicast RREP message to the Source Vehicle
ELSE
    Further broadcast RREQ to its neighbor Vehicle
END IF
```

2.4 Techniques for Selfish Vehicles Detection

There are mainly three techniques developed for detecting selfish vehicles from the network.

Credit-Based Schemes

In proposed scheme, each node has to pay some money to use services of the network. This money is like credit which is given to the intermediate nodes [9]. Those nodes which provide the services earn the credit and in the other hand those nodes which uses the services have to pay the credit for that. The main agenda of this proposed scheme is to motivate the nodes behave reliably in the network during transmission [10].

Reputation-Based Schemes

In Reputation-based schemes, matrix of reputation of nodes established based on the each node's behavior which is observed by its neighbor in the network. This reputation matrix is used to evaluate whether node is trustworthy or not. This information is then propagated throughout the network so that the detected misbehaving node can be removed from the network [9].

Acknowledgment-Based Schemes

Proposed scheme is purely based on the reception of acknowledgement to verify the message is forwarded from the sender or not. In this type of technique monitoring of RREQ and RREP packet from neighbor nodes used to detect selfish nodes from the network. Furthermore, there are three types of module integrated: (i) Reputation module, (ii) Route discover module, and (iii) Audit module [9].

3 Related Work

Selfishness of the vehicles is a real issue of the VANET which influences the system's quality furthermore throughput. In later past year, numerous analysts have built up various methodologies to overcome issue of selfishness in different types of network.

Omar et al. [1] proposed two-phase-based quality of service optimized link state routing (QoS-OLSR) protocol which was used to detect selfish vehicle from the clustered network. Here, selfish vehicle behaves selfish during cluster formation by providing fake information regarding to vehicle's speed. Incentive was used to detect selfish vehicle during cluster formation. But still selfish vehicle have benefits to behave selfishly after cluster formed. To solve this problem, in second phase cooperative watchdog model-based Dempster-Shafer model used. Younes et al. [7] proposed a Secure COngestion contrOL (SCOOL) protocol. This protocol provides integrity and authenticity of transmitted data and also provide to preserve the privacy of the cooperative vehicles and drivers. Yang et al. [11] Proposed dynamic three—layer Reputation Evidence Decision System (REDS). In which, to discriminate selfish vehicles, Dempster-Shafer evidence integration mechanism used. This proposed scheme provide rapid solution for selfish node detection. Song et al. [12] proposed privacy-preserving distance-based incentive scheme to detect location privacy and behavior of selfish nodes. Zhou et al. [13] proposed trust-based approach named security authentication method. Here two types of trust evaluation technique used. First was direct trust, in which the security vector model was set up taking into account the security practices of the new vehicle node. And second was indirect trust, in which the trust degree is computed in light of the proposal trust vectors from the vehicle nodes in the system. Khan et al. [14] proposed DMN—Detection of Malicious Nodes is improved version of DMV algorithm in terms of selection of selfish nodes. Jesudoss et al. [15] proposed payment punishment scheme (PPS) which not only detect selfish vehicles but also used to encourage selfish vehicles cooperatively after cluster formation. Here watchdog model used to monitor the vehicles and modified Extended Dempster–Shafer model used to discourage the vehicle from selfish behavior.

4 Existing and Proposed Approach

4.1 Existing Approach

During creation of a path, source vehicle checks its routing table first. If the destination vehicle is not present, then it broadcast the RREQ message to all neighbor vehicles. The cooperative neighbor vehicle will again rebroadcast it to their one-hop neighbor vehicles and this process continues until it reaches destination vehicle [16]. Meanwhile source vehicle is also a one-hop neighbor node of each of these nodes, so it will receive the same. So each time vehicle monitor its neighbor vehicles' character based on receiving back RREQ packet from neighbors. If it has not received the same RREQ from one of its neighbor vehicles within a prefixed timeout, then that node will be marked as potential misbehaving vehicle. This process continues repeatedly. For each potential misbehaving node, a threshold value is maintained. If the number of times a vehicle is marked as a potential misbehaving node beats this threshold limit, then that vehicle will be declared as selfish vehicle and this information will be sent to all other vehicles of the network [16]. After detecting selfish vehicles in the network, new path established through game theory, which help to find shortest path from the source to destination.

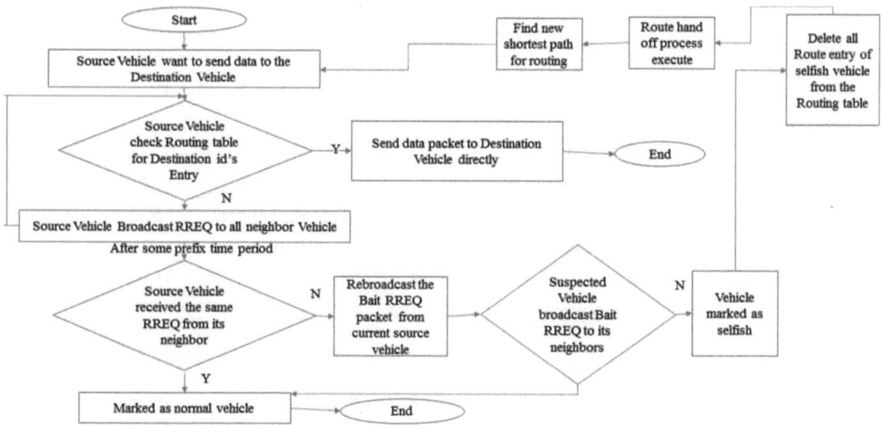

Fig. 3 Flowchart of proposed approach

4.2 *Proposed Approach*

To improve the existing approach here for detection of selfish vehicles bait method used in which whenever any vehicle drop the packet, bait mechanism used for detection of selfish vehicles.

During creation of a path, source vehicle checks its routing table first. If the destination vehicle is not present, then it broadcast the RREQ message to all neighbor vehicles. The cooperative neighbor vehicle will again rebroadcast it to their one-hop neighbor vehicles and this process continues until it reaches destination vehicle. Meanwhile source vehicle is also a one-hop neighbor node of each of these nodes so it will receive the same. So each time vehicle monitor its neighbor vehicles' character based on receiving back RREQ packet from neighbors. If it has not received the same RREQ from one of its neighbor nodes within a prefixed timeout, then current source node send the bait RREQ to its suspicious neighbor. If suspicious vehicle drop that bait RREQ packet, then that vehicle will be marked as selfish vehicle, otherwise, if it broadcast that bait RREQ packet then it will be marked as cooperative vehicle. After that all path entry related to that selfish vehicle will be deleted and route hand off process done for new path. Among all possibility shortest path will be select for further process (Fig. 3).

5 Simulation Results

The simulation is carried out on NS-2 and SUMO simulator. Scenario for the VANET generated using OSM. Here simulation is carried out in 1000 × 1000 m area of simulator, where AODV, SVAODV, and HPAODV were executed. The

Table 1 Simulation parameters

Parameters	Value	Parameters	Value
Simulator	NS 2.34	Pause time	5 s
Routing protocol	AODV, SVAODV, HPAODV	Maximum speed	5, 10, 15, 20, 25 m/s
Scenario size	1000 × 1000 m	Initial energy	300 J
Number of vehicles	20, 40, 50, 60, 80, 100	Transmit power	1.65 W
Selfish vehicles	0, 2, 4, 6, 8	Receive power	1.4 W
Simulation time	240 s	Idle power	1.15 W
Traffic types	Constant bit rate (CBR)/UDP	Sleep power	0.045 W
Number of connection	5	Transition power	2.3 W
Packet rate	4 packets/s	Transition time	800 μs

Fig. 4 VANET scenario

performance of these three routing protocol measured in four different metrics: PDR, Normalized Routing Overhead, End to End Delay and Average Consumed Energy. The major simulation parameters are listed in below Table 1. Furthermore VANET scenario of the network in SUMO is shown in Fig. 4.

5.1 Test 1: Varying Number of Vehicles

As shown in Fig. 5a, in the absence of selfish vehicles, AODV gives more than 98% PDR for all network sizes.In the presence of selfish vehicles, PDR decrease up to 50%. When proposed approach is applied on the system, it improve result in

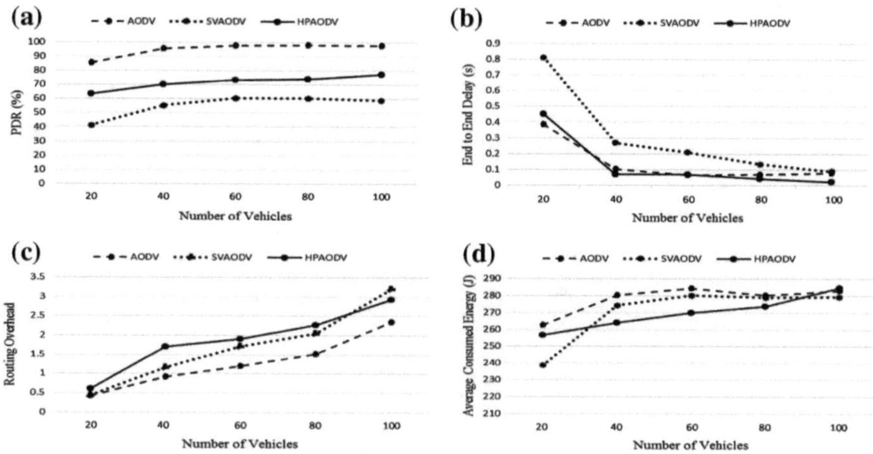

Fig. 5 **a** PDR. **b** End to end delay. **c** Normalized routing overhead. **d** Average consumed energy —when number of vehicles vary

terms of PDR up to 20–25% than selfish environment. Furthermore, proposed approach reduces up to 30–50% end to end delay during transmission. Because proposed approach takes quick action after getting suspicious vehicle in the network, so it provides the robust solution (As shown in Fig. 5b). Despite this beneficial results, proposed scheme increases normalized routing overhead (Refer Fig. 5c). Which is drawback of the proposed work. As shown in Fig. 5d, Energy consumption became low in selfish environment.

5.2 *Test 2: Varying Number of Selfish Vehicles*

As shown in Fig. 6a–d, as per selfish vehicles were increased PDR, End to End Delay, Normalized Routing Overhead and Average Consumed Energy were decrease, increase, increase and decrease respectively. In this test scenario, 0 number of selfish vehicles present normal AODV protocol scenario. When proposed honeypot approach was applied, it increases the PDR up to 30%, decrease end to end delay up to 20%. During increasing of selfish vehicles in the network, the PDR of the HPAODV protocol decreased. In terms of normalized routing overhead, proposed approach does not provide better result. Energy consumption of proposed approach changes with respect to number of selfish vehicles increase in the network.

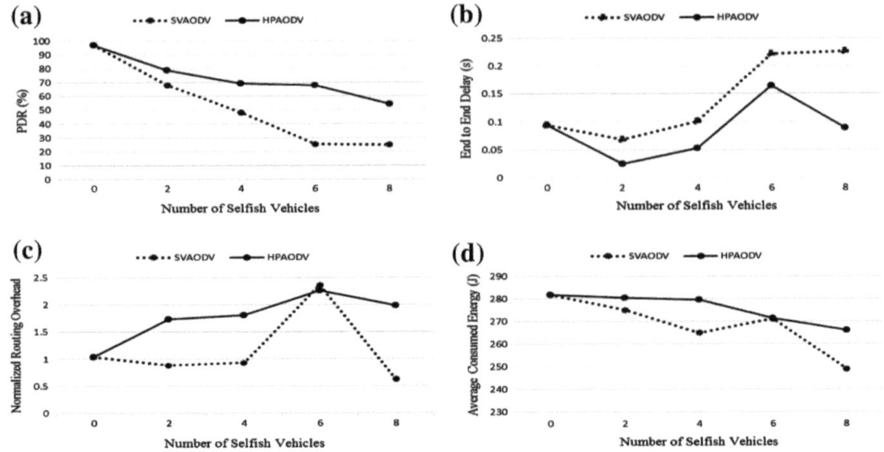

Fig. 6 **a** PDR. **b** End to end delay. **c** Normalized routing overhead. **d** Average consumed energy —when number of selfish vehicles vary

5.3 *Test 3: Varying Number of Mobility*

As shown in Fig. 7a, in the absence of selfish vehicles, AODV gives more than 98% PDR for all network sizes. In the presence of selfish vehicles PDR decrease up to 30–40%. When the proposed scheme was applied, it improves the result of packet delivery ratio up to 20–25% efficiently. As mobility increased, drastic increment occurred in terms of end to end delay in network. When the proposed approach was applied on the network, it was able to reduce the unexpected delay of the network. As shown in Fig. 7b, routing overhead increased in both scenario (i.e., in normal AODV and in SVAODV). Here, proposed approach fails to reduce the routing overhead of the network. Moreover, when the proposed approach was applied on the network, it was able to reduce average energy consumption by vehicles, which make it more reliable in terms of lifetime of the network.

5.4 *Comparison of Existing Approach with Proposed Approach*

In existing approach, detection of selfish vehicle done by comparing the threshold value of potentially misbehaving vehicle with predefined threshold value. When in the proposed approach based on Bait detection scheme in which if vehicle drop the packet then current source vehicle send fake message packet to vehicle. If target vehicle response to that message, then it consider as selfish node.

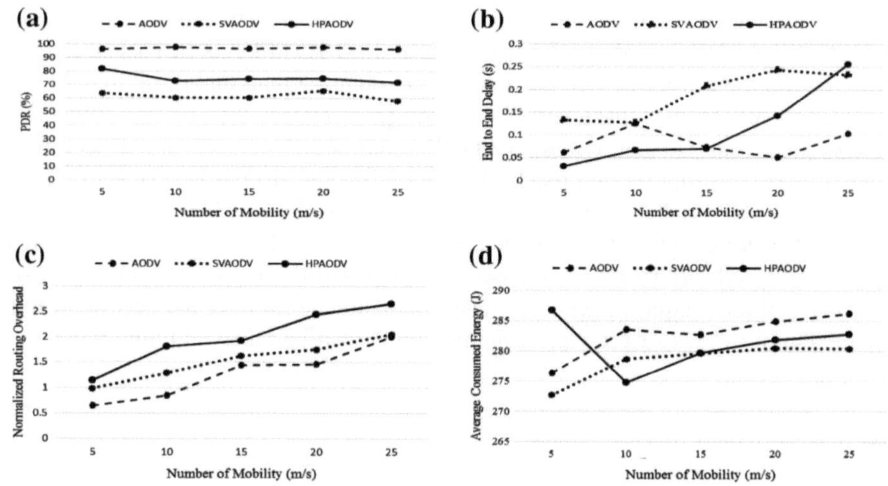

Fig. 7 **a** PDR. **b** End to end delay. **c** Normalized routing overhead. **d** Average consumed energy —when mobility vary

Table 2 Comparison of existing approach and proposed approach

Properties	Existing approach	Proposed approach
Detection technique	Based on threshold value	Based on bait mechanism
Accuracy of detection of selfish node	Less/average (expected)	More
Processing time for detection selfish node	More	Less
PDR	Less (expected)	More

In existing scheme, after some amount of packet dropping behavior existing scheme take action. Where in proposed scheme at the time when packet dropping behavior recognized by the node, it take action to detect selfish vehicle (Table 2).

6 Conclusion

Vehicular ad hoc network has been an active research area in recent years due to its ubiquitous nature and need of intelligent transportation systems for developing smart cities. This is advantageous on the one hand, while proves to be disadvantageous when selfish vehicles start misbehaving. Selfishness is a serious issue in VANET, which directly affects PDR of the network. To solve selfishness problem, here we introduced honeypot selfish vehicle detection scheme. Proposed scheme based on bait RREQ packet, which manipulate selfish vehicle to behave selfishly again so that it can be detected.

The simulation results show that performance of the AODV protocol in presence of selfish vehicles degrade performance up to 30–50% in terms of packet delivery rate. Furthermore, routing overhead and delay increase than that of normal AODV protocol. Due to the characteristics of selfish vehicles, the average consumed energy slips below to that of normal AODV where selfish vehicles are present in the network. Here, proposed approach HPAODV is able to improve packet delivery ratio up to 30%, and it also degrade the end to end delay. These make proposed approach more effective in terms of network lifetime. However, due to proposed approach, overhead will be increased in the network because of multiple forwardness of fake request packets to detect selfish vehicles over the network. Here, this is the drawback of proposed scheme, which can be taken as future work.

References

1. Wahab OA, Otrok H, Mourad A (2014) A cooperative watchdog model based on Dempster–Shafer for detecting misbehaving vehicles. Comput Commun 41:43–54
2. Jhaveri RH, Patel NM (2015) A sequence number based bait detection scheme to thwart grayhole attack in mobile ad hoc networks. Wirel Netw 21(8):2781–2798
3. Lipiński B, Mazurczyk W, Szczypiorski K, Śmietanka P (2015) Towards effective security framework for vehicular ad-hoc networks. J Adv Comput Netw 3(2)
4. Liang W, Li Z, Zhang H, Wang S, Bie R (2014) Vehicular ad hoc networks: architectures, research issues, methodologies, challenges, and trends. Int J Distrib Sens Netw
5. Ostroukh AV, Elhadi H (2015) Comparative study of routing protocols in vehicular ad-hoc networks (VANETS). Int J Adv Stud 4(2):9–14
6. Helen D, Arivazhagan D (2014) Applications, advantages and challenges of ad hoc networks. JAIR 2(8):453–457
7. Younes MB, Boukerche A (2015) Scool: a secure traffic congestion control protocol for VANETs. In: Wireless communications and networking conference (WCNC). IEEE, pp 1960–1965
8. http://eprints.utm.my/33353/2/MahmoudAhmadSalemMFSKSM2012CHAP1.pdf
9. Agarwal D, Rout RR, Ravichandra S (2015) Detection of node-misbehavior using overhearing and autonomous agents in wireless ad-hoc networks. In: Applications and innovations in mobile computing (AIMoC). IEEE, pp 152–157
10. Mittal S (2015) Identification technique for all passive selfish node attacks in a mobile network. Int J 3(4)
11. Yang Y, Gao Z, Qiu X, Liu Q, Hao Y, Zheng J (2015) A hierarchical reputation evidence decision system (REDS) in VANETs. Int J Distrib Sens Netw 501:341579
12. Song J, He CJ, Yang F, Zhang HG (2015) A privacy-preserving distance-based incentive scheme in opportunistic VANETs. Secur Commun Netw
13. Zhou A, Li J, Sun Q, Fan C, Lei T, Yang F (2015) A security authentication method based on trust evaluation in VANETs. EURASIP J Wirel Commun Netw 2015(1):1–8
14. Khan U, Agrawal S, Silakari S (2015) Detection of malicious nodes (DMN) in vehicular ad-hoc networks. Proc Comput Sci 46:965–972
15. JesudossA, Raja SVK, SulaimanA (2015) Stimulating truth-telling and cooperation among nodes in VANETs through payment and punishment scheme. Ad Hoc Netw 24:250–263

16. Das D, Majumder K, Dasgupta A (2015) Selfish node detection and low cost data transmission in MANET using game theory. Proc Comput Sci 54:92–101
17. Altayeb M, Mahgoub Imad (2013) A survey of vehicular ad hoc networks routing protocols. Int J Innov Appl Stud 3(3):829–846

Secret Information Sharing Using Extended Color Visual Cryptography

Shivam Sharma, Shivam Modi and Akanksha Sharma

Abstract Digitization of personal data is resulting in nefarious activities encircling them which in return perpetually call for immediate implementation of data security method, cryptography being one of them wherein sender can securely transmit the data over a secure platform. Sender can encrypt the data by encoding the text in an image, and receiver on another part can decode the message by collecting the respective shares. Generations of shares, where slight changes are performed accordingly, are implemented using XNOR operation. Until and unless all the shares are received by the receiver, the secret cannot be decoded, which provides an immense scope of security. Also alteration of the shares formed is not possible, and this gives a great perception of security issue being solved.

Keywords Decryption · Encryption · Visual cryptography · XNOR

1 Introduction

With the exponential growth in transmission of data [1] from one end to another end, probability of increased defect in security measures posed to be one of the most serious muddles, and developers find themselves trapped in. One has to preserve the data for not letting it go in intruder's hands, which in turn asks for

S. Sharma (✉) · S. Modi · A. Sharma
Department of Information Technology, Vellore Institute of Technology,
Vellore, Tamil Nadu, India
e-mail: shivamvitp@gmail.com

S. Modi
e-mail: shivammodi95@gmail.com

A. Sharma
e-mail: sharmakanksha007@gmail.com

H.R. Vishwakarma and S. Akashe (eds.), *Computing and Network Sustainability*,
Lecture Notes in Networks and Systems 12, DOI 10.1007/978-981-10-3935-5_41

cryptography. With the complex network topology [2] converging over, it has now become really easy for intruder to hack over the communication. If worst case is considered, hackers will not be able to categorize the image retrieved by them as an important one, guaranteeing the security issue to be preserved. The platform, on which cryptography was performed earlier [3], was wholly based on text encryption and decryption. With the era of digitization hovering up the technology, there was a slight swift to image-based encryption [4]. The term was coined by MoniNaor and Adi Shamir as visual cryptography [5]. Transmission of data becomes easier and flexible when worked on this platform. At sender side, text is encoded in image. A chain of steps is followed to generate shares, if viewed by naked eyes, and the virtue of change cannot be visualized. The shares along with the original image are networked across. Appropriate percent of manipulations is performed on the shares to conserve the property of security. The next phase of cryptography includes decoding of the text that was encrypted in image [6], which shall not be performed until the receiver has all of its components by its own, in this case, shares. The process of decryption can be decoded by visual cryptography itself and cannot be seen by human visual system.

2 Related Works

Naor and Shamir [5] in 1994 coined the term visual cryptography, which combines secret sharing and traditional cryptography. Their major drawback of the proposed algorithm was that it had random patterns of white and black pixels provoking the intruders to be suspicious of data.

After the use of gray scale image in the studies of Naor and Shamir, the era of color images has drawn its attention to the area where a lot of research is still being carried.

2.1 Non-expanded Visual Cryptography Scheme with Authentication

Huang and Chang [7] used an extended visual cryptography combined with authentication. Authentication with block encoding scheme was integrated to securely transmit the data over a network. They aim to solve the characteristic feature of image distortion during the process of image expansion, with making use of non-extended block encoding. Herein, the image is divided into four regions with a chain of sequence to generate region share, which can be achieved using block code method. The original image is divided into a number of blocks containing

2 × 2 pixels each in a block region. The regions altogether undergoes through three phases. First one is to take the original image into number of blocks. Second phase is sharing wherein same pixels generate group, and each group in return generates combinations of share blocks. The last step followed by sharing is superimposing the combination of each share block, which constructs block of confidential data. Eventually, share images can be obtained. If it is made to be shifted a certain unit, secret message can be viewed.

2.2 An Approach for Secret Sharing Using Randomized Visual Secret Sharing

Dixit et al. [8] proposed to go for randomization and pixel reversal using predefined sequence of mathematical computation, using which matrix can be formed. Once the matrix has been formed, original image can be obtained at decryption. They have gone for pixel reversal at first and then for randomization. This method is applied to one secret gray scale image.

2.3 Embedded Extended Visual Cryptography Scheme for Color Image Using ABC Algorithm

Deepa and Benziger [9] had embedded extended visual cryptography scheme for color image. Earlier, textual data were used to be encoded in an image. Here, image will be encoded in image, that's why it is being referred to as 'embedded.' For improving the quality of the reproduced image, they have used ABC algorithm. For generating shares, pixels are divided into subpixels. ABC helps in deciding which pixel to change in each block. Also, in place of RGB values, they have used CMY (Cyan, magenta, yellow) as they can produce new range of colors, as described in subtractive model.

2.4 Extended Visual Cryptography for Color Images Using Coding Tables

Kamath et al. [10] used color image for visual cryptography. With the implementation of Jarvis error filter key and a coding table, they have implemented the concept of dithering on four cover images and one secret image. A color halftone image is generated by combining the three constituent planes of red, green, and blue. Error diffusion [2] method is used in halftoning of the image, which in return uses Jarvis error filter key. Cover table and secret table were used in encoding the secret image to cover image generating shares. The secret image is obtained only when the receiver stacks the present shares. Steps such as key generation, cover image encoding, and secret image encoding are used for the generation of keys. By stacking of two or more shares along with the key image, construction of secret image can be performed at decryption. The efficiency of their proposed scheme is 50%. Mean square error (MSE), peak signal-to-noise ratio (PSNR), and normalized correlation (NC) metrics are used to compare the original secret image and the image obtained after the process of decryption at receiver's side.

2.5 A Novel Multisecret Sharing Scheme with MSB Extraction Using EVCS

With the evolution of cryptography, the need of VC was prominent. Rajput [11] proposed their scheme on VC using extended visual cryptography in which two color images are shared into n number of meaningful shares by using bit plane coding so that it provides a better contrast in the recovered image. Sharing of more than one gray scale or color secret images converted into 'n' meaningful shares is possible here. They have used a mathematical function wherein they collect coefficient of pixels from each pair of coordinates. Two cover images and two secret images are used for encryption. The process of encoding is done using the higher order four bit planes from secret images and cover images so that it generates a new gray scale image of 8 bit. This is done using EVCS. EVCS is also performed with cover images for every plane of the new combined secret image so as to generate the two shares for each bit plane. First, it creates the bit planes of the secret image followed by generating shares for each plane. Finally, combining all the bit planes into a single bit image is done with which the process of encryption is implemented. The image can be decoded using those bit planes. Also decryption is possible here using two methods: decryption using human vision and decryption with a low computation.

3 Proposed Work

Color image is taken as input for the work. Textual data are encoded in the image which will not be visible from naked eyes.

1. The text is encoded, and alphabetic characters are converted into their standard ASCII code.
2. ASCII codes are then converted into binary digits, that is, 0 and 1.

3. 2-D arrays are created with random sets which on which if XNOR operation is performed 0 or 1 can be yield, depending on the set. The array is computed by two factors, i.e., a[i][j], where ‘i’ is number of letters encoded in image and ‘j’ is 8-bit code format.
4. Take the first bit of the binary code format, and check its corresponding randomly generated set. In a given set of numbers for 0 and 1, each set has four elements in it.

3.1 Share Generation

The concept of shares was introduced so as to increase the percentage of security over a network in transmission. For example, if we transmit only one image through the network that is that image which has the data encoded in it, there are higher chances of activities related to intruders seeking into your personal data. Instead of sending only one image, ‘N’ number of shares are made to transmit across the network with some minute changes inscribed in it. These changes cannot be seen with human visual system, predominantly, resulting in making the data invulnerable to threats.

The original image is reproduced four times. Currently, we have five images including the four duplicates and one original.

Let us assume ‘0’ to be the first bit and the randomly generates set be {1, 0, 1, 1}.

Replicas of original image are taken, and their first pixel is changed with respect to the set obtained.

Four images are obtained with almost no difference in parameter of viewing from naked eyes as compared to the original one since changes in a unit of pixels cannot be detected with human visual system.

3.2 Encryption

Secure data transmission can be preserved by using the method of encryption wherein certain changes are made such that intruders cannot think about the possibility of some important message hidden over there.

Encryption is done at sender’s side to secure the transmission.

The image is segmented into 20×20 dimensional blocks.

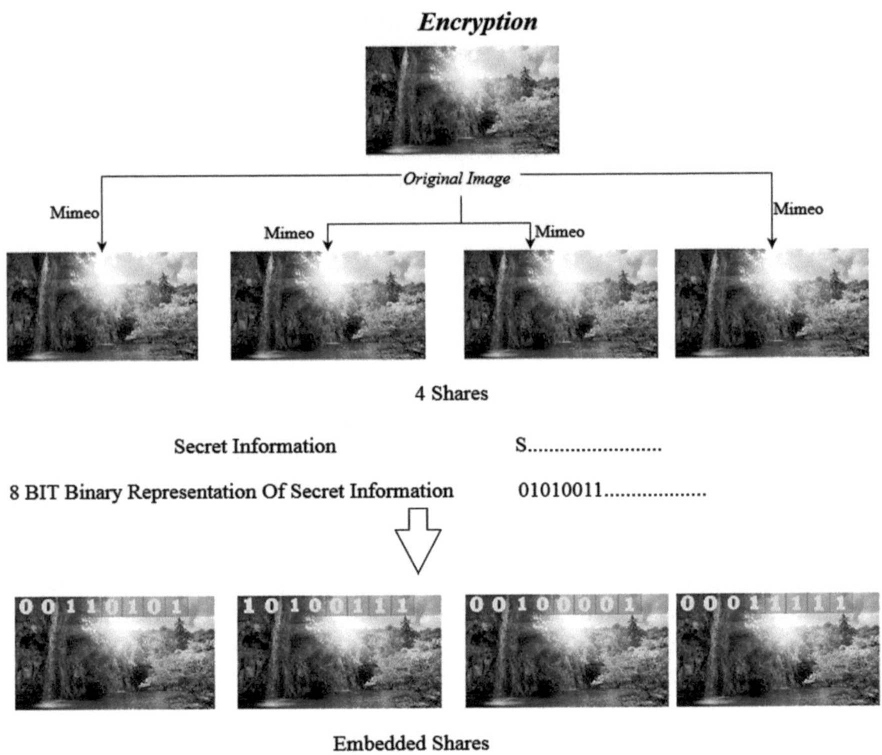

Fig. 1 Encryption process

A randomly generated pixel will be changed leaving no difference when seen with naked eyes. This operation is performed within each block for a unique pixel generated randomly.

These changes in pixels will be significant leaving intruders blank about any message in data.

Once the data are off to communication line from sender to receiver, the process of decryption occurs (Fig. 1).

3.3 Decryption

Once the message is received from sender's side, user at the other end will try to decode the message by finding out the same key through the same key (Fig. 2).

The receiver will have five images including four encrypted shares and one original image.

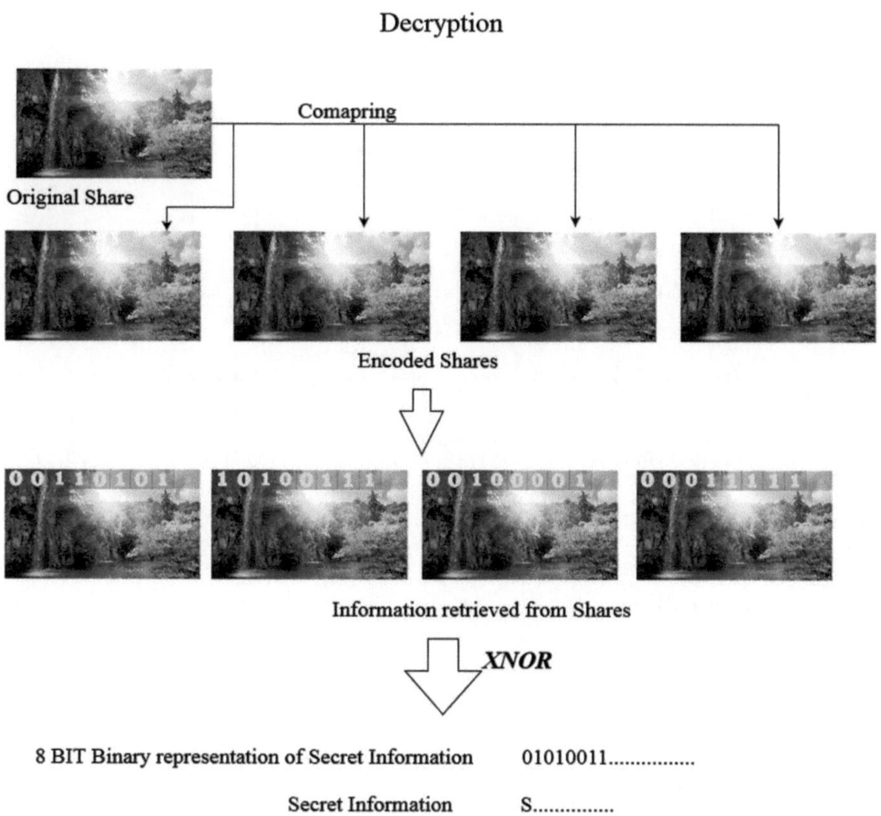

Fig. 2 Decryption process

By comparing pixel by pixel of original image to that of the shares will yield them the binary number sender has changed during the process of encryption. XNOR operation will be performed with those set of binary numbers obtained from those blocks whose pixels were changed during encryption. 8-bit binary number will be obtained after these steps which in return can be changed into ASCII characters. Hence, the message can be decoded.

Also, decoders will be needing magnifying glass to see the message since this factor could have caused a bad drift in intruder's eyes, making that data vulnerable.

Pseudo Code

```
{the following encode is utilised to encode a set of data}
public class encoder
{
      Take the value of the string from the user

int b[]= new int[c.length];
for(i=0;i<c.length;i++)
      {
Convert the charters of the string to ascii characters
      }
Make a 2 dimensional array which will store the binary for all ascii values

 Run a for loop till the length of the string
      {
            Indivisually select every character of the arrayand convert o 8
bit binary nos
      }
These two are the data set of 4 bit xnors
xnor0={{0,0,0,1},{0,0,1,0},{0,1,0,0},{0,1,1,1},{1,0,0,0},{1,0,1,1},{1,1,0,1},
{1,1,1,0}};
int[][]
xnor1={{0,0,0,0},{0,0,1,1},{0,1,0,1},{0,1,1,0},{1,0,0,1},{1,0,1,0},{1,1,0,0},
{1,1,1,1}};
{the following is the code for making the 4 2-d arrays with the  xnor
datasets}
for(i=0;i<t.length();i++)
      {
for(int j=0;j<8;j++)
          {

If the value of the element is equal to 1 then randomly select an clement
from the xnor1 data sets and store the elements in 4 different arrays
Or
Randomly select an clement from the xnor0 data sets and store the elements in
4 different arrays

          }
     {all the data in each of the arrays is being encoded in the image}
{this will be repeated 4 times for each of the shares}
Run a for loop till 20 * length of word
{
      Run a for loop till 20 * no of bits

    {
            Randomly select f1 pixel from the 20*20 block
            change their rgb values according to the array assigned to those
      elements
            add 3 if element is 1
            else subtract 3
         }
      }
```

```
{The following set of code is used to decode the following data}
public class decoder
{
      all the shares are loaded here
      repeat for all shares and store in f1,f2,f3,f4
form 2-d arrays for each of these images

{used to decode and get the data stored in the specific pixel}
{repeat the following  for all the 4 shares}
{
Start comparing every pixel of the 4 shares with that of the original image
separately in the 20*20 segments
As soon as we get a different pixel we chech for the difference in values
If difference is positive then store a 1 in the respective block of the
respective 2d array
Else
Add one to it
}
{doxnor of the given values and find the data for the given ascii values}
for(i=0;i<10;i++)
       {
                 Find  xnor  of  every  single  element  separately  with
respective elements of the 2 d array
       }
convert binary to ascii values and then to character
print the decoded string
```

4 Result

Results show that after giving color image as an input and changing the intensity of the image, the patterns cannot be seen as this was a major drawback in Naor and Adi Shamir [5]. Random patterns were recorded in the scheme proposed by them which posed as a threat, when data security is considered. Also, by visualizing the generated shares, no one can say that any important message is hidden within it which gives the sender a relaxation to not care about the hacking activities. Since the changes are made in units of pixel, it does not differ in quality as that of the original image.

5 Conclusion

We have shown that how a color image with a text encoded in it can be safely transmitted across a network from one end to another end by using extended visual cryptography. Shares are generated within which pixels are made to change based on the encoding done in the image and upon their ASCII code. Their XNOR equivalent is found from the set defined, and four elements are taken accordingly. A pixel of each block is made to change according to the elements found in the set. At encoding, a random pixel is changed and this is followed in each 20 $\times$ 20 block which is again not visible with naked eyes. The same process is used at decryption but in reverse way. User will at first compare each share with the original image and will get a binary number which will then be converted into ASCII character revealing the message.

The dimensions of retrieved image are same as those of the original image which is desirable. The quality of image is also not distorted in the reproduced image. Because of randomly changed pixels, ease of finding patterns is not an option here which is a measure to show concern. There is a vast scope of research on this topic.

References

1. Hunter P (2012) Chipping away at long-distance data. Eng Technol 7(8):78–81. doi:10.1049/et.2012.0815
2. Li C, Wu A (2010) Simple topology estimating approaches for complex networks. In: 2010 international conference on intelligent system design and engineering application. IEEE
3. Vignesh R, Sudharssun SS, Jegadish Kumar KJ (2009) Limitations of quantum and the versatility of classical cryptography: a comparative study. In: ICECS'09. Second international conference on IEEE environmental and computer science
4. Ameena MN, Binish MC (2014) Image encryption based on diffusion process and multiple chaotic maps. In: 2014 international conference on power signals control and computations (EPSCICON). IEEE
5. Naor M, Shamir A (1995) Visual cryptography. Advances in cryptology—EUROCRYPT'94. Springer, Berlin
6. Das S et al (2012) A secured key-based digital text passing system through color image pixels. In: 2012 international conference on advances in engineering, science and management (ICAESM). IEEE
7. Huang Y-J, Chang J-D (2013) Non-expanded visual cryptography scheme with authentication. In: 2013 IEEE international symposium on next-generation electronics (ISNE). IEEE
8. Dixit S, Jain DK, Saxena A (2014) An approach for secret sharing using randomised visual secret sharing. In: 2014 fourth international conference on communication systems and network technologies (CSNT). IEEE
9. Deepa AK, Benziger B (2014) Embedded extended visual cryptography scheme for color image using ABC algorithm. In: 2014 12th international conference on signal processing (ICSP). IEEE
10. Kamath M et al (2012) Extended visual cryptography for color images using coding tables. In: 2012 international conference on communication, information and computing technology (ICCICT). IEEE
11. Rajput PK (2013) A novel multi secret sharing scheme with MSB extraction using EVCS. In: 2013 sixth international conference on contemporary computing (IC3). IEEE

Particle Swarm Optimization for Disconnected Wireless Sensor Networks

Ramya Sharma and Virender Ranga

Abstract Wireless sensor networks which are formed by sensor nodes are used widely for sensing the environment and observing useful information from the data gathered. Due to harsh environmental conditions, the network can be disconnected. Restoring the network's lost connectivity is crucial for future functioning of the network and is done by placing relay nodes, which are small devices used for transmission of sensed data. This problem of reconnecting the network in an optimal manner is shown to be NP-hard; therefore, we practice meta-heuristics to this problem. In this paper, we propose Federating Network using Particle Swarm Optimization (FN-JPSO), which can be applied efficiently for restoring the lost connectivity. Our proposed approach first finds the representative node for each disconnected segment and then creates Steiner points for reconnection. These Steiner points are further used to create random spanning trees, which are used as particles in FN-JPSO to provide an optimal interconnected network.

Keywords Steiner nodes · Relay nodes · Particle Swarm Optimization · Cost

1 Introduction

In the previous years, the applications of wireless sensor networks (WSNs) have faced an uprising in their numbers [1]. For applications such as space exploration, forest fire detection, combat field reconnaissance, and machine health monitoring, in order to observe a covered area and keeping certain activities in check, a number of sensor nodes are placed in the network. By positioning sensors to control unattended surroundings, the danger to human life may be avoided and the envi-

R. Sharma (✉) · V. Ranga
Department of Computer Engineering, National Institute of Technology, Kurukshetra, India
e-mail: sharma.ramya268@gmail.com

V. Ranga
e-mail: virender.ranga@nitkkr.ac.in

H.R. Vishwakarma and S. Akashe (eds.), *Computing and Network Sustainability*, Lecture Notes in Networks and Systems 12, DOI 10.1007/978-981-10-3935-5_42

ronment will continue to be monitored. These applications use sensor nodes (SNs) which have small battery life, with limited processing and communication capabilities. After deployment, the sensor nodes set up a network with the objective of data sharing and synchronizing the actions performed. To facilitate such collaboration, the said nodes must be accessible to each other. Long distance communication for sensor nodes will be costly and would exhaust them very quickly (as energy transmission is proportional to the distance between the two nodes). Thus, relay nodes (RNs) are introduced, which are used to transmit the sensed data from sensor nodes to a base station (BS). This problem of introducing minimum relay nodes in the setting so that the entire network is interconnected is a NP-hard problem and is called relay node placement (RNP) problem. This RNP problem takes two separate architectures into account: one-tiered WSN and two-tiered WSN. In one-tiered WSN, both sensor nodes and relay nodes contribute to the routing. While in two-tiered WSN, only relay nodes are used in routing procedure. The limitation of a vast network is that the wireless devices may fail, leaving the WSN disconnected [2]. Due to harsh and unpredictable environmental conditions such as landsides, floods, or forest fires, the network faces disconnection, due to breaking of the network into smaller intra-connected segments. To restore the connectivity, a set of relay nodes can be positioned along the minimum cost path. The connectivity between every minimal cost path can be represented as a minimal cost tree.

A Steiner minimum tree (SMT) is a tree formed by interconnecting a given set of vertices with the help of some extra points (Steiner points) to gain low cost connectivity. In triangle geometry, the Steiner point is a specific point associated with a plane triangle. This point has the shortest distance from all the triangle vertices. Thus, Steiner points are used for connectivity of a graph. Creating a SMT is a NP-complete problem; thus, we pursue meta-heuristics to tackle this problem. Jumping Particle Swarm Optimization (JPSO) is a discrete variation of Particle Swarm Optimization. We propose JPSO, Federating Network using Particle Swarm Optimization (FN-JPSO), to solve the SMT problem to regain lost connectivity in the network. The structure of the paper is divided into five parts beginning with the introduction section. Related work is described in Sect. 2. Section 3 shows our contribution to this paper. Simulation results are shown in Sect. 4, and the paper is concluded in Sect. 5.

2 Related Work

In the approach suggested by Senel et al. [3], the authors considered the problem of relay node placement which uses Spider web deployment strategy, placing relay nodes in the inward direction so that the network has better connectivity and coverage. Major shortcoming of this method is the slight increment in relay node count. The proposed approach of Lee and Younis [4] for network fragmentation is an optimized relay node placement algorithm with the help of a minimum Steiner tree on the convex hull (ORC). ORC attempts to recognize Steiner points

(SPs) which are used to populate relays such that the disconnected segments will be linked with very less number of relays. ORC positions RNs inwardly from the border of the area identified by the convex hull. The authors propose a distributed cell-based optimized relay node placement (CORP) in approach by Lee and Younis [5], which practices greedy heuristics and chooses to decrease the number of relays necessary for setting up a connected inter-segment topology. CORP models the area as a grid of equal-sized cells and identifies the best neighboring cell of a segment Seg_i which lay on the shortest path that joins Seg_i to the other segments, functioning in rounds. This method attempts to minimize the relay node count.

3 Proposed Solution

3.1 Detection of the Network Fragmentation

The network detects the disconnection when multiple relay nodes do not receive the transmitted data or beacon messages. These relay nodes send the beacon messages to all the other neighboring nodes, thus discovering disconnection, if present. The relay nodes which have disconnected neighbors act as the boundary nodes in the segment.

3.2 Reconnection

Selecting a representative node for each segment

For reconnection, every segment must be represented by some node which further processes the reconnection phase. This node is called the representative node of a segment. Thus, for an *i*th segment Seg_i, the representative node rep_rn_{ij} is selected from a set of $boundary\_rn_{ij}$, where i stands for ith segment and j stands for jth boundary node (a node which has experienced disconnection of its links with other nodes).

For selecting the representative node, first, we find out the center of mass of each segment. Then, for every segment we find that boundary relay node which is closest to all the segments' center of mass, i.e., which has the value $\sum_{k \neq j} dist(boundary_{rn_{ij}}, com_k)$ minimum for that segment, where *dist()* calculates distance between two nodes.

Finding out the Steiner points

In triangle geometry, a Steiner point (SP) is described as an extra point introduced, which when connected to the tree vertices of the triangle incurs minimum cost; i.e., the distance between the SP and the vertices is minimum. Thus, for the reconnection of the segments, we use Steiner points.

In our algorithm, Steiner point of a triangle is calculated to be the circumcenter of the triangle. We randomly select three representative nodes from the network and find the Steiner point of the said triangle. We observe $(number\ of\ segments - 1)$ Steiner points (*Spoint*) in our algorithm. As these Steiner points are potential relay nodes location, they are stored for later use. The circumcenter of a triangle is calculated in Eq. 1.

$$x = \frac{((m_{pAB} \times mid_{AB}.x) - mid_{AB}.y) - ((m_{pAC} \times mid_{AC}.x) - mid_{AC}.y))}{(m_{pAB} - m_{pAC})}. \quad (1)$$

where

m_{pAB} is the slope of AB
m_{pAC} is the slope of AC
$mid_{AB}.x$ is the x coordinate of midpoint of line AB
$mid_{AB}.y$ is the y coordinate of midpoint of line AB
$mid_{AC}.x$ is the x coordinate of midpoint of line AC
$mid_{AC}.y$ is the y coordinate of midpoint of line AC

Constructing a fully connected graph and finding out the spanning trees

To ensure connectivity in the network, we construct a fully connected graph, consisting both representative relay nodes and Steiner nodes. Out of this fully connected graph, a set of n random spanning trees is generated. We can choose n accordingly, as it represents the swarm size. Here, n is considered to be 200.

Our aim is to generate a Steiner minimum tree. Thus, we use a modified version of Particle Swarm Optimization to find the best spanning tree.

Implementing the Jumping Particle Swarm Optimization algorithms

Originally, the PSO algorithm was introduced for continuous optimization problems. To deal with combinatorial optimization problems, a variation of PSO called Discrete Particle Swarm Optimization (DPSO) was proposed in [5], in which particles explore a multidimensional discrete exploration space. After introduction of DPSO, many alternatives of this algorithm have been suggested. A certain DPSO approach called Jumping Particle Swarm Optimization (JPSO) algorithm has lately been presented by Consoli et al. [6] to solve combinatorial optimization problems. In JPSO algorithm, as proposed by Qu et al. [7], instead of defining the velocity of every swarm particle to change its position in the search, the particle moves (jumps) from position to position (which are solutions) in discrete search space. If a particle has good fitness in certain area, it attracts the other particles to explore the area. This particle is called an attractor.

In our proposed algorithm, we have selected three kinds of attractors:

L_BEST, which is the best position of the current particle till now;
G_BEST, which is the best position among all particles till now; and
C_BEST, which is the best position among all particles in this iteration.

$$v_{i,j+1} = c_0 v_{i,j} + c_1 r_1 \left(l\_best - x_{i,j}\right) + c_2 r_2 \left(g\_best - x_{i,j}\right) + c_3 r_3 \left(c_{best} - x_{i,j}\right). \quad (2)$$

Equation (2) presents the original PSO equation, which can be divided into two parts: self-improvement, which means exploration of current position, and the second part which encourages the particle to explore the space with the help of attractors. Thus, a particle follows the attractors for improvement.

Every attractor has some likelihood c_x, which is the probability of selecting a particular attractor or self-improvement of the particle. To explore the space, a random number *R*, which has the range [0, 1], is generated. This *R* along with likelihood c_x determines the particle move.

```
Input:  SPAN[PART]:A particle swarm of randomly
         generated spanning trees
         N:total number of nodes n
Output: G_BEST: An optimal spanning tree

FN-JPSO(SPAN[PART], N)
       Initialize all the attractors L_BEST[PART], G_BEST
       and C_BEST
       while ITER!=0
            for( each spanning tree SPAN[i])
                 R ← RAND(0,1)
            if R ≥ 0 and R ≤ 0.24
                 SELF_IMPROVEMENT(SPAN[i])
            else if R ≥ 0.25 and R ≤ 0.49
                 PATH_REPLACEMENT(SPAN[i], LBEST[i])
            else if R ≤ 0.50 and R ≥ 0.74
                 PATH_REPLACEMENT(SPAN[i], GBEST)
            else if R ≤ 0.75 and R ≥ 0.99
                 C_BEST ← FIND_CURRENT_BEST(SPAN)
                 PATH_REPLACEMENT(SPAN[i], C_BEST)
            if COST(L_BEST[i]) > COST(SPAN[i])
                 L_BEST[i] ← SPAN[i]
            if COST(G_BEST[i]) > COST(SPAN[i])
                 G_BEST[i] ← SPAN[i]
            ITER ← ITER - 1
  PATH_REPLACEMENT(SPAN[i], ATTRACTOR)
     select minimum cost path from ATTRACTOR
     replace this path in the SPAN[i]
  SELF_IMPROVEMENT(SPAN[i])
     select a Steiner node with degree 1 or 2
     remove this Steiner node, joining the nodes
     connected to Steiner node

  return G_BEST
```

Fig. 1 Federating Network using Particle Swarm Optimization

The FN-JPSO after meeting the stopping criterion (200 iterations in this case) gives *G_BEST* as the final output, as shown in the Fig. 1.

4 Results Discussion

The Steiner minimum tree generated is then populated by the relay nodes, by placing relay nodes after distance |R| is covered, where |R| is the transmission range of a relay node. The number of relay nodes can be calculated as shown in Eq. (3):

$$Number\ of\ relay\ nodes = \frac{Total\ Cost}{Transmission\ range|R|}. \tag{3}$$

Here, we have assumed the transmission range of a single relay node to be 25 units. Thus, for MST (without Steiner nodes) (Fig. 2), the number of relay nodes turns out to be 80 relay nodes, while MST (with Steiner nodes) (Fig. 3) requires 102 relay nodes. The Steiner tree generated by our algorithm (Fig. 4) takes only 73 nodes, performing better than the other two connecting trees.

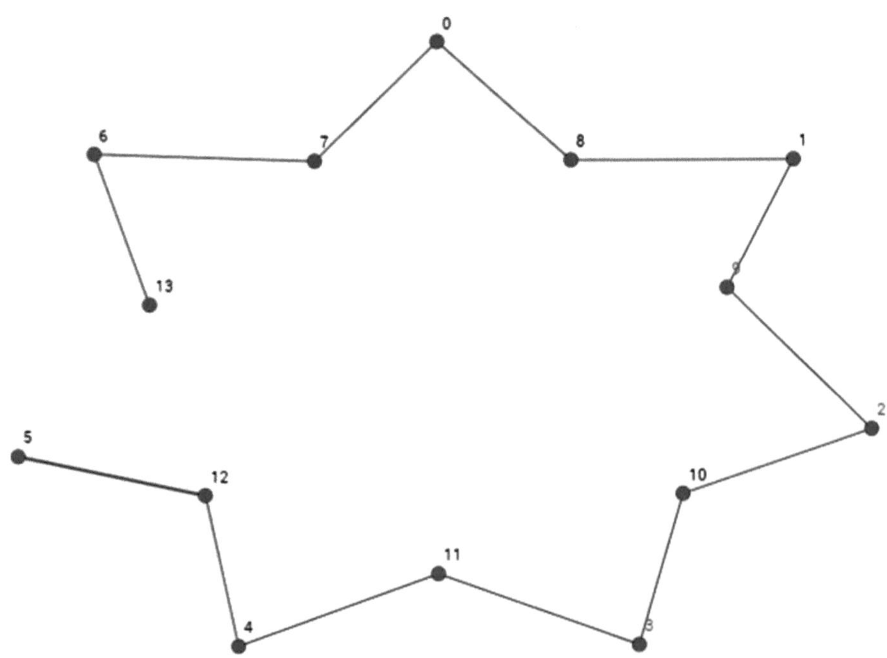

Fig. 2 MST generated by 14 segmented networks (0–13 are representative nodes of respective segments)

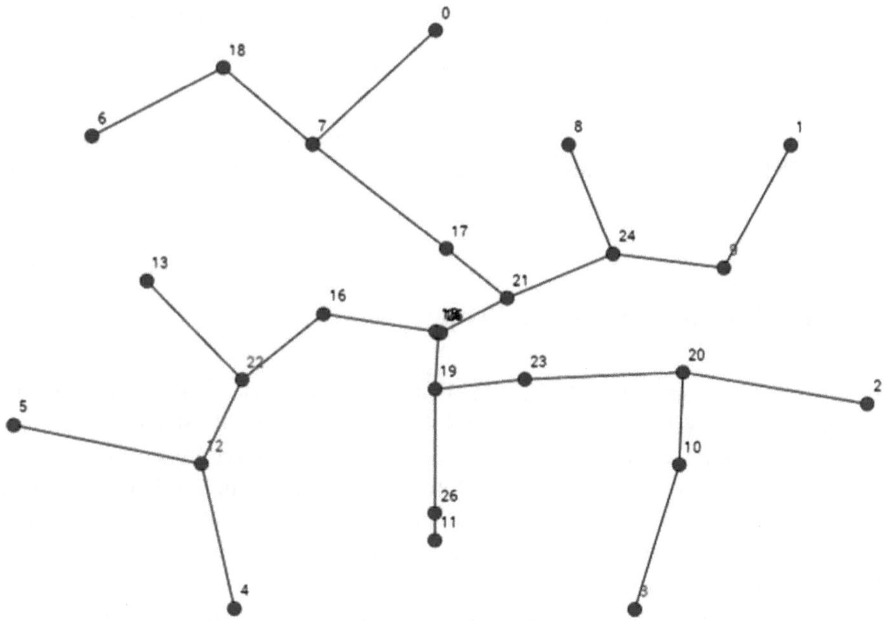

Fig. 3 Minimum spanning tree generated after introducing 13 Steiner nodes [14–27 are Steiner nodes]

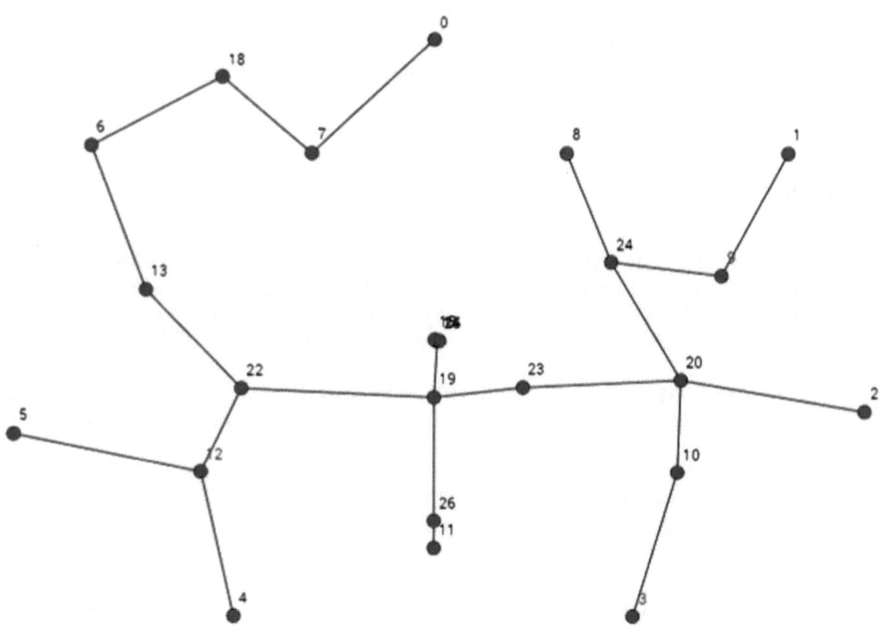

Fig. 4 Global best generated after FN-JPSO is implemented [Nodes 16, 17, and 21 are extra, hence are removed from the tree]

Time consumed during execution T can be calculated as shown in Eq. (4):

$$T = time_{steiner\,nodes} + time_{FNJPSO}. \tag{4}$$

The total time consumed for the given example was: T = 27.083 s + 8.670 s, i.e., 35.753 s, which gives an almost real-time solution for this, otherwise NP-hard problem.

5 Conclusion and Future Scope

In this paper, we present a technique, Federating Network using Particle Swarm Optimization (FN-JPSO). This methodology makes use of Steiner nodes and Steiner minimum trees and solves a NP-hard problem in polynomial time. It uses a discrete variant of Particle Swarm Optimization and Jumping Particle Swarm Optimization to generate a swarm and produce the best output, i.e., least cost Steiner minimum tree, by removing the surplus Steiner nodes and improving the particle's fitness function (cost).

As the results show, the number of relay nodes used in Steiner minimum tree generated by FN-JPSO is less than that of MST. The biggest advantage of using FN-JPSO is that it is topology independent, while the previous work in this area may fail after following heuristic techniques, and FN-JPSO will provide a better result irrespective of the placement of segments.

Future in this area is that it can be further extended for placing relay nodes or sensor nodes while establishing a new wireless sensor network.

References

1. Ranga V, Dave M, Verma AK (2006) Network partitioning recovery mechanisms in WSANs: a survey. In: Wireless personal communications, vol 72, no 2. Springer, pp 857–917
2. Ranga V, Dave M, Verma AK (2014) A hybrid timer based single node failure recovery approach for WSANs. In: Wireless personal communications, vol 77, no 3. Springer, pp 2155–2182
3. Senel F, Younis M, Akkaya K (2011) Bio-inspired relay node placement heuristics for repairing damaged wireless sensor networks. IEEE Trans Veh Technol 60(4):1835–1848. IEEE Press
4. Lee S, Younis M (2012) Optimized relay node placement for connecting disjoint wireless sensor networks. Comput Netw 56:2278–2804. Elsevier
5. Lee S, Younis M (2010) Optimized relay placement to federate segments in wireless sensor network. IEEE J Sel Area Commun Spec Issue Mission Crit Netw 28(5):742–752. IEEE Press (2010)
6. Consoli S, Moreno-Perez JA, Darby-Dowman K, Mladenovic N (2010) Discrete particle swarm optimization for the minimum labelling Steiner tree problem. Nat Comput 9(1):29–46. Springer

7. Qu R, Xu Y, Castro JP, Landa-Silva D (2013) Particle swarm optimization for the Steiner tree in graph and delay-constrained multicast routing problems. J Heuristics 19(2):317–342. ACM (2013)
8. Kennedy J, Eberhart RC (1997) A discrete binary version of the particle swarm algorithm. In: Proceedings of the world multiconference on systemics, cybernetics and informatics, Piscataway, NJ, pp 4104–4109

A Review on Cloud-Based Intelligent Traffic Controlling and Monitoring System

Swati Nigade and Anushri Kulkarni

Abstract This paper presents a cloud-based intelligent traffic controlling and monitoring system to address a major problem of traffic congestion faced by the people. Radio frequency identification (RFID) provides a cost-effective solution to implement the system. Vehicles will be deployed with RFID tags which will give vehicle unique identity. This will give vehicle density at each traffic signal and which will in turn help to control traffic congestion using a smart algorithm. Live traffic density data will be updated in the cloud-based server. Cloud data can be used by police control room (PCR) for monitoring traffic. In case, if the RFID tag belonging to a stolen car is detected, police can take special action to catch the thief. If tag belonging to the ambulance or any other special vehicle is detected, traffic can be rerouted. Android application will use cloud data to display traffic density on user demand.

Keywords Traffic · RFID · Traffic congestion · Emergency vehicle · Stolen vehicle · Android · Cloud · Sensor · Network

1 Introduction

Due to overpopulation, traffic congestion is the major problem in India. Comparatively, infrastructure growth is much slower in proportion with the number of vehicles. To address this problem, many management systems have been stated. Among those systems, wireless network-based systems have been proven cost effective and efficient. Technologies such as ZigBee, RFID, and GSM can be used in implementing traffic control solutions. RFID is a wireless technology that uses radio frequency electromagnetic energy to carry information between the RFID tag

S. Nigade (✉) · A. Kulkarni
Zeal College of Engineering and Research, Pune, Maharashtra, India
e-mail: swati.nigade92@gmail.com

A. Kulkarni
e-mail: anushri.kulkarni@zealeducation.com

H.R. Vishwakarma and S. Akashe (eds.), *Computing and Network Sustainability*,
Lecture Notes in Networks and Systems 12, DOI 10.1007/978-981-10-3935-5_43

and RFID reader. Some RFID systems will work within range inches or centimeters, while others may work for 100 m or more [1]. With improvement in Internet services in the country, this wireless sensor network system can be further extended to sensor—cloud network, wherein cloud computing is used for hefty and scalable high-execution computing when needed. By developing applications which will use the cloud as a back end, users can have easier access to the live data which can help them in deciding time of travel, which will in turn help to reduce traffic. Data stored in the cloud can also be used by police for better monitoring and deployment of police force in congested areas. Thus, it was seen that sensor—cloud network gives a scalable and secured system, with high performance.

2 Literature Survey

Due to traffic congestion, we see a number of negative effects such as delays, which may result in late arrival for employment, meetings, and education, resulting in lost business, disciplinary action, or other personal losses. Blocked traffic interferes with the passage for emergency vehicles traveling to their destinations in case of emergencies. And it eventually results in slow moving traffic, which increases the time of travel, thus stands-out as one of the major issues in metropolitan cities. In [1], intelligent traffic control system was discussed for congestion control, ambulance clearance, and stolen vehicle detection. Each individual vehicle was equipped with RFID tag placed in strategic location so that was not possible to remove or destroy. RFID reader and a microcontroller were used to count the number of vehicles that passed on a particular path during specified duration. This determined network congestion and green light duration for that path. If an RFID tag belonged to any emergency vehicle, this information was communicated to the traffic controller and the green light was turned on. If an RFID tag belonged to stolen vehicle, message will sent using GSM module to police control room.

In [2], implementation of green wave system for vehicles was discussed, and this system helps emergency vehicles to pass through the traffic without any time delay. This system also helps in synchronizing the green phase of traffic signals. With this setup, a vehicle passing through a junction will continue to receive green signals at every junction along the same road. Along with this, the system also tracks a stolen vehicle when it passes through a junction. Advantages of this system are that it helps in case of bad weather conditions and works efficiently in detecting stolen vehicles. One of the major disadvantages of the system is that the vehicle information needs to be stored and updated regularly in the system database which is connected to the traffic signal via XBee transceivers which provides low data rate and limited connectivity. This limits its performance and its evolution. Also, it is noted that when the green wave is disturbed, the disturbance can cause traffic problems which affects the synchronization and the system will fail.

In [3], road traffic control system using cloud computing was discussed. The living methods of wireless sensor network methods for traffic management are not effective as urban regions have a great stack of traffic jams. For this problem, hefty and scalable high-execution computing is needed. Cloud computing is a good engineering and provides potent and scalable computing at a low cost. Sensor cloud network—which is integrated version of wireless sensor network and cloud computing can gather, access, process, store, share, and search large number of sensor data easily.

In [4], it shows a video surveillance team for monitoring traffic in Bangalore city. It involves a manual analysis of data by the traffic management team to determine the traffic light duration at each junction. It will communicate the same to local police officers for necessary action. It can be clearly seen from this that this problem of traffic congestion can be easily taken care if we have a proper system in place.

3 Proposed System

From the existing systems, it can be seen that even after having best wireless—sensor network system, the vehicular traffic has not been reduced. To address this problem, improvement in the current wireless—sensor network was suggested using cloud computing. With various advantages of cloud computing, better algorithms can be used to solve traffic congestion. Android applications or other platform-based applications can be developed using cloud services.

We have following sections in our paper to explain the proposed system:

- Signal section (traffic load time switching using RFID technology)
- RFID-based Vehicle/Ambulance Identification
- Cloud-based Server (shortest path, traffic condition, stolen/black listed server, enquiry server)
- Android Application (traffic condition/stolen report)

3.1 Signal Section (Traffic Load Time Switching Using RFID Technology)

This section has 3 signal points. These signals will work normally under normal traffic load conditions. As soon as the traffic increases, RFID reader connected to each microcontroller will sense the increased traffic and indicate the microcontroller. The microcontroller will in turn increase the green signal time, so that the traffic congestion can be avoided. In Fig. 1, it can be seen that the number of vehicles at the junction is more whereas, in Fig. 2, the density is less. This will directly affect the time duration of the green signal (Fig. 3).

Fig. 1 Traffic congestion

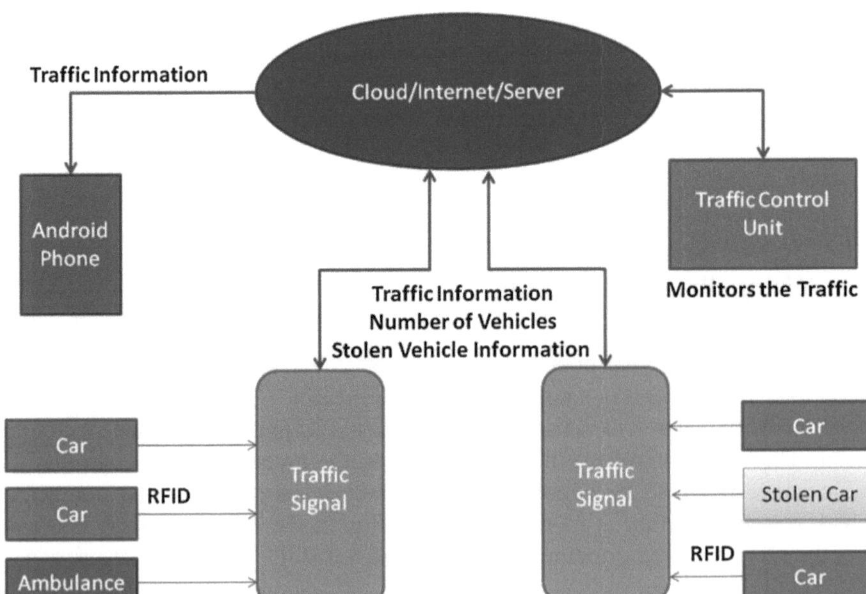

Fig. 2 Flow chart for proposed system

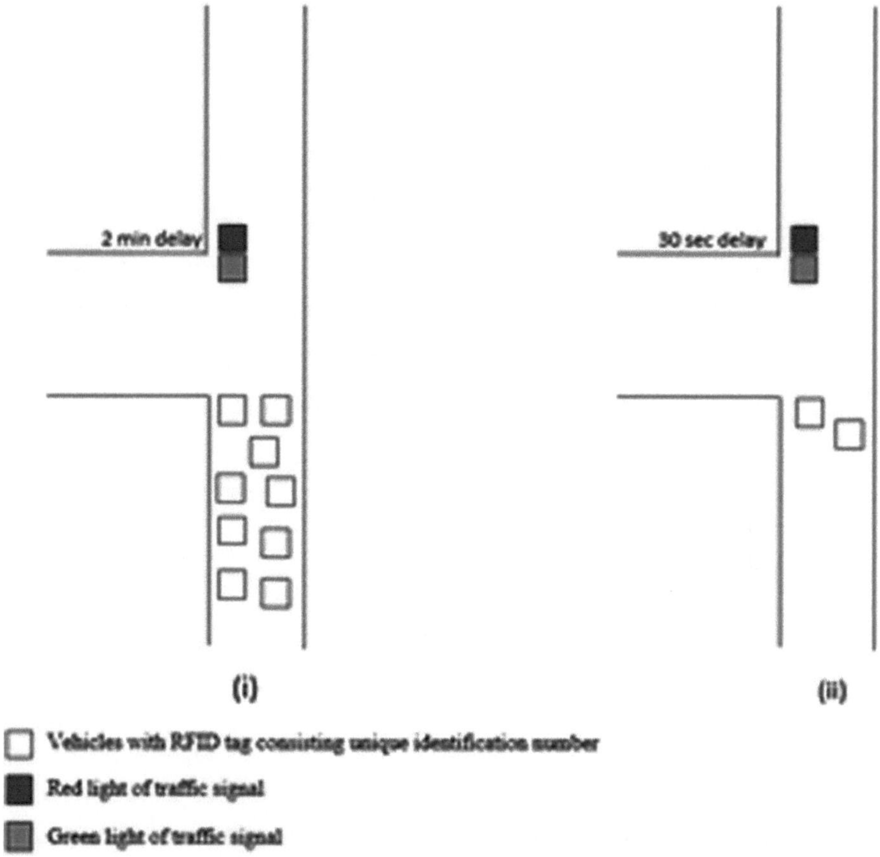

Fig. 3 Two cases for which different traffic density is observed

3.2 RFID-Based Vehicle/Ambulance Identification

The signal microcontroller is connected to an RFID reader. The reader will continuously read the RFID number of every vehicle and send the data to the cloud-based server. The RFID identification is done for two reasons: first, stolen/black listed vehicle tracking; second, for ambulance/other emergency vehicle detection. If stolen or black listed vehicles are detected, then police control room will be notified. As all the traffic signals are connected to cloud server and the police can easily track down the vehicle. To delay the vehicle at the traffic signal, they can increase the time for which red signal will be shown.

In case of emergency vehicle, as soon as any vehicle is detected the microcontroller will turn on all the red signals for 5 s indicating to all commuters that there is an ambulance. In an earlier approach, the microcontroller, after the alert to the commuters, would turn ON the green signal opposite to that of the ambulance

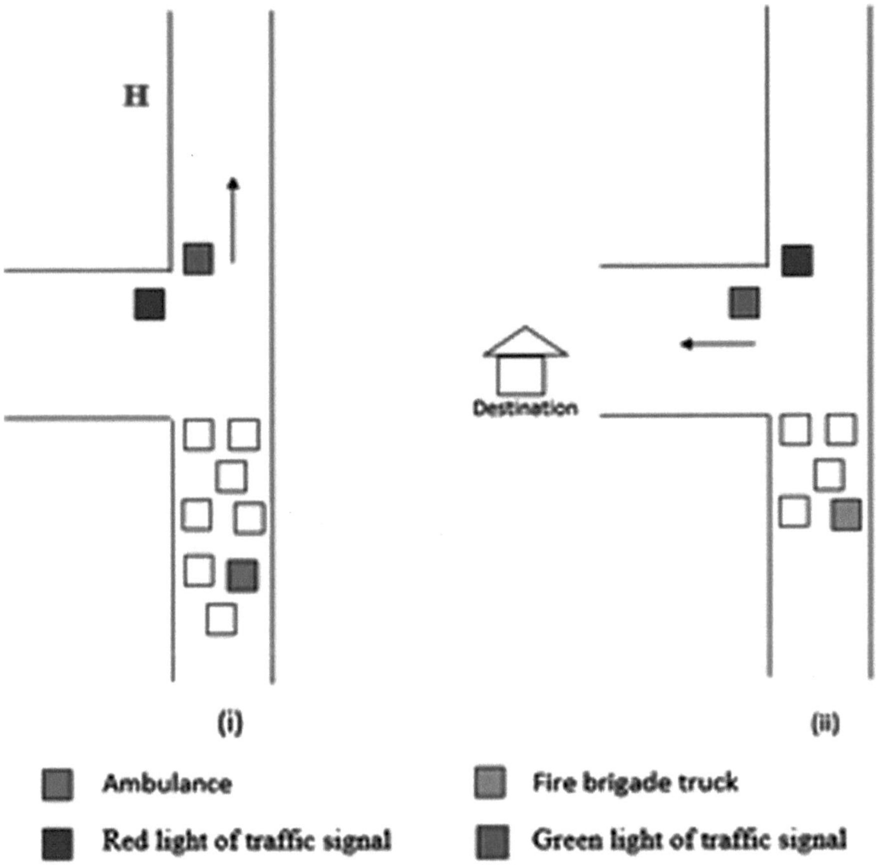

Fig. 4 Case where in special emergency vehicle is detected

so that the ambulance can pass through the traffic efficiently taking less time. But in our approach as a cloud is involved, better decisions can be made for which signal to turn green.

Consider the case of ambulance, if an ambulance is detected at the traffic signal, the microcontroller will send these details to the server, and the cloud smart algorithm will find out which is the closest and the fastest path to the hospital. The server will send these details to the microcontroller, and it will make respective signal green and not just the opposite signal. Figure 4 shows two cases wherein emergency vehicle will be given a clear path by turning required signal green.

3.3 *Cloud-Based Server (Shortest Path, Traffic Condition, Stolen/Black Listed Server, Enquiry Server)*

The server has all the data related to traffic condition with time and area information. All the data will be dynamically updated on the server by microcontrollers at the traffic signals. This data will help traffic control unit to monitor the traffic. It will also help them to deploy traffic police at locations with more traffic. The server will be back end for the android application which will help users with live traffic information and to do queries about their stolen vehicle. Further, server can be used by microcontroller to do advanced algorithms to control the traffic.

Figure 5 shows architecture followed on cloud platform, there are three users—a traffic signal, police control room, and android users. Everyone needs to login to the server and depending upon the user they will be given rights to access the data. For example, admin or the police will have all rights, microcontroller at the traffic signal will have rights to access data and update the data periodically, while android users can only view the data and have no rights to modify the data. Also android users can register a query for stolen vehicle. The system followed by the cloud platform makes it secure and reliable.

3.4 *Android Application*

An android application will be developed for giving users live traffic information. This is a user node which is used to upload as well as download the data from the

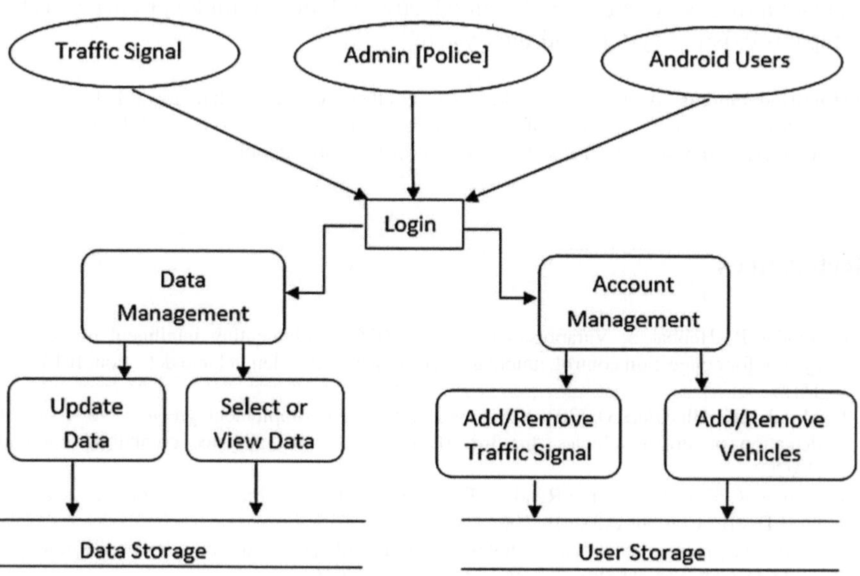

Fig. 5 Flow chart for architecture followed on cloud platform

Table 1 Comparison between existing and developed system

Existing system	Developed system
Wireless—sensor network	Cloud—sensor network
Traffic monitoring is not up to the mark	Traffic monitoring is improved. Data can be analyzed for better deployment of traffic police
Not scalable. Advanced algorithms cannot be implemented easily	Highly scalable with the help of cloud computing
Not completely secured and limited performance	Secured and high performance

server. Traffic information provided by the application will help the users to decide their time of travel or to find an alternate path to their destination in real time. Users can also upload query such as stolen vehicle along with their name, car details, time, and area name from where it was stolen. Once the query is registered with the server, the server will update all microcontrollers at the traffic signals and notify the traffic control unit so that they can monitor.

4 Conclusion

An Implementation of the system described in this paper will help to improve traffic management and reduce traffic congestion in urban area. End users will also get facilitated with user-friendly android application to get traffic updates of any area at any time due to cloud-based server. As RFID is a low-cost technology, implementation of this system provides an effective solution to traffic problems. Table 1 compares existing system and our system.

Acknowledgements Authors would like to express their sincere gratitude toward all the teaching and non-teaching staff members of E&TC department of Zeal College of Engineering and Research. Also they want to thank their parents for constant support.

References

1. Sundar R, Hebbar S, Varaprasad Golla V (2015) Implementing intelligent traffic control system for congestion control, ambulance clearance, and stolen vehicle detection. IEEE Sens J 15(2)
2. Mittal AK, Bhandari D (2013) A novel approach to implement green wave system and detection of stolen vehicles. In: 3rd, IEEE international advance computing conference (IACC)
3. Kumar K, Kaur PD (2015) Road traffic control system in cloud computing: a review. Int J Grid Distrib Comput 8(3):201–206
4. Traffic management centre. http://www.bangaloretrafficpolice.gov.in/index.php?option=com_content&view=article&id=87&btp=87. Accessed 2016

5. Traffic solution. http://phys.org/news/2013–05-physics-green-city-traffic-smoothly.html. Accessed 2016
6. Srinivas J, Reddy KVS, Qyser AM (2012) Cloud computing basics. Int J Adv Res Comput Commun Eng 1(5):343–347
7. Shruthi KR, Vinodha K (2012) Priority based traffic lights controller using wireless sensor networks. Int J Electron Signals Syst (IJESS) 1(4) ISSN: 2231–5969
8. Hegde R, Sali RR, Indira MS (2013) RFID and GPS based automatic lane clearance system for ambulance. Int J Adv Elect Electron Eng 2(3):102–107
9. Varaprasad G, Wahidabanu RSD (2010) Flexible routing algorithm for vehicular area networks. In: Proceedings of IEEE conference on intelligence, transportation system and telecommunication, Osaka, Japan, 2010, pp 30–38
10. Gokulan BP, Srinivasan D (2010) Distributed geometric fuzzy multiagent urban traffic signal control. IEEE Trans Intell Transp Syst 11(3):714–727

Author Index

H.R. Vishwakarma and S. Akashe (eds.), *Computing and Network Sustainability*,
Lecture Notes in Networks and Systems 12, DOI 10.1007/978-981-10-3935-5

MIX
Papier aus verantwortungsvollen Quellen
Paper from responsible sources
FSC® C105338

If you have any concerns about our products,
you can contact us on
ProductSafety@springernature.com

In case Publisher is established outside the EU,
the EU authorized representative is:
Springer Nature Customer Service Center GmbH
Europaplatz 3, 69115 Heidelberg, Germany

Printed by Libri Plureos GmbH
in Hamburg, Germany